Dear MyCopy Customer,

This Springer book is a monochrome print version of the eBook to which your library gives you access via SpringerLink. It is available to you at a subsidized price since your library subscribes to at least one Springer eBook subject collection.

Please note that MyCopy books are only offered to library patrons with access to at least one Springer eBook subject collection. MyCopy books are strictly for individual use only.

You may cite this book by referencing the bibliographic data and/or the DOI (Digital Object Identifier) found in the front matter. This book is an exact but monochrome copy of the print version of the eBook on SpringerLink.

Platelet Aggregation in the Pathogenesis of Cerebrovascular Disorders

Proceedings of the
Round Table Conference. Rome, October 30 – 31, 1974

Editors:
Alessandro Agnoli and Cornelio Fazio

With 150 Figures

Springer-Verlag Berlin Heidelberg GmbH

Proceedings of the Round Table Conference "Platelet Aggregation in the Pathogenesis of Cerebrovascular Disorders" Rome, October 30 – 31, 1974

Chairman: Cornelio Fazio

Secretary: Alessandro Agnoli

Edited by A. Agnoli and C. Fazio

Sponsored by Boehringer Ingelheim s.p.a. Firenze

DOI 10.1007/978-3-642-66609-4

Library of Congress Cataloging in Publication Data. Main entry under title: Platelet aggregation in the pathogenesis of cerebrovascular disorders. Includes index. 1. Cerebrovascular disease – Congresses. 2. Blood platelet aggregation – Congresses. I. Agnoli, A. II. Fazio, Cornelio, 1910- . III. Boehringer Ingelheim. RC388.5.P53. 616.8'1'071. 77-8194.

This work is subject to copyright. All rights are reserved, whether the whole or part of the material is concerned, specifically those of translation, reprinting, re-use of illustrations, broadcasting, reproduction by photocopying machine or similar means, and storage in data banks. Under § 54 of the German Copyright Law where copies are made for other than private use, a fee is payable to the publisher, the amount of the fee to be determined by agreement with the publisher.

© by Springer-Verlag Berlin Heidelberg 1977

MyCopy version of the original edition 1977

The use of general descriptive names, trade marks, etc. in this publication, even if the former are not especially identified, is not be taken as a sign that such names as understood by the Trade Marks and Merchandise Marks Act, may accordingly be used freely by anyone.

2123/3130-543210

www.springer.com/mycopy

Contents

Introductory Remarks. C. FAZIO and A. AGNOLI 1

Methodology and Physiology
Chairmen: G.V.R. BORN and K. BREDDIN

The Platelet as an Inflammatory Cell. R.L. NACHMAN.
With 2 Figures 4

Platelet Aggregation in the Pathogenesis of Cerebrovascular Disorders. G.V.R. BORN 8

The Mathematical Analysis of the Human Platelet Aggregation Mechanism and Its Clinical Application. T. ABE,
J. NISHIZAWA, and M. KAZAMA. With 16 Figures 17

Quantitative Valuation of Platelet Aggregation Curves
through the Calculation of a Numerical Index.
F. CAVALLERO. With 5 Figures 27

The Polyunsaturated Fatty Acids in Human Platelets:
Effects of Diet and Possible Functional Significance.
R. PAOLETTI. With 1 Figure 33

Enhanced Platelet Aggregation as a Risk Factor for
Progress and Complications of Vascular Disease. New
Findings with a Platelet Aggregation Test (PAT III) and
on the Dependence of Different Aggregation Tests on Morphologic Platelet Changes. K. BREDDIN, H.J. KRZYWANEK,
J. ZIEMEN, H. BAUER, and H. GRÜN. With 24 Figures 44

Pathogenesis
Chairmen: M. ANTHONY, T. ABE, E.J. ACHESON,
J.S. MEYER, J. MARSHALL, K.J. ZÜLCH

Thrombosis and Embolism as a Cause of Ischemic Cerebrovascular Disturbances. Analysis from a Series of 1000
Patients. K.J. ZÜLCH and H. von EINSIEDEL-LECHTAPE.
With 3 Figures 64

Platelet Adhesiveness and Cerebral Vascular Disease
Revisited. E.J. ACHESON 75

The Significance of Platelet Aggregation in Amaurosis
Fugax. J. MARSHALL. With 1 Figure 77

Mechanisms of Platelet 5-Hydroxytryptamine Release in
Migraine. M. ANTHONY 81

Experimental observations on Platelet Emboli in Focal
Brain Ischemia. C. FIESCHI, F. VOLANTE, N. BATTISTINI,
and E. ZANETTE. With 2 Figures 87

Observations on Platelet Aggregability in the Acute Phase
of Untreated Strokes. G.L. LENZI, F. LAGHI PASINI, and
A. VITTORIA. With 2 Figures 93

Microembolism in the Nervous System. J.M. WILLIAMS,
S. HOHMANN, N.C.R. MERRILLEES, B.L. OPPERMANN, and
P.M. ROBINSON. With 5 Figures 96

Influence of Plasma Components in the Development of Conditions of Increased Platelet Aggregation Found in a Number
of Vascular Diseases. R. BREDA, B. BIZZI, and G. LEONE.
With 4 Figures 110

Mechanism Associated with Platelet Adhesiveness in Cerebrovascular Disease. R.L. SWANK. With 3 Figures 115

Regional Intravascular Coagulation and Microthrombosis in
Traumatic Brain Lacerations in Man. S. COCCHERI and
C. TESTA. With 3 Figures 121

Platelet Aggregation in Cerebrovascular Patients.
M. PRENCIPE, F.M. PISSARRI, V. CECCONI, and A. AGNOLI.
With 2 Figures 129

In Vivo Effect of Cyclic AMP and Related Drugs on Platelet Function. F.J. PARETI and P.M. MANNUCCI 134

Platelet Hyperreactivity and Decreased Survival in Chronic
Cerebrovascular Patients. Chronic Defibrination Syndrome?
G.G. NERI SERNERI, E. SILVESTRINI, G.F. GENSINI, and
R. ABBATE-GENSINI. With 15 Figures 136

Arterial Hypertension and Platelet Aggregation in the Patho-
Pathogenesis of Cerebrovascular Diseases. S. LENTINI,
E. BOLOGNA, and C. PIRRO. With 9 Figures 152

Smoking, Cerebrovascular Diseases and Platelet Functions.
G. KAUCHTSCHISCHVILI, G. GRIGNANI, P. BO, and G. NAPPI. 167

Contribution of Platelet Aggregation and Serotonin Release
to Progressive Cerebral Infarction. J.S. MEYER,
K.M.A. WELCH, and J. BUCKINGHAM. With 6 Figures 172

Effect of Agents which Modify Platelet Aggregation and/or
Coagulation on Experimental Platelet Embolism and Intravascular Coagulation. M.G. BOUSSER, L. BARA, R.J. PROST,
and M. SAMAMA 179

Platelet Abnormalities in Cerebrovascular Diseases.
C.A. BOUVIER. With 2 Figures 191

Inhibition of Platelet Aggregation by Synthetic Organic
Acids: Quantitative Relationships Between Chemical Structures and Biological Activities. V. ČEPELÁK, M. KUCHAŘ,
B. BRŮNOVÁ, J. MUROTOVÁ, and Z. ROUBAL. With 4 Figures 197

Some Observations of Platelet Changes in Atherosclerosis
and Some Observations in the Platelet Alterations Before
and After Antiaggregant Drugs in Normal and Atherosclerotics. O.N. ULUTIN and S.B. ULUTIN. With 2 Figures ... 207

Pharmacology and Clinical Pharmacology

Chairmen: C.A. BOUVIER, W.S. FIELDS, O.N. ULUTIN, A. RASCOL

Inhibition of Platelet Thrombus Formation by Pharmacological Agents. R. KADATZ. With 2 Figures 216

A Long-Term Clinical Trial with Antiplatelet Agents in
Cerebrovascular Ischemia: Biological and Methodological
Aspects. B. GUIRAUD, B. BONEU, J. DAVID, G. GERAUD,
R. BIERME, and A. RASCOL. With 4 Figures 225

Effect of Aspirin and Dipyridamole on Platelet Function
and on Neurologic Evaluation of Patients Affected by
Stroke or Transient Ischemic Attacks. E.E. POLLI,
M. CORTELLARO, L. FRATTOLA, A. RANDAZZO, L. CANDELISE,
E. POGLIANI, A. POLITI, S. BASSI, G. SCOTTI, and
S. SANTAMBROGIO. With 11 Figures 232

The Pharmacologic Control of the Enhanced Platelet Aggregation in Preventive Neurology. F. FEDERICI, S. BIAGINI,
R. EGGER, F. MARCHIONNI, and G. PENCHINI. IN Collaboration
with A. FERRONI, E. SIGNORINI, C. TARDIOLI, and F. BAZZANELLA. With 9 Figures 243

Aspirin in Cerebral Ischemia. W.S. FIELDS, P.W. CALLEN,
and M.M. PRESLOCK 258

The Canadian Cooperative Studies of the Effect of Platelet-Suppressing Drugs in Transient Cerebral Ischemic
Attacks. D. SIMARD. With 1 Figure 262

On the Relationships Between the Activation of the Complement System and the Platelet Aggregation. T. DI PERRI,
A. AUTERI, A. VITTORIA, and F. LAGHI PASINI.
With 11 Figures 268

Discussion and Concluding Remarks

Chairmen's Considerations 282

Concluding Remarks. C. FAZIO 289

List of Contributors

ABE, T., Department of Medicine, Teikyo University, School of Medicine, Tokyo, Japan
ABBATE-GENSINI, R., Istituto di Patologia Medica II, University of Florence, Medical School, Florence, Italy
ACHESON, E.J., Department of Medicine, The Royal Infirmary, University of Manchester, Manchester, Great Britain
AGNOLI, A., Clinica Malattie Nervose e Mentali, University of L'Aquila, L'Aquila, Italy
ANTHONY, M., Division of Neurology, Prince Henry Hospital and the School of Medicine, University of South Wales, Sydney, Australia
AUTERI, A., University of Siena, Medical School, Institute of Medical Semeiotics, Siena, Italy
BARA, L., Laboratoire de Recherche sur la Thrombose Expérimentale, Service du Professeur Bilski-Pasquier et Service d'Hématologie, Pr. J. Bousser, Hôpital de l'Hotel Dieu, Paris, France
BASSI, S., Clinica Malattie Nervose e Mentali, University of Milan, Milan, Italy
BATTISTINI, N., Department of Neurology and Psychiatry, University of Siena, Siena, Italy
BAUER, H., Center of Internal Medicine, Department of Angiology, University of Frankfurt, Frankfurt/Main, W. Germany
BAZZANELLA, F., Central Laboratory, General Regional Hospital, Perugia, Italy
BIAGINI, S., Department of Nervous and Mental Diseases, University of Perugia, Medical School, Perugia, Italy
BIERME, R., Laboratoire Hémostase, Centre Transfusion Sanguine, Toulouse, France
BIZZI, B., Università Cattolica del Sacro Cuore, Istituto di Clinica Medica Generale e Terapia Medica, Rome, Italy
BO, P., Clinica Malattie Nervose e Mentali, University of Pavia, Pavia, Italy
BOLOGNA, E., Pio Istituto di S. Spirito E OO RR - VII Padiglione del Policlinico. Center for the Study and Treatment of Arterial Hypertension and Kidney, Rome, Italy
BORN, G.V.R., Department of Pharmacology, University of Cambridge, Great Britain
BOUSSER, M.G., Clinique des Maladies du Système Nerveux, Hôpital de la Salpêtrière 47, Bd. de l'Hôpital, Paris, France
BOUVIER, C.A., Unité d'Hémostase, Laboratoire Central, Geneva University Hospital, Geneva, Switzerland

BREDA, R., Università Cattolica del Sacro Cuore, Istituto di Clinica Medica Generale e Terapia Medica, Rome, Italy
BREDDIN, K., Center of Internal Medicine, Department of Angiology, University of Frankfurt, Frankfurt/Main, W. Germany
BRŮNOVÁ, B., University Hospital, Plzeň, Research Institute for Pharmacy and Biochemistry, Prague, Czechoslovakia
BUCKINGHAM, J., Department of Neurology, Baylor College of Medicine and the Baylor-Methodist Center for Cerebrovascular Research, Houston, Texas, USA
CALLEN, P.W., Neurology-Medical School, University of Texas, Health Science Center, Houston, Texas, USA
CANDELISE, L., Clinica Malattie Nervose e Mentali, University of Milan, Milan, Italy
CAVALLERO, F., Laboratori di Analisi Chimico-Cliniche e Microbiologia, Ospedale Generale Regionale, Genova, Italy
CECCONI, V., 1a Clinica Malattie Nervose e Mentali, University of Rome, Rome, Italy
ČEPELÁK, V., University Hospital, Plzeň, Research Institute for Pharmacy and Biochemistry, Prague, Czechoslovakia
COCCHERI, S., Department of Angiology and Blood Coagulation of the Regional Hospitals, Bologna, Italy
CORTELLARO, L., Clinica Medica I, University of Milan, Milan, Italy
DAVID, J., Laboratoire Hémostase, Centre Transfusion Sanguine, Toulouse, France
DI PERRI, T., Istituto di Semeiotica Medica, University of Siena, Medical School, Siena, Italy
EGGER, R., Department of Nervous and Mental Diseases, University of Perugia, Medical School, Perugia, Italy
EINSIEDEL-LECHTAPE, H., von, Max-Planck-Institut für Hirnforschung, Abteilung für Allgemeine Neurologie and the Neurological Department at the Municipal Hospital, Köln-Merheim, W. Germany
FAZIO, C., 1 Clinica Malattie Nervose e Mentali, University of Rome, Rome, Italy
FEDERICI, F., Department of Nervous and Mental Diseases, University of Perugia, Medical School, Perugia, Italy
FERRONI, A., Department of Neurophysiopathology, General Regional Hospital, Perugia, Italy
FIELDS, W.S., Neurology-Medical School, University of Texas, Health Science Center, Houston, Texas, USA
FIESCHI, C., Department of Neurology and Psychiatry, University of Siena, Siena, Italy
FRATTOLA, L., Clinica Malattie Nervose e Mentali, University of Milan, Milan, Italy
GAMBA, G., Istituto di Clinica Medica Generale e Terapia Medica, University of Pavia, Pavia, Italy
GENSINI, G.F., Istituto di Patologia Medica II, University of Florence, Medical School, Florence, Italy
GERAUD, G., Service de Neurologie, Hospital Purpan, Toulouse, France
GRIGNANI, G., Istituto di Clinica Medica Generale e Terapia Medica Generale e Terapia Medica, University of Pavia, Pavia, Italy
GRÜN, H., Center of Internal Medicine, Department of Angiology, University of Frankfurt, Frankfurt/Main, W. Germany

GUIRAUD, B., Service de Neurologie, Hospital Purpan, Toulouse, France
HOHMANN, S., Monash University, Department of Medicine and Melbourne University Anatomy, Melbourne, Australia
KADATZ, R., Pharmacological Laboratories, Dr. Karl Thomae GmbH, Biberach/Riß, W. Germany
KAUCHTSCHISCHVILI, G., Clinica Malattie Nervose e Mentali, University of Pavia, Pavia, Italy
KAZAMA, M., Department of Medicine, Teikyo University, School of Medicine, Tokyo, Japan
KRZYWANEK, H.J., Center of Internal Medicine, Department of Angiology, University of Frankfurt, Frankfurt/Main, W. Germany
KUCHAR, M., University Hospital, Plzeň, Research Institute for Pharmacy and Biochemistry, Prague, Czechoslovakia
LAGHI PASINI, F., Istituto di Semeiotica Medica, University of Siena, Medical School, Siena, Italy
LENTINI, S., Pio Istituto di S. Spirito E OO RR - VII Padiglione del Policlinico, Center for the Study and Treatment of Arterial Hypertension and Kidney, Rome, Italy
LENZE, G.L., Institute of Nervous and Mental Diseases, University of Siena, Medical School, Siena, Italy
LEONE, G., Università Cattolica del Sacro Cuore, Istituto di Clinica Medica Generale e Terapia Medica, Rome, Italy
MANNUCCI, P.M., Haemophilia and Thrombosis Centre Angelo Bianchi Bonomi, University of Milan, Milan, Italy
MARCHIONNI, F., Department of Nervous and Mental Diseases, University of Perugia, Medical School, Perugia, Italy
MARSHALL, J., Institute of Neurology, National Hospital for Nervous Diseases, London, Great Britain
MEYER, J.S., Department of Neurology, Baylor College of Medicine and the Baylor-Methodist Center for Cerebrovascular Research, Houston, Texas, USA
MERRILLEES, N.C.R., Monash University, Department of Medicine and Melbourne University Anatomy, Melbourne, Australia
MURATOVA, J., University Hospital, Plzeň, Research Institute for Pharmacy and Biochemistry, Prague, Czechoslovakia
NACHMAN, R.L., New York Hospital, Cornell Medical Center, New York, USA
NAPPI, G., Clinica Malattie Nervose e Mentali, University of Pavia, Pavia, Italy
NERI SERNERI, G.G., Istituto di Patologia Medica II, University of Florence, Medical School, Florence, Italy
NISHIZAWA, I., Department of Medicine, Teikyo University, School of Medicine, Tokyo, Japan
OPPERMANN, B.L., Monash University, Department of Medicine and Melbourne University Anatomy, Melbourne, Australia
PAOLETTI, R., Istituto di Farmacologia e Farmacognosia, University of Milan, Milan, Italy
PARETI, F.I., Haemophilia and Thrombosis Centre Angelo Bianchi Bonomi, University of Milan, Milan, Italy
PENCHINI, G., Department of Nervous and Mental Diseases, University of Perugia Medical School, Perugia, Italy
PIRRO, Cr., Pio Istituto di S. Spirito E OO RR - VII Padiglione del Policlinico, Center for the Study and Treatment of Arterial Hypertension and Kidney, Rome, Italy

PISSARI, F.M., 1a Clinica Malattie Nervose e Mentali, University of Rome, Rome, Italy
POGLIANI, E., Clinica Medica I, University of Milan, Milan, Italy
POLITI, A., Clinica Medica I, University of Milan, Milan, Italy
POLLI, E.E., Clinica Medica I, University of Milan, Milan, Italy
PRENCIPE, M., 1a Clinica Malattie Nervose e Mentali, University of Rome, Rome, Italy
PRESLOCK, M.M., Neurology-Medical School, University of Texas, Health Science Center, Houston, Texas, USA
PROST, R.J., Laboratoire de Recherche sur la Thrombose Expérimentale, Service du Professeur Bilski-Pasquier et Service d'Hématologie, Pr. J. Bousser, Hôpital de l'Hotel Dieu, Paris, France
RANDAZZO, A., Divisione Medica Urgenza, Ospedale Policlinico, Milan, Italy
RASCOL, A., Service de Neurologie, Hospital Purpan, Toulouse, France
ROBINSON, P.M., Monash University, Department of Medicine and Melbourne University Anatomy, Melbourne, Australia
ROUBAL, Z., University Hospital, Plzeň, Research Institute for Pharmacy and Biochemistry, Prague, Czechoslovakia
SAMANA, M., Laboratoire de Recherche sur la Thrombose Expérimentale, Service du Professeur Bilski-Pasquier et Service d'Hématologie, Pr. J. Bousser, Hôpital de l'Hotel Dieu, Paris, France
SANTAMBROGIO, S., Divisione Medica Urgenza, Ospedale Policlinico Milano, Milan, Italy
SCOTTI, G., Clinica Malattie Nervose e Mentali, University of Milan, Milan, Italy
SIGNORINI, E., Department of Neuroradiology, General Regional Hospitals, Perugia, Italy
SILVESTRINI, E., Istituto di Patologia Medica II, University of Florence, Medical School, Florence, Italy
SIMARD, D., Département des Sciences Neurologiques, Hôpital de l'Enfant-Jésus, Quebec, Canada
SWANK, R.L., Medical Research Foundation of Oregon and Advancement Fund of the University of Oregon, Medical School, Oregon, USA
TARDIOLO, C., Health Service Center, General Regional Hospital, Perugia, Italy
TESTA, C., I. Neurosurgical Division of the Regional Hospitals, Bologna, Italy
ULUTIN, O.N., Division of Hematology and Hemostasis Research Unit, Internal Clinic of Cerrahpasa, Medical Faculty of Istanbul University, Istanbul, Turkey
ULUTIN, S.B., Division of Hematology and Hemostasis Research Unit, Internal Clinic of Cerrahpasa, Medical Faculty of Istanbul University, Istanbul, Turkey
VITTORIA, A., Istituto di Semeiotica Medica, University of Siena, Medical School, Siena, Italy
VOLANTE, F., Department of Neurology and Psychiatry, University of Siena, Siena, Italy
WELCH, K.M.A., Department of Neurology, Baylor College of Medicine and the Baylor-Methodist Center for Cerebrovascular Research, Houston, Texas, USA

WILLIAMS, I.M., Monash University, Department of Medicine and
    Melbourne University Anatomy, Melbourne, Australia
ZANETTE, E., Department of Neurology and Psychiatry, University
    of Siena, Siena, Italy
ZIEMEN, I., Center of Internal Medicine, Department of Angiology, University of Frankfurt, Frankfurt/Main, W. Germany
ZÜLCH, K.J., Max-Planck-Institut für Gehirnforschung, Köln,
    W. Germany

Introductory Remarks

C. Fazio and A. Agnoli

Within the sphere of cerebrovascular diseases, both in the anatomo-pathologic and in the physiopathologic fields, different pathogenic interpretations have been advanced which have not always been capable of withstanding criticism.

In the therapy of the acute phase of apoplexy the results achieved have failed to live up to our expectations since, up to the present, they have not produced advantages proportionate to the amount of knowledge acquired over the last few years. We are, in fact, obliged to recognize that our main therapeutic benefits are not attributable to specific intervention on the ischemic or cerebral hemorrhagic focus but rather to refinement in the techniques of general medical and paramedical assistance, i.e., "nursing."

Thus, when the physician today employs a precise therapy - the problem being to decide whether a precise therapy does, in fact, exist - he can at most restrict the spread of the cerebral lesion without, however, being able to avoid invalidating neurologic results in the patient.

Rehabilitative therapy plays a fundamental role, partially correcting the residual effects of the acute phase; but, however, brilliant the results achieved may be, the individual will always be a person with cerebral injury with the neurologic and physiologic consequences that this implies.

It thus follows that, at present, the only valid attitude to adopt in the case of cerebrovascular disorders consists in prophylactic therapy.

Following this line of thought, numerous epidemiologic investigations have been conducted from which we have been able to single out schematically two fundamental phases in the pathogenic iter of apoplexy: an initial phase lasting over a period of many years during which predisposing factors are at work - the so-called risk factors and a second phase in which a "trigger" factor sets off the acute event.

The predisposing factors, among which we recall hypertension, diabetes, and dyslipidemia, are the same as those found in myocardial infarction and point toward a basic pathologic process: atherosclerosis.

However, it is useful to remember that in the brain the predisposing action of these factors is aggravated by the presence of congenital anomalies of the cerebrovascular tree among which we particularly recall those of the circle of Willis because of their frequency and their documented importance.

With regard to the trigger factors, hemodynamic aspects have been evaluated from time to time or mechanical events such as thrombosis and emboli.

Even though this argument is far from being settled, today we tend to integrate the two theories attributing a trigger role to embolic phenomena and evaluating the hemodynamic situation as a factor conditioning the degree and extension of the ischemic lesion.

During the last few years numerous researchers have considered the role played by platelets in atherosclerotic disease, particularly in the determinism of the microemboli from the atherosclerotic plaques.

At this point, the availability both of adequate methods of investigation and drugs capable of interfering with platelet aggregation have given rise to a series of problems, the solution of which is closely related to the "prophylaxis" of apoplexy.

The neurologist, therefore, finds himself obliged to answer a series of questions which may be listed as follows:

1. Do clinical situations exist in which there is an increase in platelet adhesiveness and platelet aggregation?

2. Is this increase determinate in the genesis of emboli and in the evolution of the clinical picture of diseases of thrombotic genesis?

3. Are the antiaggregation therapies capable of preventing cerebral ischemic episodes?

At this point the neurologist has had to ask the assistance of hematologists and biologists, the only ones capable of helping him answer these difficult questions.

In this spirit, this symposium was organized, in which specialists in different branches of medicine have met round the same table.

We are of the opinion that the findings presented here will increase our knowledge and, above all, provide us with useful practical guidelines in the preventive therapy of the cerebrovascular diseases.

Methodology and Physiology

Chairmen: G.V.R. Born and K. Breddin

The Platelet as an Inflammatory Cell

R.L. Nachman

The primary hemostatic response involving the platelet-blood vessel wall reaction in mammals resembles, to a significant extent, the basic cellular defense mechanisms of more primitive forms such as invertebrates. Mammalian primary hemostasis appears to have retained a phylogenetic vestige of primitive leukocyte behavior (Salt, 1970). In this evolutionary sense, the platelet might be thought of as a special form of leukocyte. Human platelets contain intracellular granules similar to the classic lysosomes of polymorphonuclear leukocytes (Marcus et al., 1966) and contribute to the inflammatory response accompanying tissue injury by releasing constituents which may alter blood vessel reactivity. Platelets have long been known to contain vasoactive amines which are released upon aggregation. Platelets interact with many types of particulate matter including bacteria and viruses and immune complexes which lead to the release of intracellular constituents, including serotonin (Mustard and Packham, 1970), prostaglandins (Smith and Wallis, 1971), cationic proteins (Packham et al., 1966), and various proteolytic enzymes such as cathepsin A (Nachman and Ferris, 1968), elastase (Robert et al., 1970), and collagenase (Chesney et al., 1974).

Early inflammatory changes in damaged tissues involve the release of various biologically active materials including a group of cationic proteins from polymorphonuclear leukocytes (Spitznagel and Zeya, 1964). This group of proteins mediates an entire spectrum of activities including antibacterial activity (Spitznagel and Zeya, 1964), increased vascular permeability (Seeger and Janoff, 1966), fever production (Herion et al., 1966) and anticoagulant activity (Saba et al., 1967). Cationic lysosomal proteins are important contributors to the inflammatory process. Platelet cationic proteins increase vascular permeability by several mechanisms including induction of endogenous histamine release as well as delayed chemotaxis (Nachman et al., 1970, 1972). Some rabbit platelet cationic proteins are also bactericidal (Weksler and Nachman, 1971). Platelets aggregate in intact vessels at site of altered or disturbed blood flow such as arterial bifurcations. Vascular permeability enhancing factors released from these aggregates lead to protein and lipid deposition at focal vascular sites which may eventually contribute or lead to atherosclerosis. Recent studies suggest that, in addition, platelets release growth factors which stimulate the proliferation of medial smooth muscle cells (Ross et al., 1974). Platelets also interact with the complement system. One mechanism involves the

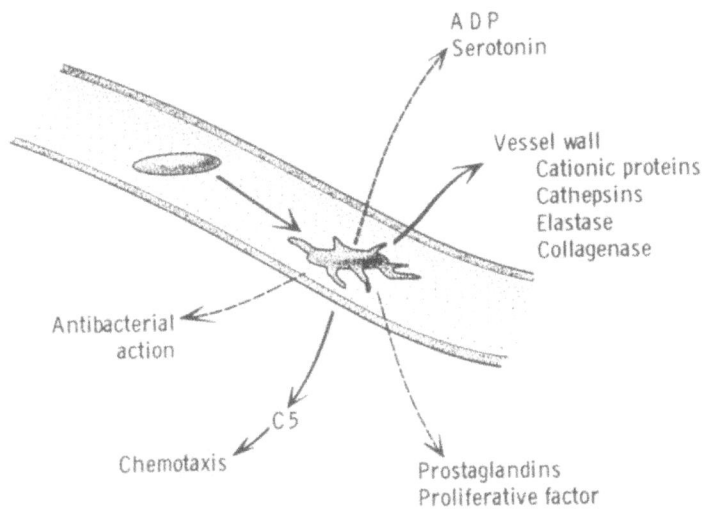

Fig. 1. Potential mechanisms whereby the platelet may contribute to inflammatous responses

release from activated platelets of an enzyme which activates C5 leading to the generation of chemotactic activity which initiates leukocyte accumulation in perivascular sites (Weksler and Coupal, 1973). Complement activation via the properdin pathway can also initiate platelet aggregation and release (Siraganian, 1972). Figure 1 summarizes in schematic fashion the potential mechanisms whereby the platelets may contribute to inflammatory responses. In a general sense, there is evidence which suggests that platelets do participate in inflammatory responses (Packham et al., 1966). These cells accumulate in blood vessels adjacent to areas of inflammation and tissue damage. Human kidney transplant rejections may be associated with the formation of platelet aggregates in renal vessels. A similar phenomenon may characterize the kidney changes seen in the generalized Schwartzman reaction. In rabbits, the several mechanisms for inducing immune vasculitis can be prevented if the animals are first rendered thrombocytopenic.

Two potential pathogenetic mechanisms (Fig. 2) should be considered which might explain the role of platelets in cerebrovascular disorders. (A) Underlying vascular lesions such as ulcerated atheromatous plaques could provide the initiating damaged surface which eventually leads to the formation of a platelet plug. Embolization from these primary sites might then lead to distal occlusion in smaller peripheral vessels. (B) In the absence of overt underlying vessel damage, various intravascular stimuli of a systemic nature could lead to circulating intravascular platelet aggregates which produce focal vascular damage secondary to the release of intracellular constituents. It is thus possible that platelets initiate as well as sustain thrombotic events in cerebral vessels. These considerations suggest a rationale for the use of antiplatelet agents in some forms of cerebrovascular disease.

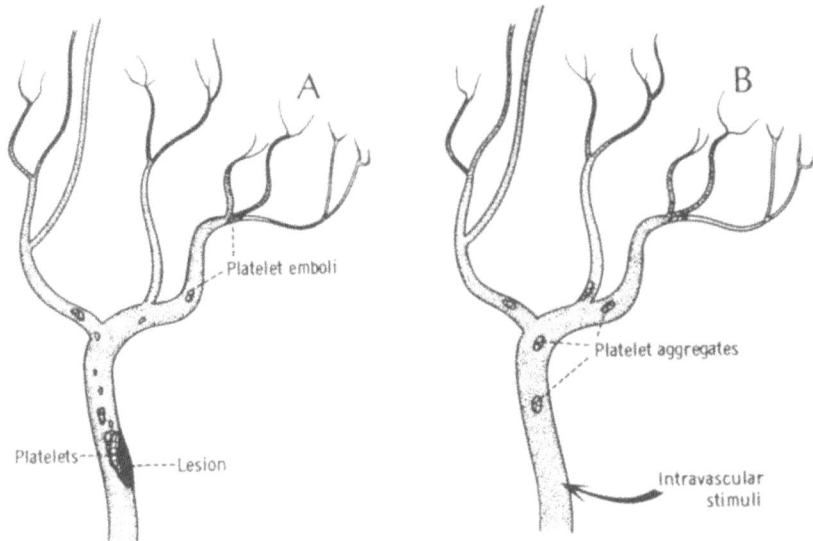

Fig. 2. Potential role of the platelet in cerebral vascular occlusive disease. A Vessel wall damage leads to platelet plug formation. B Intravascular stimuli lead to platelet aggregates which damage vessels due to release of inflammatory mediators

References

Chesney, C. McI, Harper, E., Colman, R.W.: Human platelet collagenase. J. clin. Invest. 53, 1647 (1974)

Herion, J.C., Spitznagel, J.K., Walker, R.I., Zeya, H.I.: Pyrogenecity of granulocyte lysosomes. Amer. J. Physiol. 211, 693 (1966)

Marcus, A.J., Zucker-Franklin, D., Safier, L.B., Ullman, H.L.: Studies on human platelet granules and membranes. J. clin. Invest. 45, 14 (1966)

Mustard, J.F., Packham, M.A.: Factors influencing platelet function: adhesion, release and aggregation. Pharmacol. Rev. 22, 97 (1970)

Nachman, R.L., Ferris, B.: Studies on human platelet protease activity. J. clin. Invest. 47, 2530 (1968)

Nachman, R.L., Weksler, B., Ferris, B.: Increased vascular permeability produced by human platelet granule cationic extract. J. clin. Invest. 49, 274 (1970)

Nachman, R.L., Weksler, B., Ferris, B.: Characterization of human platelet vascular permeability-enhancing activity. J. clin. Invest. 51, 549 (1972)

Packham, M.A., Anderson, E.E., Mustard, J.F.: Response of platelets to tissue injury. Biochem. Pharmacol. (Suppl.) 17, 171 (1966)

Robert, B., Szigati, M., Robert, L., Legrand, Y., Legrand, G., Caen, J.: Release of elastolytic activity from blood platelets. Nature (Lond.) 227, 1248 (1970)

Ross, R., Glomset, J., Karuja, B., Harker, L.: A platelet dependent serum factor that stimulates the proliferation of arterial smooth muscle cells in vitro. Proc. nat. Acad. Sci. (Wash.) 71, 1207 (1974)

Saba, H.I., Roberts, H.R., Herion, J.C.: The anticoagulant activity of lysosomal cationic proteins from polymorphonuclear leukocytes. J. clin. Invest. 46, 580 (1967)

Salt, G.: The cellular defense reactions of insects. In: Cambridge Monographs on Experimental Biology 68. Cambridge, England: Cambridge University Press 1970

Seeger, W., Janoff, A.: Mediators of inflammation in leukocyte lysosomes VI. J. exp. Med. 124, 833 (1966)

Siraganian, R.P.: Platelet requirement in the interaction of the complement and clotting systems. Nature (New Biol.) 239, 208 (1972)

Smith, J.B., Wallis, A.L.: Aspirin selectively inhibits prostaglandin production in human platelets. Nature (New Biol.) 231, 235 (1971)

Spitznagel, J.K., Zeya, H.I.: Basic proteins and leukocyte lysosomes as biochemical determinants of resistance to infection. Trans. Ass. Amer. Phycns. 77, 126 (1964)

Weksler, B., Nachman, R.: Rabbit platelet bactericidal protein. J. exp. Med. 134, 1114 (1971)

Weksler, B.B., Coupal, C.E.: Platelet dependent generation of chemotactic activity in serum. J. exp. Med. 137, 1419 (1973)

Platelet Aggregation in the Pathogenesis of Cerebrovascular Disorders

G.V.R. Born

Platelets in the Formation of a Haemostatic Plug

Normal spontaneous haemostasis depends primarily on the formation of a haemostatic plug which fills any gap in the continuity of an injured vessel and covers it like a capsule. At first this haemostatic plug consists only of platelets which have somehow been arrested in their circulation by the injury. A fundamental problem is, therefore, the nature of the signal from the injury site to the platelets.

Contact with a vascular lesion causes a remarkably rapid change in platelets which makes them adhere and cause other platelets chancing to touch them to adhere also. Thus, the formation of a haemostatic plug involves first *adhesion* of platelets to other tissues, followed very rapidly by the *aggregation* of platelets to each other. Initially the platelets adhere loosely to each other so that the plasma and cells continue to pass out of the vessel. Within a few minutes the platelets become packed much more closely, indeed almost as closely as is theoretically possible (Born and Hume, 1967) so that the plug becomes more effective in its haemostatic function. The mechanism of platelet aggregation is probably different in these two phases (see below); whether either mechanism is also responsible for the initial adhesion of platelets to an abnormal vessel wall is still uncertain.

Platelet Adhesion

Platelets do not adhere to normal endothelial cells but do adhere to gaps between endothelial cells, even when the gaps are produced in normal vessels by the action of pharmacologic agents (Tranzer and Baumgartner, 1967). The gaps expose vascular basement membrane to which platelets apparently tend to adhere. Whether this tendency has any physiologic or pathologic significance is not yet known. In small venules, histamine and some other vasoactive agents cause contraction of the endothelial cells with the appearance of gaps between them (Majno et al., 1969). These gaps presumably account for the great increase in the permeability of these vessels to proteins and other constituents of plasma in inflammation. The adhesion of platelets in such gaps might have the effect of diminishing vascular permeability.

When a blood vessel is injured, the exposed subendothelial tissues are covered by a layer of adhering platelets. This adhesion is so rapid (Hugues, 1953) that its mechanism provides interesting problems. The probability of a platelet adhering presumably increases with both the closeness and the duration of contact between the platelets and the site. These factors have to be related to biochemical mechanisms capable of providing sufficient attraction to hold a platelet against the shear forces of the flowing blood. No forces are known which could attract platelets toward an adhesion site over distances greater than a few hundred angstroms. That means that adhesion depends on collision and poses the problem in another way, i.e., what is the collision rate between circulating platelets and vessel wall and what fraction of the collisions is successful in making platelets stick? Measurements of successful collisions have provided estimates of the adhesiveness of other cell types in vitro (Curtis, 1967) and could be made with platelets but the in vivo situation poses much more difficult problems.

As might be expected on the basis of the foregoing, the trauma required to make platelets adhere to a vessel may be extremely slight; indeed, the question has still to be answered as to what constitutes the least abnormality for adhesion to occur. Platelets have been found to adhere in endothelial gaps in apparently normal venules situated at some distance from an experimentally injured vessel. The application of adenosine diphosphate (ADP) in small amounts by iontophoresis onto the outside of an apparently undamaged venule in the hamster cheek pouch causes the adhesion of platelets in the lumen within a few seconds (Begent and Born, 1970) in the absence of gaps between the endothelial cells.

When vessel walls are injured, or diseased, mural thrombi do not form unless the site has been denuded of endothelium. Which tissue constituents normally separated from circulating platelets by intact endothelium could bring about their adhesion when a lesion makes contact possible? Up to now, the answer depends on inferences from in vitro experiments. These have shown that the adhesiveness of platelets is markedly increased by several substances which are present in the normal vessel wall; they include collagen fibres, the catecholamines adrenaline and noradrenaline, and particularly ADP, a constituent of all cells which tends to be increased by any form of cellular injury. Another potent agent is thrombin, the formation of which may be initiated on lesions by tissue and/or plasma prothrombinase. The effects of thrombin and collagen depend at least in vitro, on ADP released from the platelets themselves. This suggests that, in vivo, when a circulating platelet is thrown into contact with collagen it induces the platelet to adhere until the ADP mechanism (see below) comes into play. In vitro, collagen produces platelet aggregation only after a lag of several seconds. In vivo, by contrast, the process seems very much faster. This discrepancy is not yet understood, although other agents such as adrenaline, which may be released or produced at the same time, may accelerate aggregation in vivo as in vitro. The effectiveness of these agents is presumably diminished, however, by the blood flow which dilutes them.

What happens then is determined in different vessels by their blood flow. In a large artery, e.g., the aorta, an area from which endothelium is removed is covered by only a single layer of platelets; these are gradually replaced by granulocytes. In a small artery similar damage to endothelium is followed by adhesion of platelets not only to the denuded area but also to each other until the aggregates may block the lumen and so arrest the blood flow altogether. This difference can be explained by assuming that in large arteries the blood flow removes agents released from the damaged wall before their concentration is sufficient to affect platelets, even those that come in contact with the lesion and that this does not happen and that in small vessels with slow flow; however, this has not yet been established.

Platelet Aggregation

The adhesion of platelets to each other is called *platelet aggregation* unless the reaction is immunological, then it is called *platelet agglutination*. After an injury to all but the largest blood vessels, adhesion of platelets to the damaged wall is followed very rapidly by the formation of platelet aggregates on the adhering layer. The initial adhesion process cannot be validly imitated outside living vessels so that the mechanism is difficult to investigate experimentally. The process of aggregation, on the other hand, can be observed in vitro by various methods in which it appears to operate much as in vivo. Much more is known, therefore, about aggregation than about adhesion.

The method which has provided most information about aggregation (Born, 1962a,b) depends on the continuous measurement of changes in optical density of a suspension of platelets either in plasma or in physiologic saline solutions. When platelets aggregate the optical density decreases and when the aggregates disperse the optical density increases. This has made possible quantitative measurements of the aggregation process and provided much information about its mechanism and about promoting and inhibiting agents (for reviews see Born, 1970a,b; Mills, 1969; Mustard and Packham, 1970).

Aggregating Agents. In vitro, human platelets are caused to aggregate by adenosine diphosphate (ADP), adrenaline, 5-hydroxytryptamine, thrombin, collagen, and certain fatty acids, as well as by several other agents less immediately relevant to haemostasis. Each agent must have the ability to react initially with some kind of receptor site on the platelet surface membrane. With most, if not with all, this primary reaction apparently induces the formation in and/or release of ADP from platelets, and apparently it is this which causes the changes in surface properties of platelets resulting in their aggregation. This conclusion is based on (1) the demonstration of the release of ADP from platelets by the other agents, (2) the inhibition of aggregation by enzymes which remove ADP from the plasma, and (3) inhibition by specific antagonists of the effect of ADP (for review see Haslam, 1968).

Because aggregation by other agents apparently involves ADP, it is convenient to describe the effects of added ADP on platelets and then to point out any differences in the effects of the other agents.

Effects of ADP. Change in Shape. When suspensions of platelets in plasma are stirred, the record of optical density shows small, nearly uniform oscillations. When ADP is added, the oscillations disappear almost immediately; at the same time there is a rapid increase in the optical density of the plasma, amounting to a decrease of a few per cent in light transmission. The optical changes indicate the first and probably the only effect of ADP itself on platelets, namely to change their shape from smooth discs to spheres with pseudopodia of varying lengths protruding from the surface (Macmillan and Oliver, 1965). This effect, already referred to when describing platelet adhesion in vivo, has some interesting properties: (1) It appears to be due to a reaction between ADP and a specific receptor on the platelet surface; the reaction does not require the cofactors calcium ions and fibrinogen which are required for subsequent aggregation. (2) It is very rapid and has an extraordinarily high temperature coefficient (Born, 1970c) suggesting an underlying reaction with a high activation energy such as, for example, polymerization changes in proteins. (3) Some, but not all, substances most closely related to ADP inhibit the effect: adenosine triphosphate (ATP) inhibits by direct competition at the ADP receptor (Macfarlane, 1974); adenosine monophosphate (AMP) does not inhibit at all whereas adenosine inhibits also but by a different mechanism (see below).

The opposing effects of ADP and ATP, as well as the inactivity of AMP, suggest that the action of these nucleotides on platelets may be similar to their actions on the contractile protein system of muscle (Szent-Gyorgyi, 1968). These similarities and the inhibition of platelet functions by an antibody against the contractile protein *thromboasthenin* in platelets (Nachman et al., 1960) is the best evidence so far that the initial reaction of ADP involves these proteins, some of which must be presumed to be localized in the surface membrane of platelets. (4) The change in shape is *not* accompanied by a change in volume of the platelets but with some rearrangement of their contents which can be seen on electromicrographs. A bundle of microtubules which encircles the platelets just beneath their outer membrane appears contracted and the cytoplasmic granules become concentrated within it (White, 1968). Changes in shape are brought about also by other aggregating agents with the exception of collagen which causes aggregation to begin only after a considerable lag period.

The pseudopodial extensions which appear during the shape change increase the effective diameter of the platelets so that the function of the shape change could well be to increase the probability of contact between platelets in the flowing blood and a lesion in the vessel wall. The negative charge density on a cell surface is decreased at the tips of thin pseudopodia or cytoplasmic extrusions, so that such extrusions diminish the effect of electrostatic repulsion between cells and increase the effect of other forces making for adhesion (Bangham, 1964).

First Phase of Aggregation. The optical effect of the shape change is followed by an effect in the opposite direction, i.e., an increase in light transmission which, for the most part, is also much larger. This part of the record is the resultant of several simultaneous processes in which single platelets adhere to each other to form small aggregates and to aggregates already formed, and in which small aggregates adhere to each other to form larger ones. So far, there is too little information for the construction of mathematical models of these events. It is known that the stages in which the aggregates are small, i.e., containing less than ten platelets, are passed through very rapidly and that throughout this phase the platelets adhere to each other rather loosely (Born and Hume, 1967). The looseness is also seen on electron microscopic pictures of small platelet aggregates adhering to injured vessels in vivo.

Disaggregation. The first phase of aggregation by ADP, just described, is completely reversible and the dispersion of the aggregates is shown by an increase in the optical density of the plasma. Aggregation of human platelets reverses spontaneously when caused by low concentrations of ADP; higher concentrations may induce the second phase of aggregation (see below) which obscures and delays disaggregation. Disaggregation is caused by the breakdown of ADP in plasma by enzymic reactions which simultaneously produce inhibitory derivatives (see below); and change in the platelets whereby the effectiveness of ADP on them is diminished.

Essential Cofactors. At least two other substances are essential for aggregation by ADP, namely ionized calcium and fibrinogen (Born and Cross, 1964; Cross, 1964). The velocity of aggregation increases with the concentration of *calcium* up to about 1 mM; higher concentrations inhibit aggregation. Therefore, plasma containing EDTA as anticoagulant cannot be used for investigating platelet aggregation, whereas citrate in the concentration usually added interferes little.

Fibrinogen also is essential for physiologic aggregation of platelets in plasma. In vitro, aggregation velocity increases with fibrinogen concentration. In patients with congenital afibrinogenaemia, platelet aggregation is greatly slowed but not abolished (Inceman et al., 1966); the reason seems to be that, even in the severest cases, some fibrinogen remains associated with the platelets themselves.

Potentiating Agents. Aggregation of platelets by ADP is potentiated by *potassium* ions (Born and Cross, 1964) but neither these nor sodium ions are essential.

The effect of ADP is greatly increased by *adrenaline* (Ardlie et al., 1966), even in very low concentrations. This potentiation shows itself both as an acceleration of primary aggregation and as a diminution in the ADP concentration required to initiate the second phase of aggregation in which aggregating substances are released from the platelets. These observations are given clinical significance by the appearance of such concentrations of adrenaline in the plasma of people during stress who seem particularly liable to suffer from thrombotic episodes. Adrenaline also potentiates aggregation by thrombin and by collagen (Thomas, 1968) but these effects may, at least in part, represent potentiation of ADP which these agents release from platelets.

Second Phase of Aggregation and the Release Reaction. The optical method resulted in the discovery that critical concentrations of ADP added to citrated plasma of man (Macmillan, 1966) or guinea pig (Constantine, 1966) at $37^{\circ}C$ cause two distinct phases of decrease in optical density. The second phase is associated with the release of ADP from the platelets themselves so that its concentration in the plasma may increase up to seven times (Mills et al., 1968). Other substances released at the same time include ATP and 5-hydroxytryptamine as well as platelet factor 3 which accelerates coagulation of plasma (Hardisty and Hutton, 1966). This release reaction can be induced also by thrombin (Grette, 1962) or adrenaline, and the latter diminishes the concentrations of other agents required to initiate the reaction.

The decrease in optical density during this phase of aggregation is caused by the contraction of aggregates already formed rather than by the formation of larger aggregates. There is evidence that this contraction also occurs in vivo where it presumably increases the effectiveness of the platelet plug as a barrier against further blood loss.

Inhibition of Platelet Aggregation

The successive response of platelets to ADP and to other agents can be inhibited by a variety of means. The photometric method has been particularly successful in the discovery of substances capable of preventing or reversing aggregation. Further research may provide a drug effective against platelet thrombosis without, at the same time, interfering with the haemostatis function of platelets. That this can be achieved in principle was shown by infusing adenosine intravenously into rabbits when it accelerated the dispersal of intravascular platelet thrombi without prolonging the bleeding time (Born et al., 1964).

From the proceding descriptions it is clear that inhibition of the initial shape change should also prevent both phases of aggregation; inhibition of the first phase should also prevent the second phase; and that inhibition of the second phase without inhibition of the first ought to arrest the growth of platelet aggregates. All these possibilities have been demonstrated experimentally.

Platelet aggregation can be inhibited irreversibly or reversibly. Irreversible inhibition is produced by exposing platelets to grossly unphysiologic conditions or to inhibitors of metabolism; these effects are of little interest. More interesting is the irreversible inhibition produced by acetylsalicylic acid; this will be discussed later.

Reversible inhibition can be achieved by conditions which cause no significant damage to the platelets, shown by normal functions in vitro and/or by normal survival in vivo. Thus, platelet aggregation is inhibited reversibly by (1) lowering the temperature; (2) lowering the pH by not more than about one unit; (3) removing or inactivating the cofactors in plasma; (4) eliminating ADP or the other aggregating agents from plasma; (5) antagonizing the effects of ADP or of the other aggregating agents by specific inhibitors; and (6) the inactivation or deficiency of specific receptors on the platelets. The last two are those of greatest potential interest for clinical medicine.

Specific Inhibitors. The change in shape and first phase of aggregation caused by ADP are inhibited by a number of related nucleotides and nucleosides (Born and Cross, 1963) including ATP (Macfarlane, 1974).

Particularly interesting is the inhibition by adenosine and several closely related synthetic nucleosides, including 2-chloroadenosine which is about ten times more potent than adenosine itself (Born, 1964). Inhibition by these substances increases considerably during a short period after their addition to platelet-rich plasma; the reason for this is not yet clear. The uptake of adenosine by platelets can be almost completely prevented by certain other drugs without diminishing its inhibitory action on aggregation (Born and Mills, 1969). This observation is not compatible with the hypothesis (Rozenberg and Holmsen, 1968) that inhibition by adenosine depends on its phosphorylation by platelet ATP, unless that ATP is situated in the outer membrane itself. This suggestion (Born, 1970c) implies that membrane ATP is essential for aggregation, presumably through involvement in the initial shape change.

The adenosine analogues that inhibit platelet aggregation produce vasodilation causing hypotension (Born et al., 1965). The nucleotide 2-methylthio-adenosine 5'-monophosphate has no noticeable vasodilator effect; this reopens the possibility that a compound related to ADP may possess the properties required from an inhibitor for clinical use.

Both the shape change and the first phase of aggregation are strongly inhibited by prostaglandin E_1 (Kloeze, 1969). This has provided another clue to the mechanism of inhibition. The prostaglandin accelerates the formation in platelets of cyclic 3', 5'-AMP (Zieve and Greenough, 1969) by activating adenylate cyclase in the cell membrane. The resulting activation of phosphokinase(s) results in the phosphorylation of protein(s) (Born, 1972) from ATP and so provides another basis for the suggestion that these potent inhibitors act by diverting platelet ATP from its function in aggregation.

The second phase of aggregation and the concomitant release reaction are powerfully inhibited by certain drugs which have no effect on primary aggregation. They are phenothiazine and imipramine derivatives (Mills and Roberts, 1967) and nonsteroidal anti-inflammatory drugs including phenylbutazone (Evans et al., 1967) and acetylsalicylic acid or aspirin. The mechanism of these inhibitions is not known. Aspirin in small doses inactivates platelets irreversibly, possibly by an acetylation reaction (Al-Mondhiry et al., 1969). This effect of aspirin may provide a new therapeutic use for it in the prevention of platelet thrombosis. On the other hand, the same effect may account at last for the well-established observation that aspirin treatment causes gastrointestinal bleeding, presumably by preventing platelets from contributing to haemostasis in mucosal erosions.

References

Al-Mondhiry, H., Marcus, A., Spaet, T.H.: Acetylation of human platelets by aspirin. Fed. Proc. 28, 576 (1969)

Ardlie, N.G., Glew, G., Schwartz, C.J.: Influence of catecholamines on nucleotide-induced platelet aggregation. Nature (Lond.) 212, 415 (1966)

Bangham, A.D.: The adhesiveness of leucocytes with special reference to the zeta potential. Ann. N.Y. Acad. Sci. 116, 945 (1964)

Begent, N., Born, G.V.R.: Growth rate in vivo of platelet thrombi produced by iontophoresis of adenosine diphosphate as a function of mean blood flow velocity. Nature (Lond.) (in press, 1976)

5. Born, G.V.R.: Quantitative investigations into the aggregation of blood platelets. J. Physiol. (Lond.) 162, 67 (1962a)

Born, G.V.R.: Aggregation of blood platelets by adenosine diphosphate and its reversal. Nature (Lond.) 194, 927 (1962b)

Born, G.V.R.: The functional physiology of blood platelets. Symp. zool. Soc. Lond. 27, 75 (1970a)

Born, G.V.R.: Platelet pharmacology in relation to thrombosis. In: Thrombosis and Coronary Heart Disease, First Paavo Nurmi Symposium, Porvoo. Basel: Karger 1970b

Born, G.V.R.: Observations on the change in shape of blood platelets brought about by adenosine diphosphate. J. Physiol. (Lond.) 209, 481-511 (1970c)

Born, G.V.R.: Platelet aggregation and cyclic AMP. From "Effects of Drugs on Cellular Control Mechanisms". Rabin, B.R., Fredman, R.B. (eds.). London: Macmillan 1972, pp. 237-257

Born, G.V.R., Cross, M.J.: The aggregation of blood platelets. J. Physiol. (Lond.) 168, 178 (1963a)

Born, G.V.R., Cross, M.J.: Effect of adenosine diphosphate on the concentration of platelets in circulating blood. Nature (Lond.) 196, 974 (1963b)

Born, G.V.R., Cross, M.J.: Effects of inorganic ions and of blood platelets by adenosine diphosphate. J. Physiol. (Lond.) 170, 397 (1964)

Born, G.V.R., Haslam, R.J., Goldman, M., Lowe, R.D.: Comparative effectiveness of adenosine analogues as inhibitors of blood platelet aggregation and as vasodilators in man. Nature (Lond.) 205, 678 (1965)

Born, G.V.R., Honour, A.J., Mitchell, J.R.A.: Inhibition by adenosine and by 2-chloroadenosine of the formation and embolization of platelet thrombi. Nature (Lond.) 202, 761 (1964)

Born, G.V.R., Hume, M.: Effects of the numbers and sizes of platelet aggregates on the optical density of plasma. Nature (Lond.) 215, 1027 (1967)

Born, G.V.R., Mills, D.C.B.: Potentiation of the inhibitory effect of adenosine on platelet aggregation by drugs that prevent its uptake. J. Physiol. (Lond.) 202, 41 (1969)

Constantine, J.W.: Aggregation of guinea-pig platelets by adenosine diphosphate. Nature (Lond.) 210, 162 (1966)

Cross, M.J.: Effect of fibrinogen on the aggregation of platelets by adenosine diphosphate. Thromb. Diath. Haemorrh. 12, 524 (1964)

Curtis, A.S.G.: The Cell Surface: Its Molecular Role in Morphogenesis. London: Academic Press Inc. Ltd. 1967

Evans, G., Packham, M.A., Nishizawa, E.E., Mustard, J.F.: The effect of platelet-collagen reaction and blood coagulation on hemostasis. J. clin. Invest. 46, 1053 (1967)

Grette, K.: Studies on the mechanism of thrombin-catalyzed haemostatic reactions in blood platelets. Acta physiol. scand. 56, Suppl. 195, 5 (1962)

Hardisty, R.M., Hutton, R.A.: Platelet aggregation and the availability of platelet factor 3. Brit. J. Haemat. 12, 764 (1966)

Haslam, R.J.: Biochemical aspects of platelet functions. In: Proc. 12th Congr. int. Soc. Haemat. N.Y. 198 (1968)

Hugues, J.: Contribution à l'étude des facteurs vasculaire et sanguins dans l'hémostase spontanée. Arch. int. Physiol. 61, 565 (1953)

Inceman, S., Caen, J., Bernard, J.: Aggregation, adhesion and viscous metamorphosis of platelets in congenital fibrinogen deficiences. J. Lab. clin. Med. 68, 21 (1966)

Kloeze, J.: Influence of prostaglandins on ADP-induced platelet aggregation. Acta physiol. pharmacol. neerl. 15, 50 (1969)

Macfarlane, D.E.: ATP specifically inhibits ADP effects on blood platelets. Fed. Proc. 33, 269 (1974)

Macmillan, D.C.: Secondary clumping effect in human citrated platelet-rich plasma produced by adenosine diphosphate and adrenaline. Nature (Lond.) 211, 140 (1966)

Macmillan, D.C., Oliver, M.F.: The initial changes in platelet morphology following the addition of adenosine diphosphate. J. Atheroscl. Res. 5, 440 (1965)

Majno, G., Shea, S.M., Leventhal, M.: Endothelial contraction induced by histamine-type mediators. J. Cell Biol. 3, 647 (1969)

Mills, D.C.B.: Platelet aggregation. In: The Biological Basis of Medicine. Bittar, E.E., Bitter, N. (eds.). Vol. 3. London: Academic Press 1969

Mills, D.C.B., Robb, I.A., Roberts, G.C.K.: The release of nucleotides, 5-hydroxy-tryptoamine and enzymes from human blood platelets during aggregation. J. Physiol. (Lond.) 195, 715 (1968)

Mills, D.C.B., Roberts, C.G.K.: Membrane active drugs and the aggregation of human blood platelets. Nature (Lond.) 214, 35 (1967)

Mustard, J.F., Packham, M.A.: Factors influencing platelet function: adhesion, release, and aggregation. Pharmacol. Rev. 22, 97 (1970)

Nachman, R.L., Marcus, A.J., Safier, L.B.: Platelet thrombasthenin: subcellular localisation and function. J. clin. Invest. 46, 1380 (1967)

Rozenberg, M.C., Holmsen, H.: Adenine nucleotide metabolism of blood platelets. II. Uptake of adenosine and inhibition of ADP-induced platelet aggregation. Biochim. biophys. Acta 155, 342 (1968)

Szent-Gyorgyi, A.G.: The role of actin-myosin interaction in contraction. In: Aspects of Cell Motility. London and Cambridge: Cambridge University Press 1968, p. 17

Thomas, D.P.: The role of platelet catecholamines in the aggregation of platelets by collagen and thrombin. Exp. Biol. Med. 3, 129 (1968)

Tranzer, J.P., Baumgartner, H.R.: Filling gaps in the vascular endothelium with blood platelets. Nature (Lond.) 216, 1126 (1967)

White, J.G.: Fine structural alterations induced in platelets by adenosine diphosphate. Blood 31, 604 (1968)

Zieve, P.D., Greenough, III, W.B.: Adenyl cyclase in human platelets: activity and responsiveness. Biochem. biophys. Res. Commun. 35, 462 (1969)

The Mathematical Analysis of the Human Platelet Aggregation Mechanism and Its Clinical Application

T. Abe, J. Nishizawa, and M. Kazama

Studies on platelet aggregation have been promoted remarkably by the aggregometer, which has provided graphic data on the process of this phenomenon. But the curves were modified variously by aggregation-inductor or inhibitory reagents and their concentration, platelet incubation time with inhibitory substance, coexisting protein or peptide, pH, and the ionic strength of the suspension medium as well as the type of apparatus. The analysis and evaluation of the curves also varied depending on the understanding of this process.

We have tried to find some standard criteria to analyze curves on an objective common measure and treat them mathematically in accordance with morphologic observation. So far, some investigators have already proposed certain mathematical equations concerning this process and to some extent, applicable to actual curves; but at times the equations have apparently deviated from the real values, suggesting that some fundamental view of the aggregation mechanism might differ from the actual conditions. From this point of view we proposed a new equation and proved its applicability on real aggregation curves.

Methods and Materials

In this experiment aggregometers of EEL type 169, made by Evans Company, England, ADP of Sigma Company, USA, and collagen of Nippon Hikaku Company, Japan, were used. ADP and collagen were employed at final concentrations of 2.12-21.2 µM/ml for ADP and 0.01-0.1 mg/ml for collagen as noted on each figure. As the standard concentrations of these agents, 2.12 µM/ml and 0.1 mg/ml were selected, respectively, to produce no disaggregation or diphasic pattern.

For blind value of transparency, a basic line was drawn at the condition in which the cuvette hole was plugged with a nontransparent rod, and the transparency of platelet-rich (PRP) and -poor plasma (PPP) was also plotted on the aggregogram paper where PPP transparency was set as 100.

The shape and count of platelets were examined with phase contrast and ordinary microscopes and model DIPS-1 computer of Nippon Telephone Telegram Public Cooperation was employed for calculation.

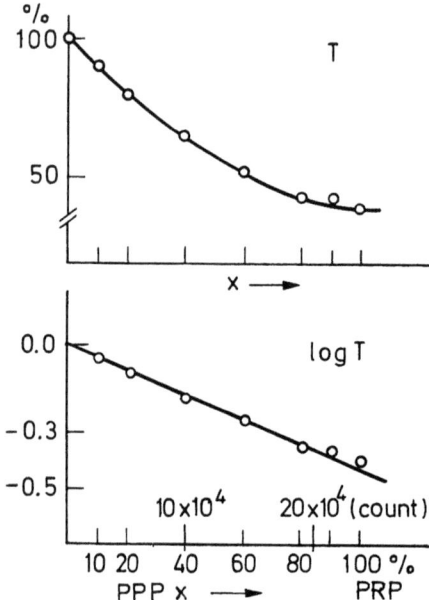

Fig. 1. Correlation between transparency (T) and platelet count in PRP

RA233 of Boehringer Sohn Co. was used in final concentrations of 10 and 50 µg/ml and Flurbiprofen of Nippon Kakenyakukako Co., in concentrations of 0.61, 1.0, 6.1, and 61 µg/ml.

Tested platelets were obtained from human subjects having various diseases such as arteriosclerosis, nephropathies, and hemorrhagic disorders as well as normal adults.

Results

Proposal of a Mathematic Equation for Platelet Aggregation Curves

The availability of Lambert-Beer's law for the transparency T of platelet suspension was examined to find that logT had good correspondence with the number of platelets within the range under 400,000 of its count (Fig. 1). However, when the shift of transparency of PRP was followed after the addition of aggregating agents, the values of logT and log $\frac{T\infty}{T}$ drew concave curves to the upper side, log $\frac{T\infty}{T}$ was selected in order to again be treated logarithmically so that a straight line could be drawn in the form of: log log $\frac{T\infty}{T}$ = -At + B, where T was the highest transparency of the test PRP, A, the slope of this straight line, and B, the constant (Fig. 2).

Proposal of Reaction Mode of Platelet

There are two probable modes of platelet aggregation imagined and the first one is shown in Figure 3, in which each platelet

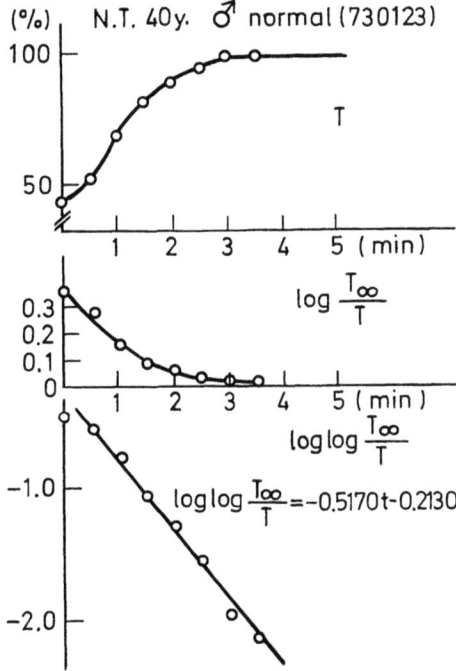

Fig. 2. Analysis of platelet aggregation curve

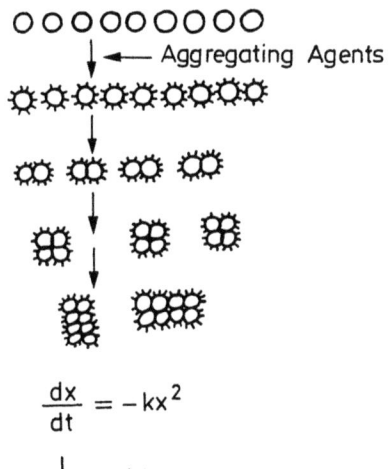

$$\frac{dx}{dt} = -kx^2$$

$$\frac{1}{x} = kt + c$$

Fig. 3. Probable mode of platelet aggregation (Type A) and supposed equation of the reaction

acquires aggregability through ADP and gathers each other at the same rate as expressed with an equation of $\frac{dx}{dt} = -kx^2$, that is, $\frac{1}{x} = kt + c$ (second order reaction).

The second mode imagined (Fig. 4), in which only a few platelets can start to aggregate although each one has aggregability and accelerate themselves so rapidly that very scarce but large aggregates are formed before other platelets can get gathered as

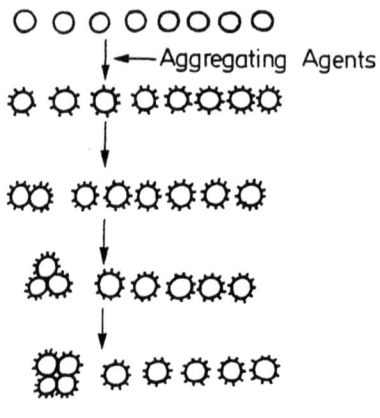

$$\frac{dx}{dt} = kx$$
$$\log x = -kt + c$$

Fig. 4. Probable mode of platelet aggregation (Type B) and supposed equation of the reaction

Fig. 5. Platelets before addition of ADP

expressed with an equation of $\frac{dx}{dt} = kx$; this is, $\log x = -kt + c$ (first order reaction).

Microscopic Examination

The sample in the cuvette was taken before and 5 min after addition of ADP and was put into counting chambers to observe the aggregation state of platelets and to count the number of particles. In the sample before addition of ADP, individual platelets were scattered (Fig. 5), but at 5 min after addition of ADP platelets made only a few large aggregates, some remained scattered (Fig. 6). From these findings the validity of the first order reaction mode was suggested and our double logarithmic equation was also understandable.

Aggregation Curves and Equations with ADP and Collagen

ADP and collagen were added to the same platelet suspension at the final concentrations of 5 µM/ml and 0.5 mg/ml, respectively, and from aggregation curves transparency values were measured

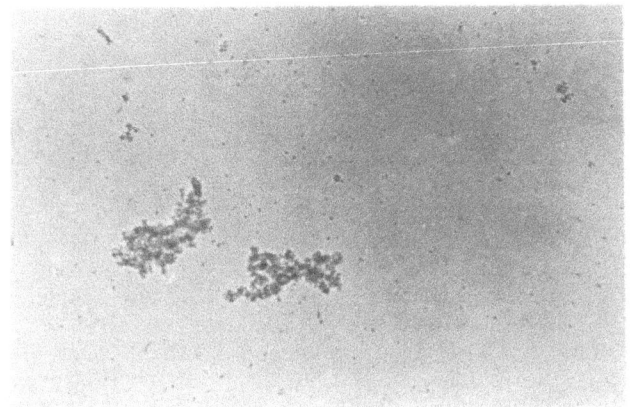

Fig. 6. Platelets 5 min after addition of ADP

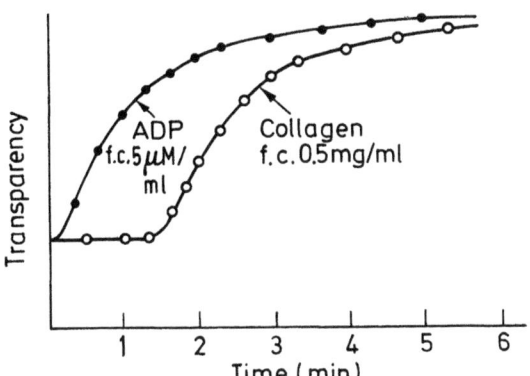

Fig. 7. Platelet aggregation profiles with ADP and collagen (I)

every 30 s until 5 min for ADP and also every 30 s from 2-7 min for collagen after addition of these agents (Fig. 7). These values were put into a computer which was previously fed our double logarithmic equation, having slope A and constant B.

From this calculation two lines were obtained and the collagen curve had a certain refractory period, its intersecting point with the ordinate was negative whereas that of ADP curve was positive, and the slope of the latter was steeper than the former (Fig. 8).

Aggregation Curves of Different Count of Platelets

Platelet aggregation curves of PRP with different counts (10, 15, 26, and 34 × 10^4) of platelets were drawn to give different transparency, and although the higher count of platelets gave the lower B value in the calculation of these curves, the difference of slope of the lines was within the range of measuring error (Fig. 9). This meant that the shift of transparency was changed by the number of platelets, but its changing ratio to the initial number of platelets was similar to each other in the course of aggregation; in other words, slope A showed some proper aggregating character of the majority of platelets.

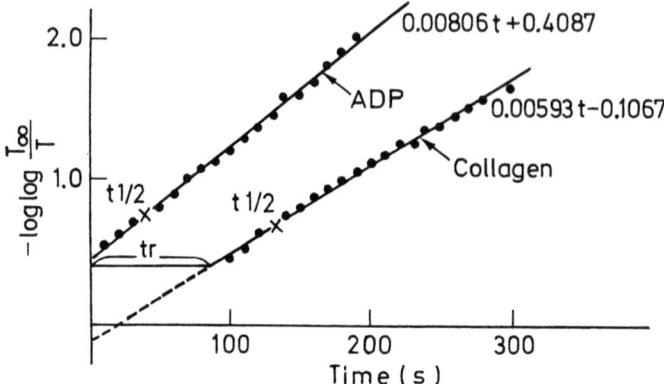

Fig. 8. Platelet aggregation profiles with ADP and collagen (II)

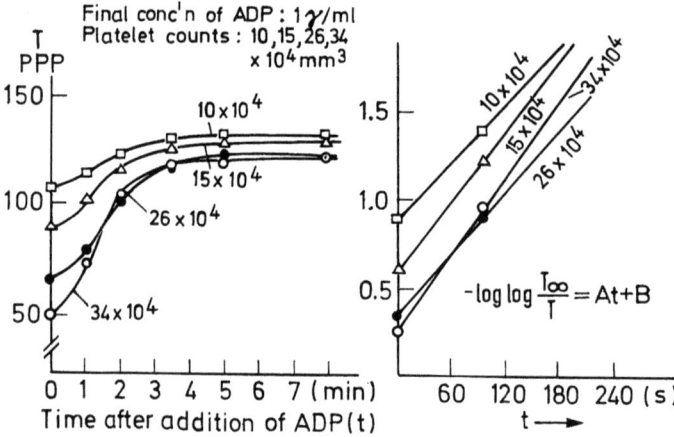

Fig. 9. Platelet aggregation profiles at different platelet counts in PRP and their conversion into double logarithmic equation (741002)

Aggregation Curves with Different Concentration of ADP

When ADP solutions with different concentrations were added to the same PRP, the higher concentration of ADP gave a greater shift of transparency and the slopes were changed according to ADP concentration, although they did not show special correlation to it (Fig. 10).

Effect of Flurbiprofen on Platelet Aggregation

The effect of flurbiprofen on platelet aggregation with ADP was examined, changing its concentration. It was found that the higher concentration of this material gave the lower maximal transparency and the larger slope A (Fig. 11).

Aggregation Curves with Different Concentration of Collagen

When collagen solutions with different concentrations were added to the same PRP, the higher concentration of collagen gave a greater shift of transparency and shorter refractory time as well as the larger A and smaller B, and it was found that our double logarithmic equation was available for this case also (Fig. 12).

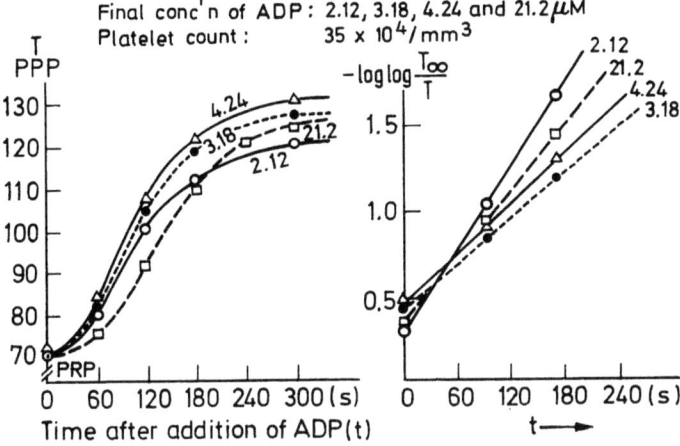

Fig. 10. Platelet aggregation profiles induced with ADP at different concentrations and their conversion into double logarithmic equation (740610)

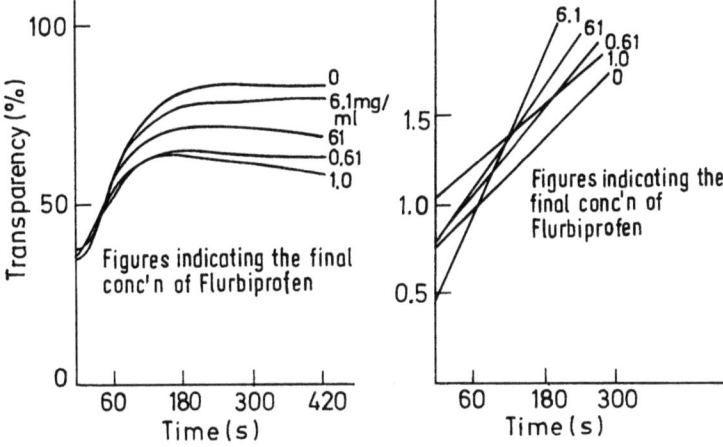

Fig. 11. Effect of Flurbiprofen on platelet aggregation with ADP and the conversion of the aggregation profiles on double logarithmic equation

Effect of RA 233 on Platelet Aggregation

The effect of RA 233 on platelet aggregation with collagen was examined changing its concentration. It was found that the higher concentration this substance displayed the greater the inhibition on aggregation and the shorter the refractory time (Fig. 13).

Distribution of A and B Calculated from Human Platelet Aggregation

Slope A and constant B were calculated from aggregation curves which were drawn with platelets of normal human subjects and patients with various diseases such as arteriosclerosis, nephropathies, and hemorrhagic disorders, and it was found that normal platelets gave rather large A and smaller B values (Figs. 14 and 15), A and B also had some negative correlation coefficients with the exception of nephropathies (Fig. 16).

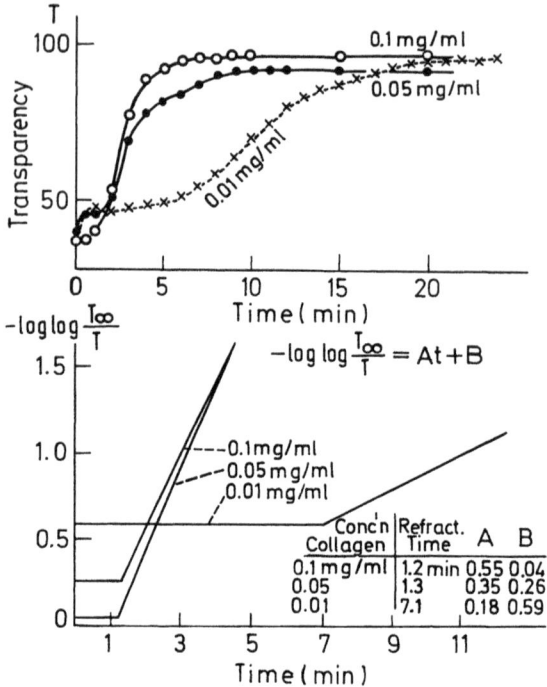

Fig. 12. Platelet aggregation profiles with different concentration of collagen and their conversion on double logarithmic equation (740925)

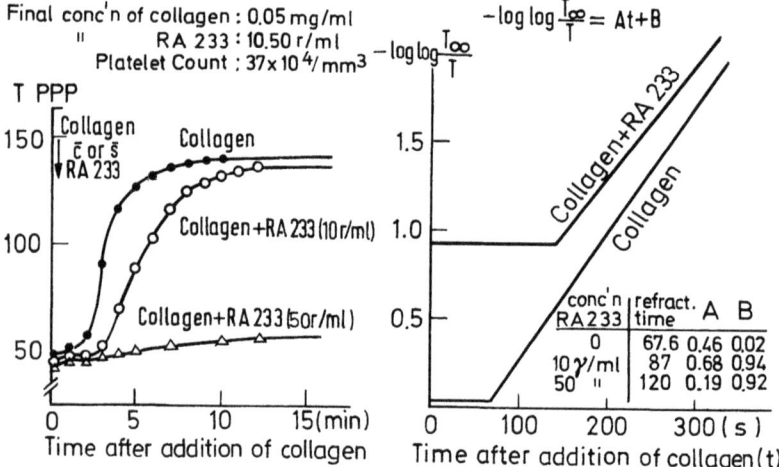

Fig. 13. Effect of RA 233 on platelet aggregation profiles induced with collagen and the conversion of these profiles in double logarithmic equation

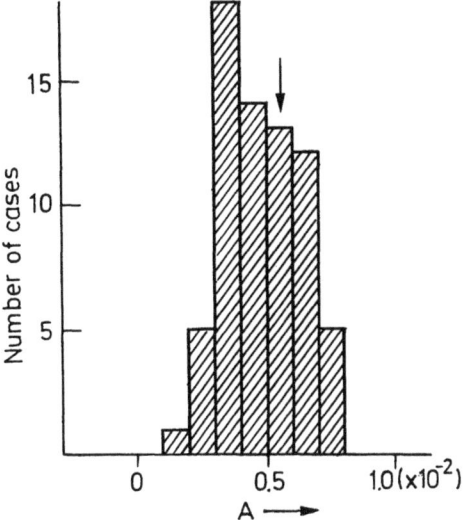

Fig. 14. Distribution of A in platelet aggregation (double logarithmic) equation of normal subjects and patients with various disorders

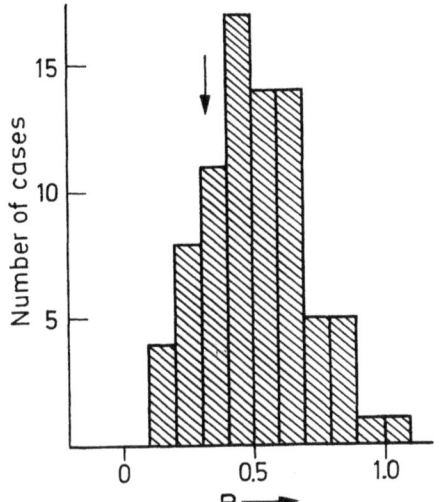

Fig. 15. Distribution of B in platelet aggregation (double logarithmic) equation of normal subjects and patients with various disorders

Discussion and Conclusion

So far the estimation of the conversion of platelet aggregation curve into mathematic equations has not been clearly understood and some difficulties remain in applying this calculation, particularly in cases of disaggregation or of diphasic pattern. However, our double logarithmic equation showed a pleasing correspondence with the actual values of aggregation curves even beyond the calculated region, and gave us some objective criteria for comparison of different curves.

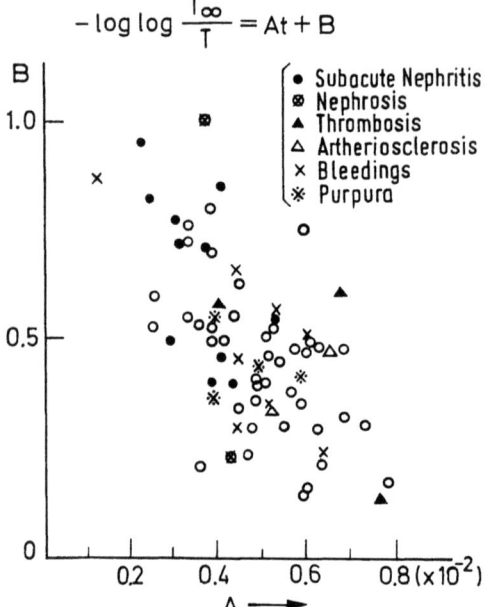

Fig. 16. Correlation between A and B in platelet aggregation (double logarithmic) equation of normal subjects and patients with various disorders

References

Born, G.V.R.: Aggregation of blood platelets by adenosine and its reversal. Nature (Lond.) 195, 927-929 (1962)

Born, G.V.R., Hume, M.: Effects of the numbers and sizes of platelet aggregates on the optical density of plasma. Nature (Lond.) 215, 1027-1029 (1967)

Praga, C.A., Pogliani, E.M.: Effect of temperature on ADP-induced platelet aggregation: Its significance in studying anti-aggregating drugs. Thromb. Diath. haemorrh. (Stuttg.) 29, 183-189 (1973)

Quantitative Valuation of Platelet Aggregation Curves through the Calculation of a Numerical Index

F. Cavallero

According to Born, the platelet aggregation test (PAT) during the last years has been employed by clinicians and researchers interested in the study of thrombotic processes.

The curves obtained by recording changes in light transmission during the aggregation process show different shapes depending upon the type and the concentration of the aggregating agent employed.

So far particular attention has been paid to calculation of the angle of inclination in the steepest slope of the curve and to the highest percentage of aggregation, the presence of a latency time, of two-phase waves, and of irreversible aggregation, or disaggregation.

An improvement has been reached by determining for each patient the threshold dose of the aggregating agent needed to induce two-phase waves. This method is very useful for evaluation of the platelet aggregation tendency, but it requires that at least five tests be performed for each sample, each time using an increasing concentration of aggregating agent. This procedure reduces a large utilization of PAT in the clinical field.

To simplify a quantitative evaluation of the platelet aggregation curves and to allow comparison of different shaped curves, I tried to build up a procedure of evaluation of the curves, which may furnish a numerical index.

With this method, curves of different shape and intensity, obtainable through the use of the same aggregating agent, utilized in a singular convenient concentration, may be compared and valued in a quantitative way. My way of proceeding consists of transforming, with sufficient care, a parabolic curve into a line artificially broken in segments, each of which is equivalent to a given time and to a certain amount of percentage of aggregation (Fig. 1).

In this way an aggregation curve is composed of one or more segments with a different ratio of time and percentage of aggregation.

The aspect of these curves, or segments of curves, may be schematized in five fundamental types:

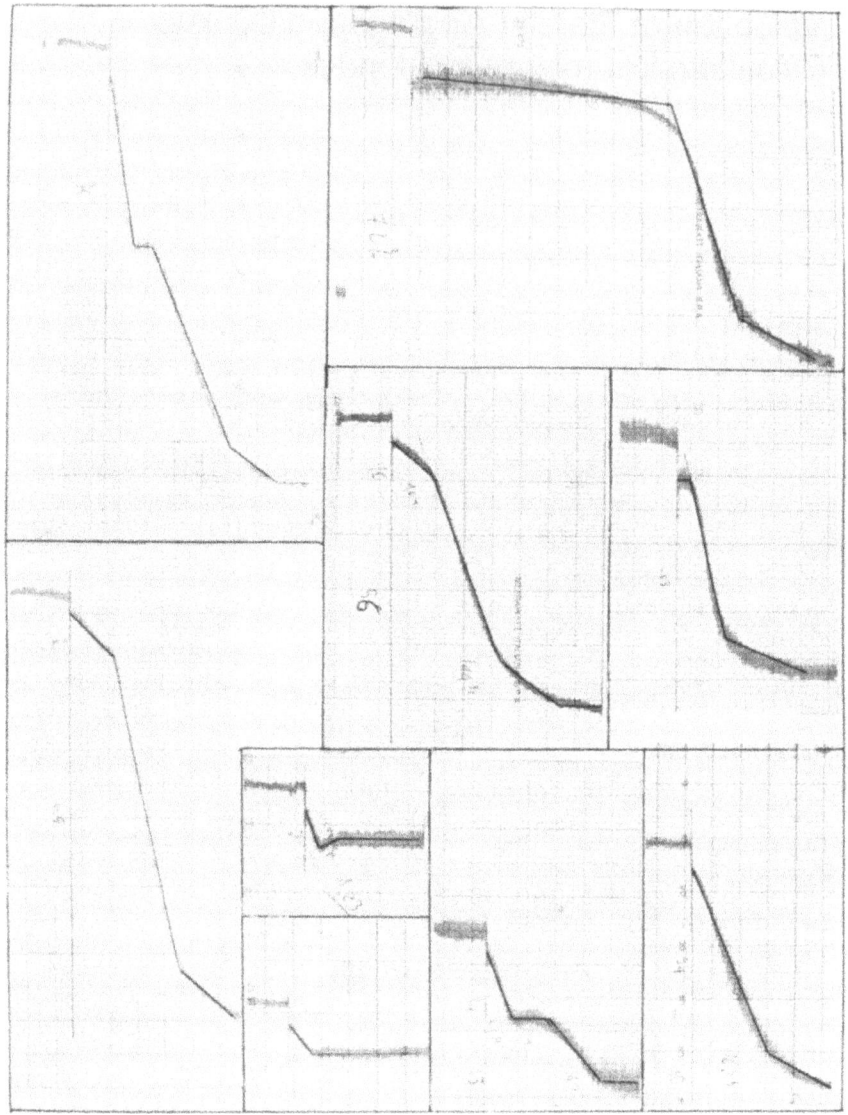

Fig. 1

1. A unique slope without disaggregation (Fig. 2I)
2. Two or more segments of slope with different decreasing speed (Fig. 2II)
3. A segment of slope after a latency time (Fig. 2III)
4. Two segments with increasing speed (Fig. 2IV)

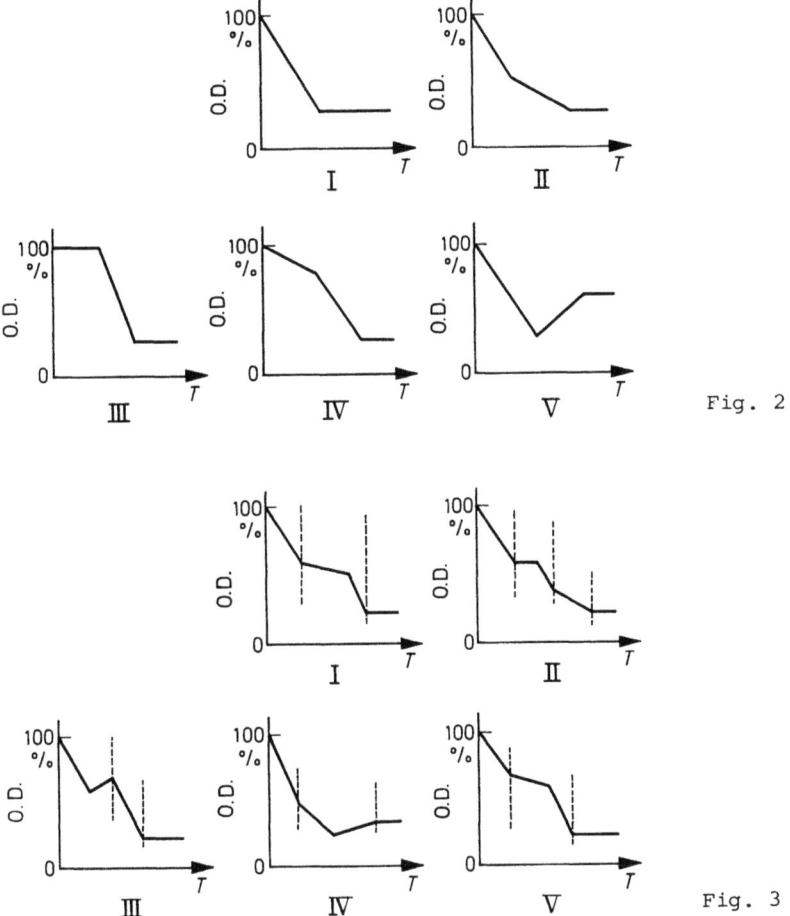

Fig. 2

Fig. 3

5. A falling slope followed by rising one; that is to say a phase of aggregation followed by a disaggregation (Fig. 2V).

These various types of segments mentioned may be variously combined in a singular curve (Fig. 3).

To evaluate a curve it is necessary to examine separately the various segments of which it is composed, to calculate and sum up each one of them.

To obtain the numerical index in the fundamental types of curves, the following formulae were applied:

As far as the first type is concerned, schematized in Fig. 4I, calling b the aggregation percentage and c the time in seconds, the formula is $\frac{b^2}{c}$. Therefore, $\frac{b}{c}$ indicates the speed and b the intensity of aggregation; so that $b \times \frac{b}{c} = \frac{b^2}{c}$.

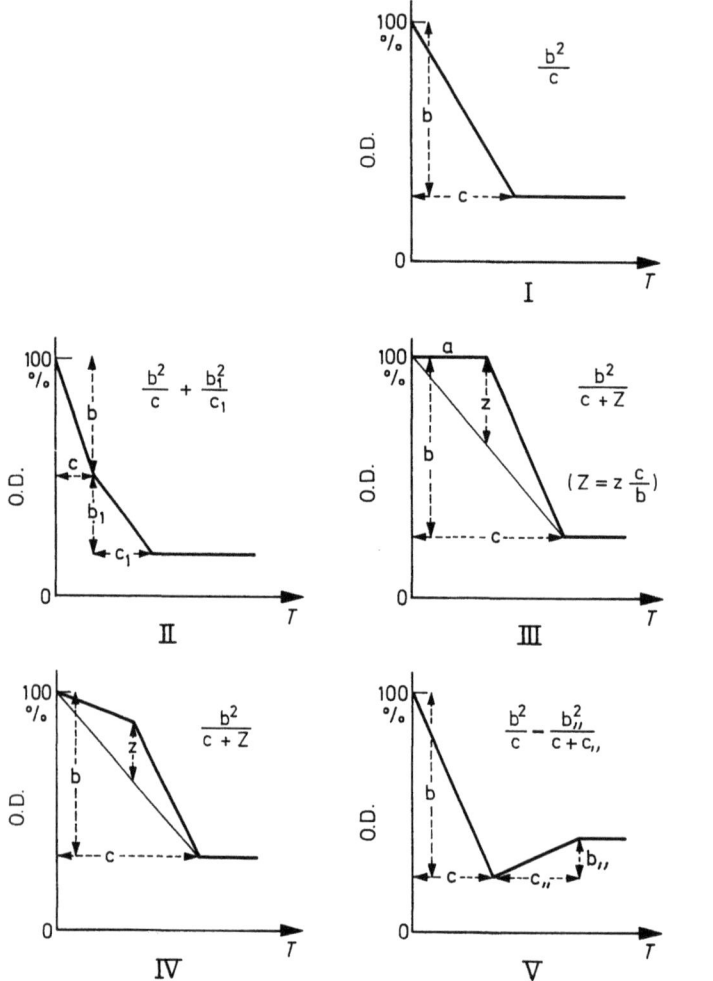

Fig. 4

In the second type (Fig. 4II) each segment is analyzed separately and the results are summed up, so that $\frac{b^2}{c} + \frac{b_1^2}{c^1}$.

In the third type (Fig. 4III) the formula $\frac{b^2}{c + Z}$ allows for the latency time (called a) and the steepness of the curve which follows it. Therefore, while a is the time which has taken place from the beginning of the curve and the initial flexion of same, $z = \frac{a \times b}{c}$, that is to say, it varies in relation to the ratio $\frac{a}{c}$ and proportionally to b. To value z like time (Z) and to add it to the time c of the curve, z would have to be multiplied by the ratio $\frac{c}{b}$. Therefore $\frac{z}{b} = \frac{Z}{c}$. Consequently time $Z = z\frac{c}{b}$. In this particular case time Z is equal to a. Therefore $Z = \frac{a \times b}{c} \times \frac{c}{b} = a$.

As far as the fourth type of curve considered (Fig. 4IV) the preceding formula is still valid ($\frac{b^2}{c+z}$), therefore even in this case a is equal to the time between the beginning and the flexion point.

It must be stressed that if z is not directly measured on the recording, but acquired from $z = \frac{a \times b}{c}$, it is necessary to change the percentage of variation in optical density in the place of b. Also, in this case the time of z is equal to $z \times \frac{c}{b}$. The time of z in this fourth type of curve will always be shorter than a.

In the fifth type of curve (Fig. 4V), which is a phase of disaggregation, the proposed formula is $\frac{b^2}{c} - \frac{b^2}{c+c''}$.

Valuing the phase of disaggregation, I think it is necessary to value c together with c", because it appears logical to consider that disaggregation begins before its appearance in the recording. The disaggregation must already be present during the aggregation process and it shows itself only when it overcomes the aggregation.

In Figure 5 are recorded (as an example) the data obtained, with mentioned formulae, from curves which, in the same time of performance, show different shapes, although having the same optical density variation. Note the growing increase of values from the curve with latency time to the steepest one.

The numerical index, obtainable by applying the mentioned formulae, have a meaning when we wish to value and to compare curves obtained from the same type of aggregating agent, as long as it is always employed at the same concentration.

The tests have to be performed always using plasma samples at the same number of platelets (about 250,000/ml). To show the results of PAT, it may be useful to report if the test produced a unique aggregative curve, or a two-phase curve either with disaggregation or with initial latency, and possibly reporting the index obtained in each segment, than concluding with the global numerical index.

It may be also useful to show the number of platelets contained in 1 ml of blood. Lastly, it should be useful to standardize the final concentration of each aggregating agent to be utilized.

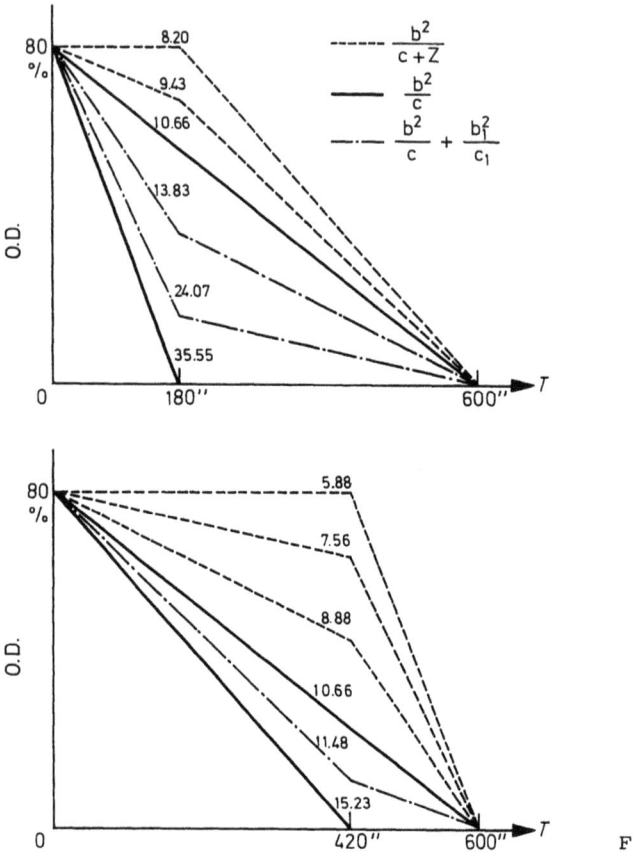

Fig. 5

Reference

Born, G.V.R.: Aggregation of blood platelets by adenosine and its reversal. Nature (Lond.) 195, 927-929 (1962)

The Polyunsaturated Fatty Acids in Human Platelets: Effects of Diet and Possible Functional Significance

R. Paoletti

A diet rich in saturated fats plus the addition of cholesterol, thiouracil, and cholic acid causes thrombosis in the rat (Thomas and Hartroft, 1959; Gresham and Howard, 1960; Renaud and Allard, 1962) and the thrombogenic effect of adenosine diphosphate is intensified by propylthiouracil added to a diet rich in saturated fats and cholesterol (Nordbøy and Chandler, 1964). A thrombogenic syndrome may be observed (Mustard et al., 1963) in pigs on a diet rich in saturated fats and cholesterol.

Research on human platelets has shown a diminution in adhesiveness in persons on a fat-poor diet (McDonald and Edgill, 1958) while hyperlipidic diets reduce the survival of the platelets (McDonald and Edgill, 1958; Mustard and Murphy, 1962).

Thrombogenic diets - rich in saturated fatty acids - cause an increase in platelet response to thrombin-induced aggregation while the response to ADP and collagen-induced aggregation diminishes (Renaud et al., 1920). It is, therefore, evident that platelet function may be greatly affected by the diet even in conditions where these elements of the blood are washed and thus deprived of their contact with the plasma of the subjects from whom the blood was drawn. We must, therefore, deduce that the lipids in the diet can modify the structural components of the platelets, probably the fatty acid composition of the phospholipids in the platelet membranes.

In in vitro systems, phosphatidylserine and other phospholipids reach their maximum activity in coagulation (Therriault et al., 1958; Slotta, 1960; Marcus et al., 1962) and may thus be identified with platelet factor 3 (PF-3) which, according to Marcus, is a phospholipid or combination of phospholipids (Marcus, 1969). The possible control of the diet on factor PF-3 has, therefore, been the subject of recent investigations. Renaud (Renaud et al., 1970) analyzed total lipids in the platelets of rats and reported an increase in the saturated fatty acids + monounsaturated/polyunsaturated fatty acid ratio after administration of a butter- or stearic acid-rich diet. A considerable increase in the incidence of thrombosis coincides with these variations. Nordøy and Rødset (1971) suggested that the fatty acid composition of the phospholipids in human platelets is related to the activity of the PF-3 factor in platelet-rich plasma. The linoleic acid content of the phosphoglyceride in human platelets increases after 3 weeks of a soya seed oil-rich diet. Other investigations

in rats also emphasize (Gautheron and Renaud, 1972) that the
variations induced by the diet (rich in cocoa butter) in the
saturated + monounsaturated/polyunsaturated ratio in PS or in
PS + phosphatidylinositol (PI) are closely related to the variations in PF-3 activity.

This data from medical literature thus implies the importance
of studying the effects of the diet on the fatty acid composition of the phospholipids in the platelets of man, also.

Diets and Fatty Acids in the Phospholipids of Human Platelets

Research conducted in our laboratory in cooperation with Dr.
James M. Iacono, Division of Clinical Nutrition, Department of
Internal Medicine, College of Medicine, University of Cincinnati,
Ohio, USA, led to a comparative study on the phospholipid composition of platelets and red corpuscles in healty subjects
whose age ranges were 20-29, 30-39, 40-49 years, residing in
Cincinnati, Milan, and three mountain communities in western
Sicily.

This choice was made with the objective of studying the effect
of diets widely different in the percentage content of lipids
and their fatty acid composition on the composition of platelets. The methods by which platelets and red corpuscles were obtained for extraction and analysis of the phospholipids and respective fatty acids are described in detail by Iacono et al.
(1973).

The results of investigations on the lipid composition of the
platelets of the various groups (age and geographical origin)
are shown in Table 1.

The PE content in the platelets obtained from the Sicilian population is considerably higher when compared with the levels in
Milan and Cincinnati, whereas the PI content is higher in the
platelets obtained in Milan and Sicily. In contrast no significant variation is observed in the three age groups studied.
Tables 2-6 show the analytical data of the saturated, monounsaturated, and polyunsaturated fatty acid composition of the phospholipids for the three population groups. Monounsaturated fatty
acids of the total fatty acids of PE, PC, and sphingomyelin are
higher in Milan and in Sicily than in Cincinnati. In the β position of PE (Table 4) the polyunsaturated fatty acids are predominant. With regard to the PE, the plasmalogen fatty acid, palmitaldehyde (16:0 DMA) and stearaldehyde (10:0 DMA) are present only
in the α position (Table 3) while arachidonic acid is present in
high concentrations in the β position (75-76%) (Table 4).

The major fatty acids of PC are palmitic, stearic, oleic, linoleic and arachidonic acids (Table 5). Palmitic, stearic and oleic
acids predominate in the α position (Table 6).

Considering the fatty acids in the position of the PC (Table 7)
we observe a higher percentage of linoleic acid and arachidonic

Table 1. Lipid composition of the platelets

	Area[a]			Age[b]		
	Milan	Cincinnati	Sicily	20-29 years	30-39 years	40-49 years
Proteins[c]	58.1 ± 1.4	58.1 ± 1.8	57.3 ± 5.2	59.7 ± 0.6	58.4 ± 2.4	55.4 ± 3.5
Total lipids[c]	19.2 ± 4.7	19.5 ± 1.9	19.4 ± 0.9	20.9 ± 3.6	18.7 ± 1.9	18.4 ± 2.6
Cholesterol[d]	24.2 ± 2.8	24.2 ± 2.7	20.4 ± 3.0	20.7 ± 3.1	22.7 ± 3.3	25.0 ± 1.8
Phospholipids[d]	65.3 ± 6.2	62.2 ± 4.2	63.5 ± 7.3	59.7 ± 3.3	63.1 ± 3.7	68.3 ± 3.2
Phosphatidylinositol (PI)	6.8 ± 0.5	11.2 ± 1.6	6.0 ± 0.8[f]	8.0 ± 2.0	8.9 ± 2.9	7.1 ± 1.9
Sphingomyelin (SPH)	13.7 ± 3.2	19.0 ± 5.4	16.1 ± 0.3	17.1 ± 1.8	16.7 ± 5.8	14.9 ± 1.1
Phosphatidylcholine (PC)	28.7 ± 1.0	29.3 ± 4.8	32.4 ± 1.3	31.4 ± 1.7	28.0 ± 3.6	31.0 ± 1.6
Phosphatidylserine (PS)	11.1 ± 4.1	12.1 ± 1.9	10.5 ± 2.6	10.3 ± 2.6	11.4 ± 2.9	12.0 ± 1.6
Phosphatidyl-ethanolamine (PE)	18.0 ± 2.3	16.7 ± 0.9	23.5 ± 0.8[f]	20.2 ± 2.2	19.5 ± 3.7	18.6 ± 3.2
Phosphatidic acid (PA)	3.9 ± 2.0	6.0 ± 0.4	2.3 ± 1.0	4.3 ± 2.1	3.5 ± 1.6	4.4 ± 1.8
Cardiolipin	3.5 ± 2.1	5.9 ± 1.0	3.1 ± 0.3	4.5 ± 0.8	3.5 ± 2.2	4.5 ± 1.5

[a] Average of all samples collected in the area
[b] Average of the samples of all three areas belonging to one of the three age groups
[c] Percent of total solid residue
[d] Percent of total lipids
[e] Percent phospholipidic total
[f] $p < 0.01$

Table 2. Total phosphatidylethanolamine (PE) fatty acids in platelets

Fatty acids	Area			Age		
	Milan	Cincinnati	Sicily	20-29 years	30-39 years	40-49 years
16:0 DMA	5.2 ± 2.7	5.3 ± 2.3	6.5 ± 2.0	3.1 ± 1.0	6.5 ± 0.8	7.4 ± 0.9
16:0	5.0 ± 1.9	8.0 ± 3.2	6.7 ± 1.2	8.8 ± 2.3	5.9 ± 2.3	5.0 ± 0.5
16:1	1.7 ± 1.2	2.2 ± 2.7	1.4 ± 0.7	3.5 ± 1.5	0.7 ± 0.6	1.2 ± 0.3
18:0 DMA	9.9 ± 6.4	6.2 ± 4.2	7.2 ± 2.4	2.9 ± 1.7	10.2 ± 2.9	10.4 ± 4.0
18:0	21.7 ± 3.7	18.2 ± 3.6	23.8 ± 7.0	23.8 ± 9.0	20.3 ± 1.1	19.4 ± 1.3
18:1	10.9 ± 2.9	9.0 ± 1.0	12.0 ± 1.2	12.6 ± 2.1	9.9 ± 1.5	9.5 ± 1.5
18:2	2.8 ± 1.0	2.9 ± 1.0	3.9 ± 1.1	3.7 ± 0.9	2.1 ± 0.5	3.5 ± 1.3
20:4	42.1 ± 3.0	48.0 ± 3.6	38.4 ± 6.0	40.8 ± 10.5	44.3 ± 2.1	43.4 ± 2.2

Table 3. Fatty acids in the β position of platelet PE

Fatty acids	Area			Age		
	Milan	Cincinnati	Sicily	20-29 years	30-39 years	40-49 years
16:0 DMA	7.2 ± 2.2	10.7 ± 3.1	11.9 ± 3.4	6.2 ± 1.4	13.0 ± 1.8	14.8 ± 4.8
16:0	12.2 ± 0.7	15.9 ± 5.5	9.9 ± 2.3	11.3 ± 6.9	7.1 ± 1.4	6.2 ± 1.4
16:1	3.0 ± 2.1	3.3 ± 5.7	1.7 ± 0.4	2.9 ± 4.2	1.0 ± 1.0	0.5 ± 1.6
18:0 DMA	11.3 ± 5.2	11.9 ± 6.7	13.0 ± 5.8	3.8 ± 2.8	20.4 ± 3.8	20.8 ± 3.4
18:0	39.4 ± 8.8	43.3 ± 7.4	42.9 ± 7.4	41.3 ± 8.6	36.2 ± 4.8	33.2 ± 8.2
18:1	14.4 ± 1.1	13.2 ± 2.4	11.6 ± 1.5	12.9 ± 1.4	9.5 ± 0.8	10.5 ± 1.5
18:2	1.4 ± 1.2	tr.	2.0 ± 0.4	1.4 ± 1.2	0.5 ± 0.9	3.7 ± 1.3
20:4	8.4 ± 6.3	1.6 2.8	6.9 ± 5.0	15.0 ± 2.0	15.2 ± 5.5	11.2 ± 8.6

Table 4. Fatty acids in the β position of platelet PE

Fatty acids	Area			Age		
	Milan	Cincinnati	Sicily	20-29 years	30-39 years	40-49 years
16:0 DMA						
16:0	7.1 ± 2.3	2.1 ± 1.9	5.5 ± 2.0	6.3 ± 3.0	4.7 ± 4.3	3.8 ± 0.6
16:1	2.2 ± 0.7	0.9 ± 1.6	2.8 ± 0.7	4.1 ± 1.7	1.4 ± 1.2	2.1 ± 0.4
18:0 DMA						
18:0	9.0 ± 4.4	1.1 ± 1.0	6.3 ± 5.0	6.3 ± 4.9	4.4 ± 4.0	5.6 ± 7.1
18:1	11.2 ± 3.0	6.9 ± 2.2	13.0 ± 2.9	12.3 ± 3.6	10.3 ± 4.2	8.5 ± 3.3
18:2	6.8 ± 4.5	4.1 ± 1.0	5.5 ± 1.2	5.9 ± 1.3	7.2 ± 3.7	3.3 ± 1.2
20:4	63.6 ± 6.4	84.5 ± 5.1	81.4 ± 11.8	66.6 ± 13.0	73.4 ± 5.7	75.6 ± 7.5

Table 5. Composition of total fatty acids of platelet phosphatidylcholine (PC)

Fatty acids	Area			Age		
	Milan	Cincinnati	Sicily	20-29 years	30-39 years	40-49 years
16:0	27.9 ± 4.4	30.4 ± 7.1	31.2 ± 5.2	28.5 ± 2.7	30.0 ± 8.3	31.7 ± 6.6
16:1	3.0 ± 0.8	tr.	1.5 ± 0.1	1.7 ± 1.8	0.7 ± 1.0	1.3 ± 1.2
18:0	19.2 ± 3.9	14.4 ± 2.2	17.3 ± 4.5	12.9 ± 10.2	16.8 ± 1.1	15.5 ± 5.6
18:1	31.6 ± 3.7	27.5 ± 1.3	29.7 ± 1.2	28.4 ± 0.6	28.4 ± 3.2	30.9 ± 2.9
18:2	7.0 ± 0.2	9.3 ± 1.5	7.1 ± 2.2	7.7 ± 0.8	7.8 ± 4.5	8.1 ± 0.9
20:0	0.6 ± 0.8	tr.	1.1 ± 2.1	0.9 ± 0.8	0.4 ± 0.6	0.3 ± 0.6
20:4	10.6 ± 1.1	17.9 ± 5.8	11.9 ± 2.6	14.8 ± 4.1	15.8 ± 9.8	11.4 ± 1.6

Table 6. Fatty acids in the position of platelet phosphatidylcholine (PC)

Fatty acids	Area			Age		
	Milan	Cincinnati	Sicily	20-29 years	30-39 years	40-49 years
16:0	34.7 ± 6.1	49.1 ± 4.6	55.3 ± 7.8	45.3 ± 6.2	48.9 ± 1.8	49.8 ± 17.7
16:1	3.1 ± 3.5	tr.	1.0 ± 0.1	2.2 ± 3.0	0.5 ± 0.7	0.6 ± 0.6
18:0	35.6 ± 6.7	30.5 ± 2.7	27.7 ± 5.7	29.9 ± 2.1	31.4 ± 1.8	31.0 ± 9.6
18:1	21.4 ± 6.0	16.8 ± 5.8	12.7 ± 2.6	17.0 ± 3.8	17.3 ± 3.2	15.3 ± 9.0
18:2	2.0 ± 0.1	2.5 ± 2.2	1.4 ± 0.6	2.5 ± 0.8	0.5 ± 0.7	2.4 ± 1.5
20:0	1.0 ± 2.1	tr.	1.8 ± 0.1	1.5 ± 2.1	0.9 ± 1.3	0.6 ± 1.0
20:4	1.3 ± 1.8	tr.	0.5 ± 0.4	1.1 ± 1.4	0.4 ± 0.6	0.2 ± 1.0

Table 7. Fatty acids in the β position of platelet phosphatidylcholine (PC)

Fatty acids	Area			Age		
	Milan	Cincinnati	Sicily	20-29 years	30-39 years	40-49 years
16:0	21.0 ± 2.5	19.3 ± 2.1	24.0 ± 7.2	20.2 ± 2.2	26.9 ± 7.6	19.1 ± 1.8
16:1	4.1 ± 0.2	2.7 ± 0.5	3.7 ± 1.0	3.7 ± 1.0	3.6 ± 0.7	3.0 ± 1.1
18:0	4.6 ± 1.8	1.3 ± 0.7	4.8 ± 3.1	4.9 ± 3.4	3.6 ± 2.2	1.9 ± 1.4
18:1	42.2 ± 0.5	37.2 ± 5.6	40.1 ± 8.4	40.9 ± 3.1	31.0 ± 0.8	43.8 ± 2.1
18:2	10.0 ± 1.5	14.2 ± 1.6	11.2 ± 1.3[a]	11.3 ± 3.3	12.0 ± 0.2	12.6 ± 2.2
20:0	0.3 ± 0.4	tr.	1.2 ± 2.0	1.2 ± 2.0	0.3 ± 0.4	tr.
20:4	18.1 ± 2.1	25.2 ± 3.6	15.0 ± 2.5[b]	17.8 ± 6.3	22.4 ± 9.4	19.6 ± 2.4

[a] $p < 0.01$
[b] $p < 0.05$

acid (20:4) in the Cincinnati group. Arachidonic acid is typically contained in animal tissues and is, therefore, prevalent in a meat-rich diet.

The major fatty acids in PS were stearic, oleic and arachidonic acids (Table 8). The Cincinnati samples contained the highest levels of stearic acid, whereas the percentages of palmitic acid were highest in the samples from Milan and Sicily.

In the sphingomyelin (Table 9), the Cincinnati group had a lower level of oleic and erucic acids and a higher level of arachidic acid. These variations also may be related to dietetic differences such as the absence, in the United States, of rapeseed oil, rich in erucic acid, which was present in the Italian diet during the period of this study.

Discussion

A number of differences in platelet lipids as related to geographical location and age have been observed in the study. Some of these observations can be associated with current thoughts on the influence of dietary fat in the pathogenesis of experimental thrombosis. In fact it has been shown that PF-3 required for blood coagulation is dependent on the type of fatty acids in PS or PS + PI (Nordøy and Rødset, 1971). It has also been postulated that there is a correlation of the phospholipid clotting time with the dietary modification of the fatty acids of PS or PS - PI (Gautheron and Renaud, 1972). An increase of stearic and oleic acids and a decrease in polyunsaturated fatty acids seem responsible for an increased clotting time.

The observation of higher levels of stearic and oleic acids in the PS and of higher levels of PI as percentage of phospholipids suggests an increase in the phospholipid clotting time of the Cincinnati platelets, according to Gautheron and Renaud (1972).

These data obtained in man confirm that obtained from the investigations conducted in our laboratory (Andreoli and Miras, 1971), which indicated that an olive oil- and carbohydrate-rich diet in the pig induces considerable variations in the composition of fatty acids in the platelets.

The specific accumulation of arachidonic acid in the β position of the PE (Tables 3 and 4) is also remarkable in comparison to the α position. In the Cincinnati sample 95% of total arachidonic acid was in this position.

Age does not seem an important factor as far as fatty acid composition or the percentage of phospholipids present is concerned.

The modifications observed may be related to the dietetic conditions present in the three regions studied. We should also consider, however, that platelets are metabolically active with regard to the incorporation of long chain fatty acids (Cohen et al., 1970; Deykin and Desser, 1968), the biosynthesis of fatty

Table 8. Fatty acids of platelet phosphatidylserine (PS)

Fatty acids	Area			Age		
	Milan	Cincinnati	Sicily	20-29 years	30-39 years	40-49 years
14:0	0.8 ± 1.0	tr.	0.6 ± 0.5	0.3 ± 0.6	0.7 ± 1.1	0.4 ± 0.4
16:0	4.4 ± 1.0	0.2 ± 0.4	15.0 ± 4.9[a]	8.3 ± 9.2	6.8 ± 8.2	4.7 ± 3.8
16:1	1.1 ± 0.7	tr.	1.4 ± 0.4[a]	1.2 ± 1.1	0.6 ± 0.7	0.8 ± 0.7
18:0	42.2 ± 1.5	45.6 ± 4.7	32.6 ± 5.4[a]	37.7 ± 7.4	42.5 ± 10.6	41.4 ± 6.6
18:1	25.8 ± 2.9	34.9 ± 1.3	26.1 ± 5.1[a]	26.7 ± 9.7	29.2 ± 5.1	29.8 ± 6.0
18:2	1.1 ± 1.2	tr.	2.7 ± 1.0	1.5 ± 2.0	0.8 ± 1.6	1.4 ± 1.2
20:0	1.1 ± 1.1	tr.	1.6 ± 0.6	0.9 ± 0.8	1.0 ± 1.0	0.8 ± 1.4
20:4	23.6 ± 2.6	19.3 ± 4.2	20.1 ± 2.6	23.5 ± 2.7	18.4 ± 1.9	20.7 ± 3.7

[a] $p < 0.05$

Table 9. Fatty acids of platelet sphingomyelin (SPH)

Fatty acids	Area			Age		
	Milan	Cincinnati	Sicily	20-29 years	30-39 years	40-49 years
14:0	1.0 ± 0.4	0.7 ± 0.7	1.6 ± 0.8	1.4 ± 0.8	1.3 ± 0.7	0.5 ± 0.5
16:0	24.2 ± 13.6	30.2 ± 4.9	23.0 ± 15.1	23.5 ± 8.6	19.7 ± 12.3	34.3 ± 8.8
16:1	1.2 ± 1.8	tr.	1.6 ± 0.5	1.7 ± 1.6	0.7 ± 1.0	0.4 ± 0.6
18:0	10.4 ± 5.7	10.1 ± 6.4	4.3 ± 2.0	4.9 ± 2.4	6.5 ± 1.7	13.4 ± 6.5[a]
18:1	9.5 ± 6.0	2.0 ± 1.1	5.5 ± 2.3[a]	3.2 ± 2.3	5.0 ± 3.0	9.0 ± 6.9[a]
18:1	1.2 ± 1.1	tr.	0.7 ± 0.7	0.1 ± 0.2	0.6 ± 0.5	1.3 ± 1.2
20:0	5.0 ± 2.6	7.8 ± 0.8	4.5 ± 1.2[a]	6.0 ± 2.1	7.0 ± 1.3	4.2 ± 2.3
22:0	19.5 ± 9.2	24.9 ± 1.8	18.9 ± 4.2	21.5 ± 4.7	25.0 ± 3.3	16.6 ± 6.5
22:1	2.8 ± 1.3	tr.	0.9 ± 0.9[a]	0.9 ± 0.7	1.8 ± 1.8	1.2 ± 2.1
24:0	7.9 ± 5.1	9.5 ± 1.2	14.0 ± 7.5	13.9 ± 7.6	10.6 ± 2.5	7.0 ± 3.9
24:1	17.4 ± 9.3	15.0 ± 1.4	25.0 ± 7.6	23.1 ± 7.7	21.9 ± 6.2	12.2 ± 4.8

[a] $p < 0.05$

acids (Marks et al., 1960; Deykin and Desser, 1968; Lewis and Majerus, 1969; Cohen et al., 1970), and their elongation (Schoene and Iacono, 1973). In particular, the re-acylation of the platelet phospholipids has been demonstrated (Cohen et al., 1970). All these factors must be borne in mind when evaluating the significance of fats in the diet in the regulation of the lipid composition of platelets and their behavior in coagulation or thrombosis. Both in laboratory animals and in man, modifications in the fatty acid composition of platelet phospholipids and increase in platelet aggregation coincide with the administration of lipids in the diet.

The Role of Polyunsaturated Fatty Acids in Platelet Phospholipids

Arachidonic acid in platelet phospholipids (particularly in the β position of the PC and PE) is the essential precursor for the synthesis of prostaglandin and thromboxane according to the folfowing sequence.

While in numerous organs and tissues (central nervous system, kidney, seminal vesicles) $F_{2\alpha}$- and E_2-type synthesis of the prostaglandins seems to predominate, in the platelets it has recently been demonstrated that the thromboxanes represent a high percentage of the products of arachidonic acid metabolism. The thromboxanes are responsible for platelet aggregation and also for contraction of the vessel walls (representing the active component of the so-called RACS (rabbit aorta contracting substance)). The extremely short half-life (a few tenths of a second) has made the identification impossible up to recently (Andreoli and Miras, 1971).

The formation of thromboxanes from arachidonic acid emphasizes the importance of that variations in polyunsaturated fatty acids in the phospholipids of the platelet membrane may have in the function of these hematic components in their interaction with the vessel walls.

References

Andreoli, V.M., Miras, C.J.: Life Sci. **10**, 481 (1971)
Cohen, P., Derksen, A., Bosh, H., Van den: J. Clin. Invest. **49**, 128-129 (1970)
Deykin, D., Desser, R.K.: J. Clin. Invest. **47**, 1590-1602 (1968)
Gautheron, R., Renaud, S.: Thrombosis Res. **1**, 353-370 (1972)
Gresham, G.A., Howard, A.N.: Brit. J. Exp. Path. **41**, 395-402 (1960)
Iacono, J.M., Zellner, D.C., Paoletti, R., Ishikawa, T., Frigeni, V., Fumagalli, R.: Haemostasis **2**, 141-162 (1973)
Lewis, N., Majerus, P.W.: J. Clin. Invest. **48**, 2114-2123 (1969)
Marcus, A.J., Ullman, H.L., Safier, L.B., Ballard, H.S.: J. Clin. Invest. **41**, 2198-2212 (1962)
Marcus, A.J.: New Engl. J. Med. **280**, 1330-1335 (1969)
Marks, P.A., Gellhorn, A., Kidson, C.J.: Biol. Chem. **235**, 2579-2583 (1960)
McDonald, L., Edgill, M.: Lancet **I**, 996-998 (1958)
Mustard, J.F., Murphy, E.A.: Brit. Med. J. **1**, 1651-1655 (1962)
Mustard, J.F., Roswell, H.C., Murphy, E.A., Downie, H.G.: J. Clin. Invest. **42**, 1783 (1963)
Nordøy, A., Chandler, A.B.: Scand. J. Haemat. **1**, 202-211 (1964)
Nordøy, A., Rødset, J.M.: Acta Med. Scand. **190**, 27-34 (1971)
Renaud, S., Kinlough, R., Mustard, J.F.: Lab. Invest. **22**, 339-343 (1920)
Renaud, S., Allard, C.: Circulation Res. **11**, 388-399 (1962)
Renaud, S., Kuba, K., Goulet, C., Lemire, Y., Allard, C.: Circulation Res. **26**, 553-564 (1970)
Schoene, N.W., Iacono, J.M.: Circulation **46**, 11-32 (1973)
Slotta, K.H.: Proc. Soc. Exp. Biol. Med. **103**, 53-56 (1960)
Therriault, D., Nichols, T., Jensen, H.: J. Biol. Chem. **233**, 1061-1065 (1958)
Thomas, W.A., Hartroft, W.S.: Circulation **19**, 65-72 (1959)

Enhanced Platelet Aggregation as a Risk Factor for Progress and Complications of Vascular Disease. New Findings with a Platelet Aggregation Test (PAT III) and on the Dependence of Different Aggregation Tests on Morphologic Platelet Changes

K. Breddin, H.J. Krzywanek, J. Ziemen, H. Bauer, and H. Grün

Our interest in the measurement of enhanced platelet aggregation resulted from the idea that it may be a risk factor for the development of thrombosis and for the progression of atherosclerosis. For the development and for the control of antiaggregating drugs reliable methods for the measurement of platelet aggregation also are needed, which reflect at least partially in vivo conditions. With different methods, contradictory results have been published on this subject. We present some new findings on the interdependence of different aggregation tests and on the changes of induced and "spontaneous" aggregation found parallel to morphologic platelet changes which generally occur in platelet-rich plasma. Using platelet aggregation test I (PAT I), in which platelet aggregates are evaluated after 10 min rotation at 37°C in a glass flask we demonstrated enhanced platelet aggregation in patients with diabetes, venous thrombosis, recent myocardial infarction, and progressive atherosclerosis.

For clinical investigations and for the further identification of the underlying reactions we developed a new measuring system, platelet aggregation test III (PAT III), for the continuous registration of the aggregation process.

Principle

A small quantity of PRP (0.6 ml) is rotated in a disc-shaped plastic cuvette at 37°C. The light beam of a photometer is directed through the plasma sample so that the changes in optical density, which occur by the formation of platelet aggregates, can be continuously registered with a suitable recorder (Figs.1,2). The aggregometer was developed in collaboration with B. Braun-Melsungen A.G., 3508 Melsungen. It was designed as supplementary equipment to the commercial Eppendorf photometer.

The following parameters can be measured:

1. Angle α_1, between the horizontal line and the tangent of the transmission curve from the start of rotation.

2. Angle α_2, between the horizontal and the tangent of the steepest part of the aggregation curve is a measure for the maximum speed of aggregation.

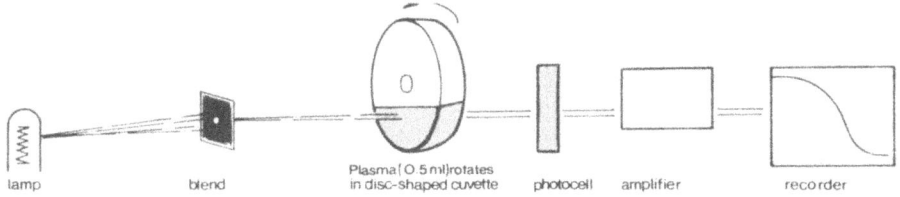

Fig. 1. Principle of PAT III. A disc-shaped cuvette, containing 0,6 ml of PRP is rotating in light beam of photometer at 37°C. Aggregation leads to increase in light transmission, which is registered on a recorder

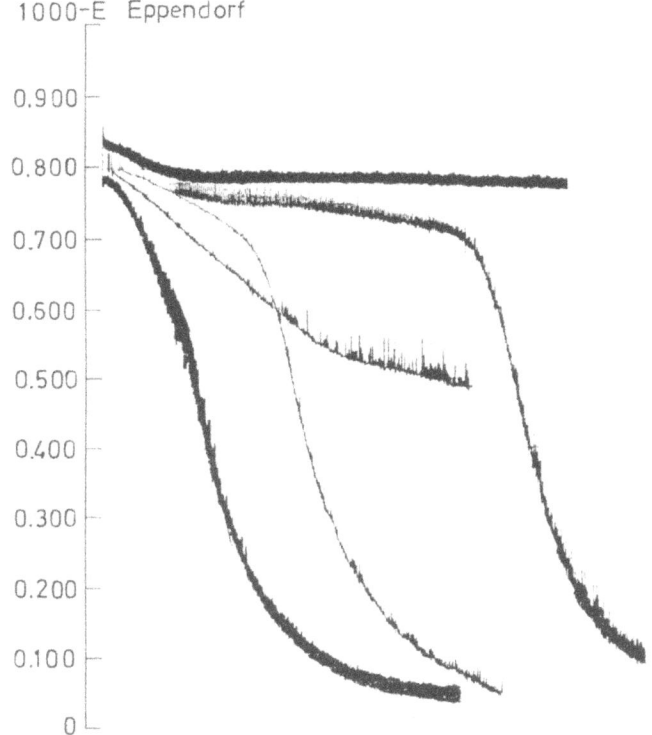

Fig. 2. Examples of PAT III curves: 1. no aggregation; 2. curve shape falsified by lipemia; 3. enhanced aggregation with prolonged Tr time; 4. and 5. strongly enhanced aggregation

3. Reaction time Tr from the start of rotation to the onset of max. aggregation.

4. Maximal amplitude Ma, the difference between initial and maximal transmission (Fig. 3).

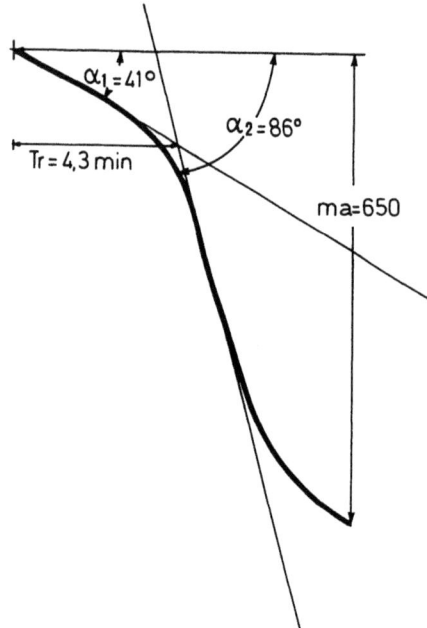

Fig. 3. Evaluation of PAT III curves. 1. Angle α_1 between the tangent of flat part of curve from the start of rotation to steepest descent of aggregation curve and the horizontal line. α_1 may be missing. 2. Angle α_2 (maximum aggregation speed) between the tangent of steepest part of curve and the horizontal line. 3. Tr is the time (minutes) from start of rotation to onset of maximum aggregation (intersection of α_1 and α_2), measured in the horizontal plane. 4. Ma (maximum amplitude) describes amplitude of aggregation curve from the starting point to end point, measured vertically

Table 1. Changes of pH in PAT III

Aggregation	pH prior to rotation	S	pH after rotation	S	Difference	N
Absent	7,83	0,14	8,56	0,11	0,73	27
Enhanced Alpha$_2$ > 30°	7.81	0,11	8,55	0,117	0,74	38

For clinical investigations Tr and α_2 are the most interesting parameters.

pH and Aggregation

During rotation plasma pH regularly rises. But we found no difference between spontaneously aggregating and nonaggregating plasma samples as to the degree of pH increase (Table 1). If the PRP was covered with a layer of air enriched with CO_2 (10-15% CO_2) ahead of rotation in the disc cuvette, no aggregation occurred and pH remained constant during rotation. The rise in pH is a conditioning factor for the aggregation in this test system as in PAT I.

Time-Dependent Changes of PAT III Results

If citrated blood is centrifuged immediately after blood sampling and PRP is rotated shortly thereafter, aggregation practically

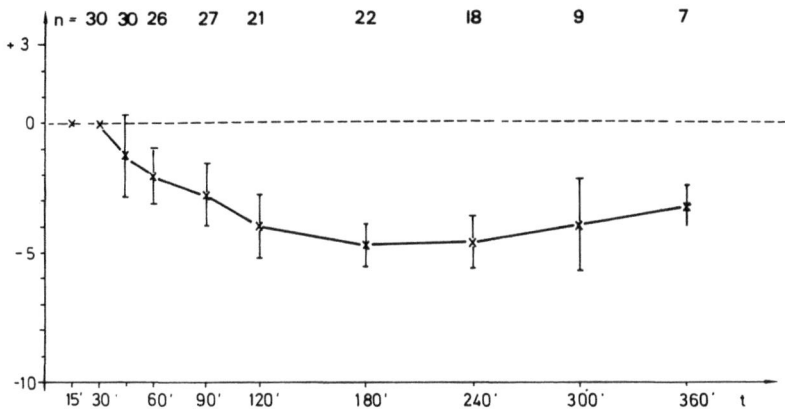

Fig. 4. Dependence of Tr on the time interval between blood sampling and rotation. Changes in relation to each preceding value and standard deviations have been calculated. Tr becomes shorter in 120 min following blood drawing and remains fairly constant for next 3 h; it is slightly prolonged thereafter. Reference (zero) value was set at 10 min

never occurs during 10-15 min rotation time. The changes of Tr in 30 test series are shown in Figure 4. Each PRP was tested at various time intervals after blood sampling. Tr is shortened during the first 90 min after blood sampling and remains rather constant thereafter. α_2 is rising during the first hour after blood withdrawal, reaching a maximum after 90 min and remains rather constant for the following 4 h (Fig. 5). Figure 6 shows an example of the shape change of the aggregation curves with the time after blood sampling.

Time-dependent changes of induced aggregation were described by Fyfe and Hamilton (1967), by Reuter et al. (1973), and by Warlow et al. (1974) for ADP- and adrenaline-induced aggregation.

Time-Dependent Changes of ADP-Induced Aggregation

ADP-induced aggregation was registered with a Braun-Aggregometer adapted for the measurement of induced aggregation, using the principle of Born's EEL-aggregometer. Citrated blood was collected from healthy volunteers and from patients with diabetes, recent myocardial infarction, or occlusive peripheral arterial disease, who had not taken acetylsalicylic acid containing drugs during the last 6 days. Citrated blood was centrifuged immediately after withdrawal for 1.5 min at 150 g at room temperature. PRP was pipetted into a fresh polystyrol tube and kept at room temperature. From each PRP a PAT III curve was recorded 60-120 min after blood sampling. 5×10^{-7}, 10^{-6}, and 5×10^{-6} of ADP were used to induce aggregation 10, 20, 30, 60, 90, 120, and 180 min after blood sampling.

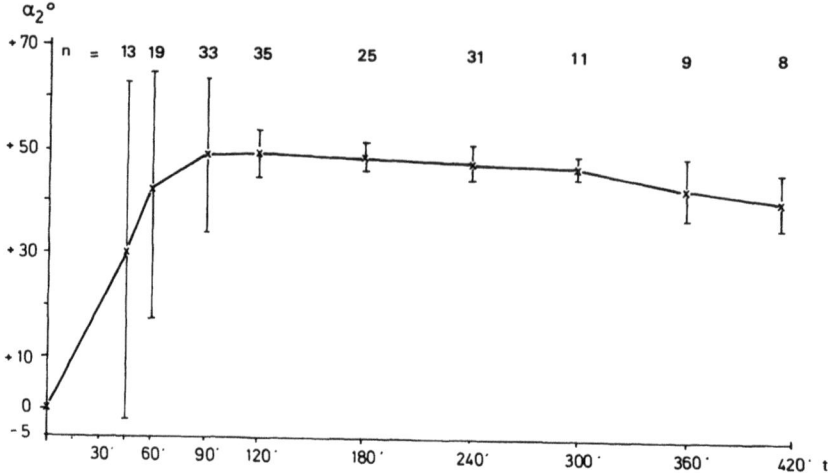

Fig. 5. Dependence of angle α_2 in comparison with respective preceding value have been registered. In 60 min following blood sampling α_2 increases rapidly. The wide variation of test results depends mainly on variation of aggregation in time rather than on variability of α_2. Between 60 min and 90 min α_2 increases slightly, it remains fairly constant for 3-4 h thereafter

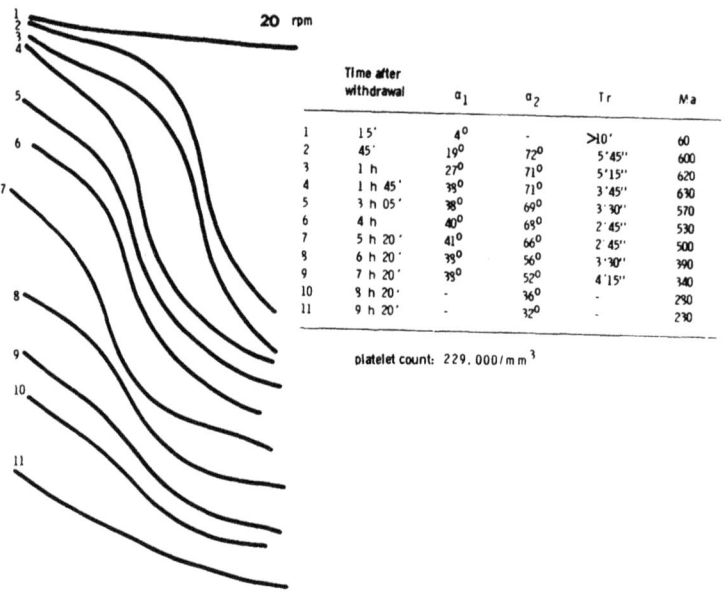

Time after withdrawal	α_1	α_2	Tr	Ma
1	15'	4°	-	>10'	60
2	45'	19°	72°	5'45"	600
3	1 h	27°	71°	5'15"	620
4	1 h 45'	39°	71°	3'45"	630
5	3 h 05'	38°	69°	3'30"	570
6	4 h	40°	69°	2'45"	530
7	5 h 20'	41°	66°	2'45"	500
8	6 h 20'	39°	56°	3'30"	390
9	7 h 20'	39°	52°	4'15"	340
10	8 h 20'	-	36°	-	280
11	9 h 20'	-	32°	-	230

platelet count: 229.000/mm^3

Fig. 6. Variability of aggregation curves at 20 rpm depending on time interval from blood sampling. No aggregation occurred 15 min after blood drawing (1). At 45-240 min α_2 remains largely constant with increasing shortening of Tr (2-6). At 320-560 min Ma is rapidly diminishing (7-11). Curves were drawn under each other to avoid superimposing them. They start always at same extinction

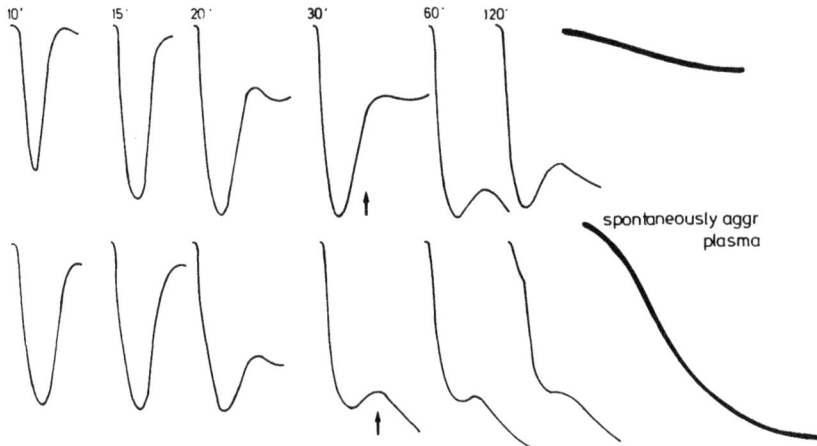

Fig. 7. Changes with time after blood sampling; ADP-induced aggregation with 10^{-6} M of ADP. The most striking change with time is the decrease of % disaggregation. Using PRP which aggregates in PAT III the decrease is reached earlier than in nonaggregating plasma

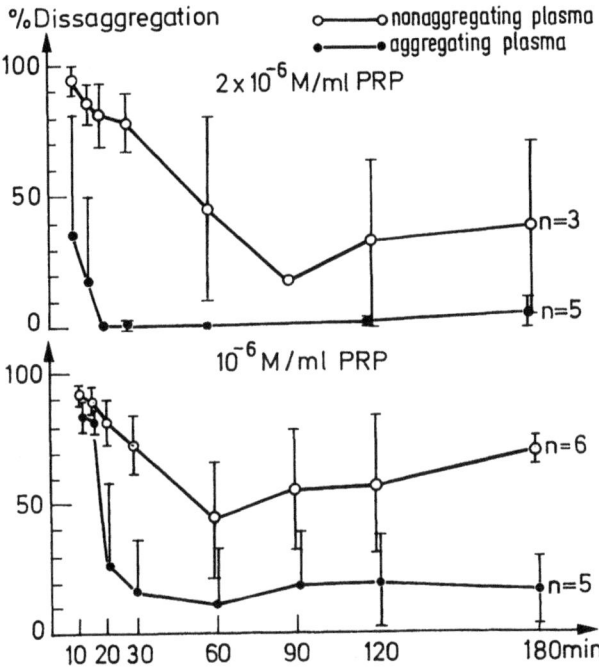

Fig. 8. Differences in % disaggregations in ADP-induced aggregation with time after blood sampling. Differences are largest 30 min after blood sampling with 2×10^{-6} and 10^{-6} M ADP

ADP generally induced prompt aggregation followed by disaggregation during the first 30 min after venipuncture. Later a second wave of aggregation often appeared. Disaggregation markedly changed with time after blood sampling and generally the degree of disaggregation was lower in PRP which showed aggregation in PAT III than in nonaggregating PRP (Ex. Figs. 7 and 8). The cor-

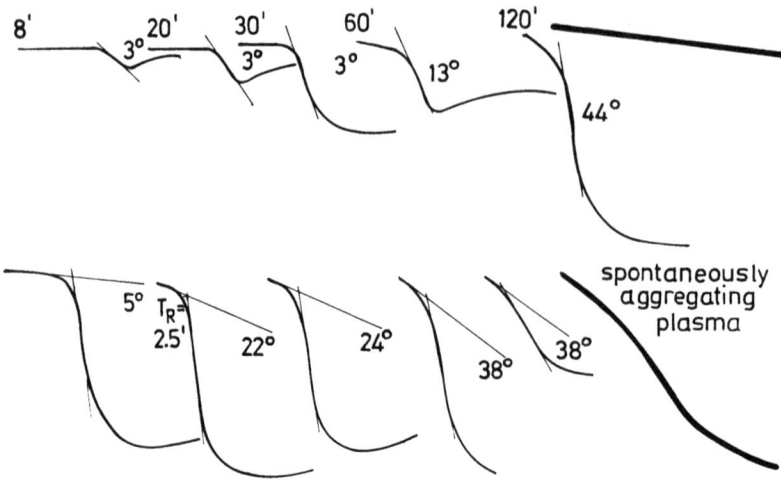

Fig. 9. Changes with time after blood sampling: collagen-induced aggregation with 1 µg/ml PRP. With time after venipuncture maximal amplitude rises, angle of primary aggregation becomes steeper, and time to onset of maximal aggregation is shortening. In PRP, which aggregates in PAT III-test, Tr is shorter and angle of primary aggregation is steeper than in non-aggregating plasma

relation (Spearman-rang) of aggregating-nonaggregating PRP (angle α_2 of PAT III) with the percentage of disaggregation was best using 10^{-6} M ADP and performing the test 30 min after blood sampling (R = 0.677, N = 53, P = 0.001). If the test was performed 20 min after blood sampling R was 0.339 (N = 47, P ~ 0.02) and 45 min after venipuncture R was 0.464 (N = 16, P < 0.1). With 5×10^{-6} or 5×10^{-6} M ADP, a lesser correlation between α_2 and disaggregation was found. In principle, ADP-induced aggregation yielded similar results as PAT III if 10^{-6} M ADP were used to induce aggregation and if the test was performed 30 min after venipuncture.

Time-Dependent Changes of Collagen-Induced Aggregation

With time after blood sampling, the angle of primary aggregation increases and the time Tr between addition of collagen and the onset of maximal aggregation shortens (Fig. 9). The maximal amplitude of the aggregation curves is a less reliable parameter. Time-dependent changes are most pronounced if small doses of collagen are used. With 1 µg collagen/ml PRP, the difference in Tr between aggregating and nonaggregating PRP in the PAT III was greatest 30 min after blood sampling (Fig. 10). So, also, collagen-induced aggregation shows marked time-dependent changes and using low doses 30 min after venipuncture results similar to PAT III could be obtained. With 5×10^{-6} M adrenaline used to induce aggregation, marked time-dependent changes could also be demonstrated.

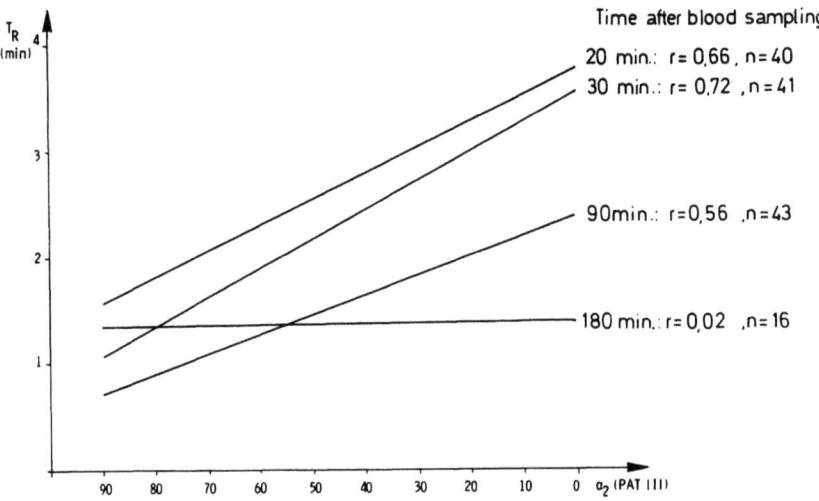

Fig. 10. Correlation of Tr in collagen-induced aggregation (1 µg/ml PRP) with angle α_2 of PAT III at different times after blood sampling. Correlation is best 30 min after venipuncture and there is no correlation at 180 min

Which changes in PRP are responsible for the rising sensitivity of the platelets to aggregation inducers observed during the first hour after blood withdrawal?

"Primary" Shape Change of Platelets

If platelets are fixated immediately at venipuncture by feeding the blood directly through a large cannula into 6% glutar-aldehyde, most platelets appear as flat discs (Fig. 11); about 20-25% show one or several processes. Seconds later many platelets in citrate blood show small protrusions and during the following minutes platelets swell to some degree and develop more and more extrusions (Fig. 12).

The number of platelets with tentacles can easily be counted using phase-contrast microscopy.

If citrate blood is fixated in 6% glutar-aldehyde at different times after blood sampling and if the blood is incubated at room temperature, the number of platelets with tentacles reaches 90% after 60 min (Fig. 13). The process of swelling, which is more difficult to evaluate using light microscopy, continues during the following hours. There seemed to be a good correlation between these "primary" shape changes platelets generally undergo after blood sampling and the changing behavior in different aggregation tests.

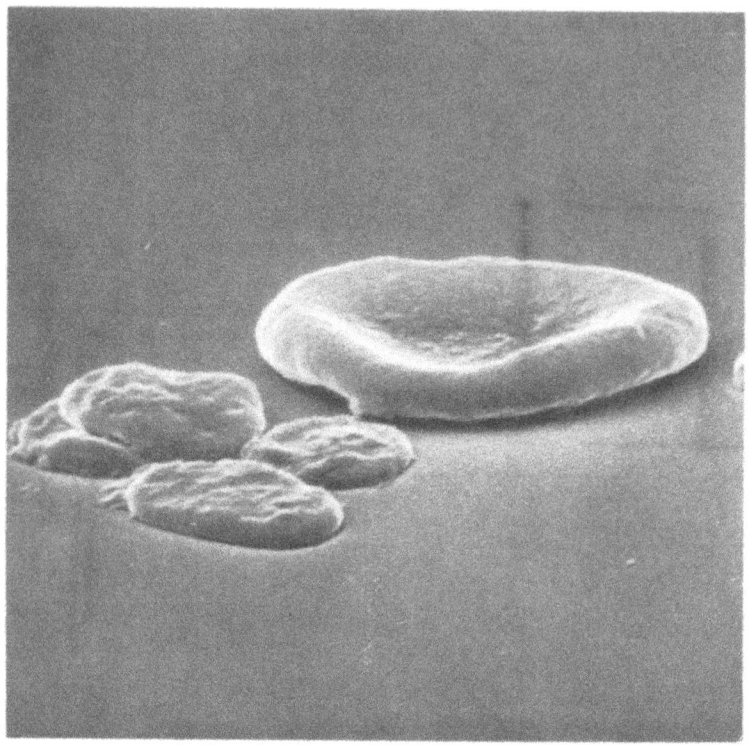

Fig. 11. Blood was fixated by directly feeding it through large cannula into 6% glutar-aldehyde. Platelets were separated by differential centrifugation. They appear as flat round discs, some of them showing hairlike processes. One platelet is swollen with many small protrusions, some show several small protrusions. Platelet in the center is lengthened. Scanning electron micrograph: magnification 5000 ×

Influence of Different Incubation Temperatures on the Time-Dependent Changes of Spontaneous and Induced Aggregation and on the Shape of Platelets after Blood Sampling

If citrate blood is centrifuged at 33°C and the PRP is thereafter incubated at 37°C, the time-dependent change of sensitivity to ADP and the behavior of spontaneously aggregating plasma (at room temperature) are strongly altered. Throughout the incubation time the PRP is less sensitive to ADP, the sensitivity decreases with time. In PAT III PRP will not aggregate at all or much later than at room temperature (Fig. 14). The formation of tentacles is markedly slowed at 37°C (Fig. 13).

If citrate blood or PRP are incubated at 4°C the formation of tentacles and the swelling of platelets are greatly enhanced (Fig. 13). At 4°C 100 percent of platelets develop extrusions within 10 min.

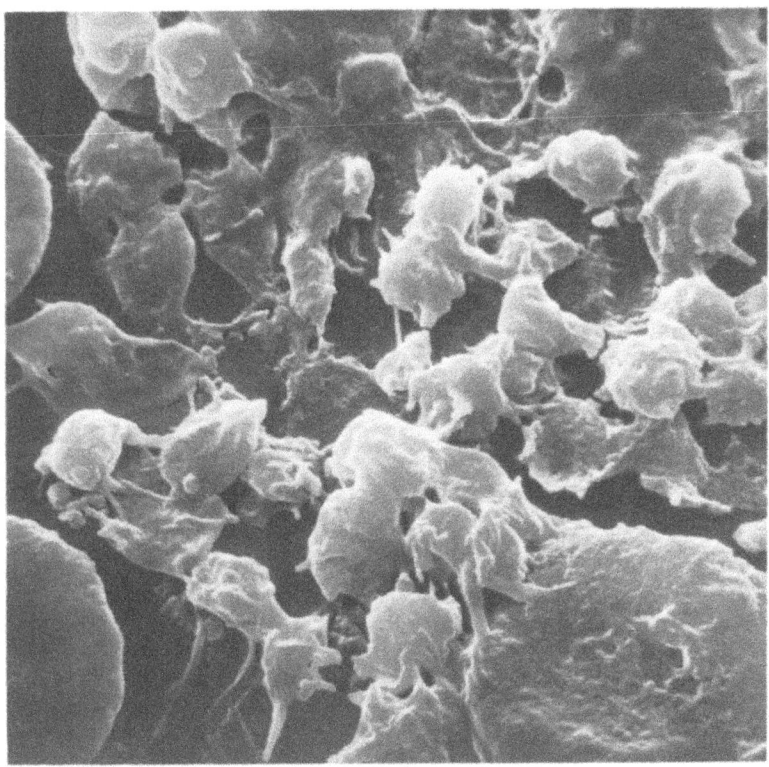

Fig. 12. Citrate blood in 6% glutar-aldehyde 30 min after venipuncture at room temperature. All platelets are swollen and show several tentacles and a very irregular surface. Scanning electron micrograph: magnification 5000 ×

The higher sensitivity of PRP to ADP after 10 min incubation at 4°C and the shape changes can be partly reversed by reincubating the PRP at 37°C. Results of such experiments are shown in Figures 15 and 16.

Conclusions

The time-dependent changes observed in PAT III, ADP-, collagen-, and adrenaline induced aggregation are well correlated with the primary shape change platelets undergo after blood sampling. This shape change, which was described in principle by Zucker and Borelli (1954), is slowed at 37°C and accelerated at lower temperatures. The shape changes and the sensitivity of the PRP to low doses of aggregation inducers are partially reversible in vitro.

PAT III is a very sensitive test system. Spontaneous aggregation only occurs if the rising pH in the PRP further enhances the formation of shape change platelets. On the other hand, neither the rising pH nor the shape change are responsible for the differences

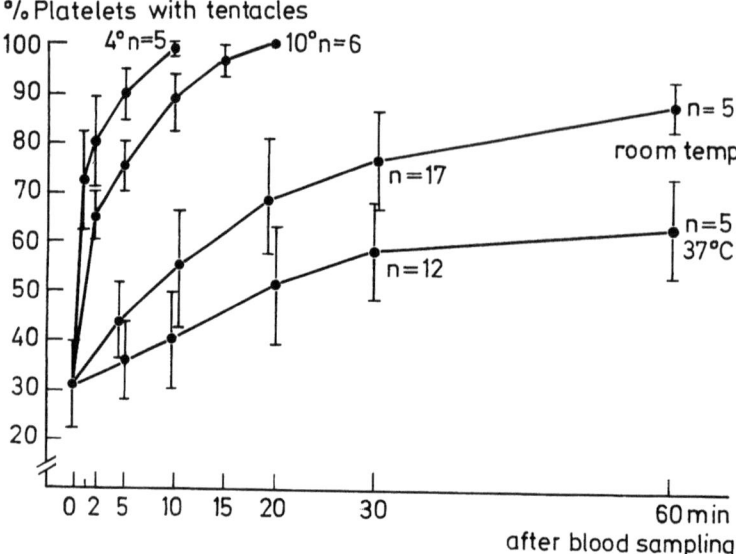

Fig. 13. Percent platelets with tentacles fixated at different times after venipuncture and incubated at different temperatures. In citrate blood kept at 4°C all platelets (100%) develop tentacles within 10 min after blood sampling. At 30 min after blood drawing and at incubation temperature of 37°C, 55% of platelets have pseudopods

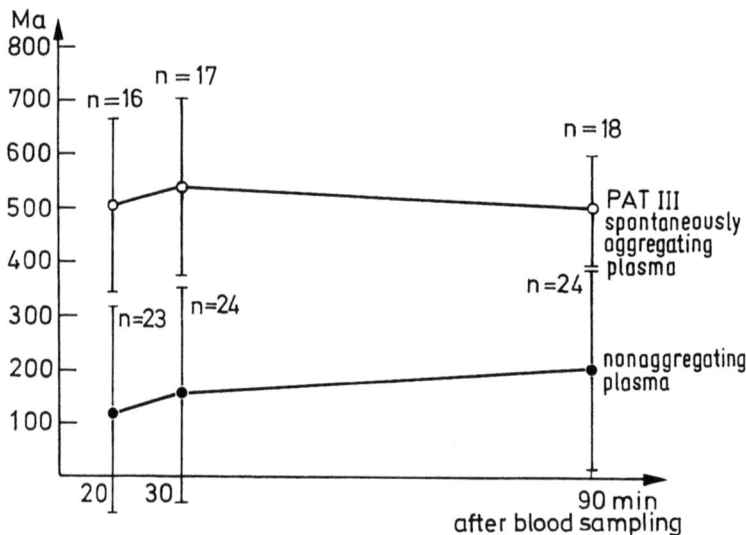

Fig. 14. Different incubation temperatures lead to different results in all aggregation tests at 37°C. Angle α_2 of PAT III is much lower than at room temperatures. At 37°C the sensitivity of PRP is less than at room temperature

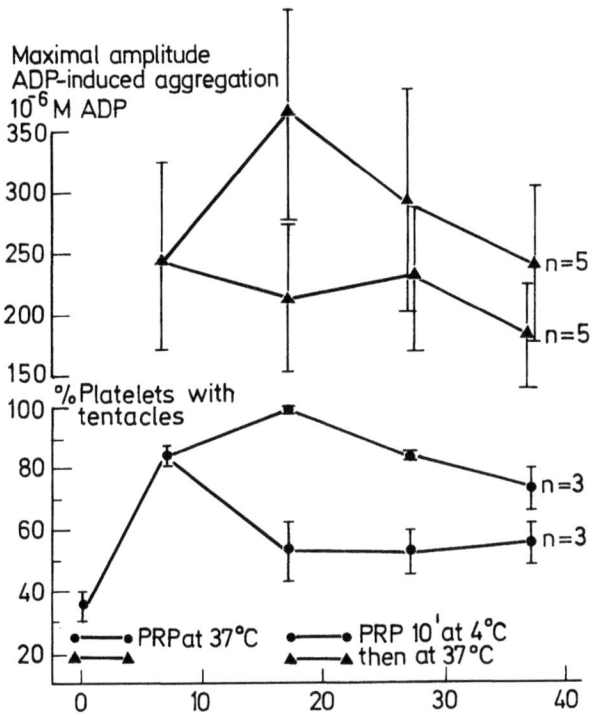

Fig. 15. Variation of incubation temperatures leads to partial reversal of the formation of platelet protrusions in PRP. Parallel to these changes in platelet morphology the maximal amplitude of aggregation curves in ADP-induced aggregation (10^{-6} M) changes: greater amplitude if PRP is kept at 4°C for 10 min, decrease of Ma after reincubation at 37°C

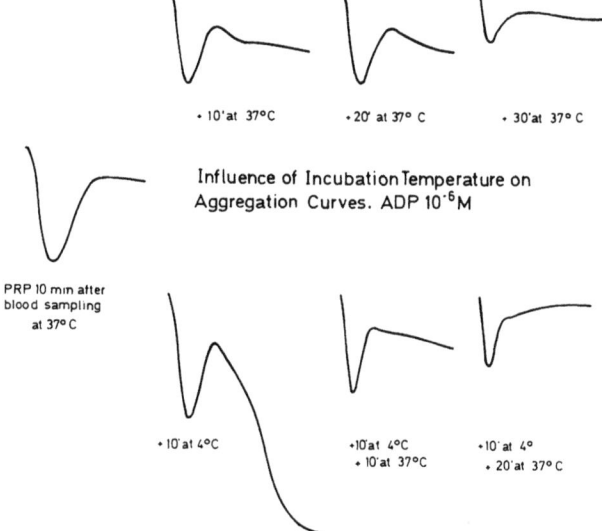

Fig. 16. Influence of incubation temperature on aggregation curves. This example shows enhanced aggregation with 10^{-6} M ADP if PRP is kept for 10 min at 4°C. After reincubation at 37°C aggregation curves resemble curves obtained with PRP, which was kept at 37°C (upper row)

found between PRP of healthy individuals and aggregating PRP. The count of tentacle formation in aggregating and nonaggregating plasma with time after blood sampling showed no significant differences.

These findings are congruent with our hypothesis that a high molecular weight plasma protein, which we named Platelet Aggregating Factor (PAF), is responsible for enhanced aggregation in PAT III. It seems likely that in vivo only irritated - that is shape changed - platelets to adhere to the vessel wall or aggregate. Raster electron microscopic investigations usually show platelets with "pseudopods" adhering to the endothelial surface.

In vitro the formation of irritated platelets is accelerated by low temperatures, by contact to foreign surfaces, by collagen, ADP, thrombin, epinephrine, and probably by many other factors and substances. It seems also likely that the formation of irritated platelets is completely or at least partially reversible also in vivo. If the term shape change is used, some workers usually refer to ADP-induced changes. These differ morphologically from the changes which we call primary shape change. ADP induces formation of large spherical extrusions from the platelets which may be three to five times larger than the original platelets and which easily disrupt. The number of platelets transforming in this way increases with increasing concentrations of ADP. The rest swell to some degree and form some tentacles.

Clinical Investigations with PAT I and III

Selection of Patients

A group of 545 "healthy" subjects included nurses, students, members of the medical staff, and other personnel of the Frankfurt University Medical Center and almost 300 policemen of the city of Frankfurt with no apparent disease. These persons stated that they felt to be in good health and had no medical treatment at the time of the study. They were not examined whether they had symptoms of manifest atherosclerosis or its complications.

In contrast to the commonly used aggregation tests (Born, O'Brien) in which aggregation is induced by addition of ADP, collagen, etc. to PRP, the PAT enables us to study the spontaneous aggregation of thrombocytes in their normal plasma environment, and the addition of inducing substances is not necessary. The other group was in- and out-clinic patients of the University Medical Center, Frankfurt/Main, with a variety of acute and chronic internal diseases.

A number of patients, who were seen for other reasons, had a thromboembolic accident in the weeks or months following the performance of an aggregation test. In this way data were obtained of 148 patients "prior to a thromboembolic event."

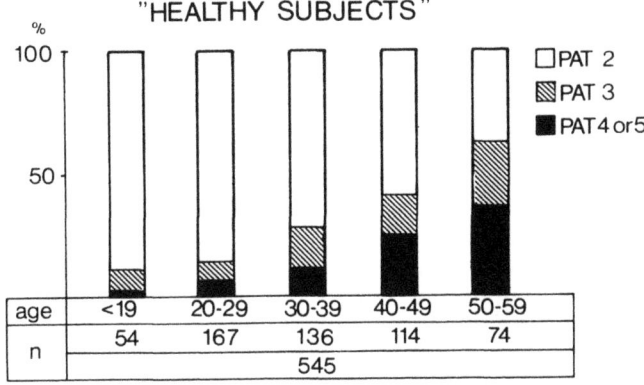

Fig. 17

Figs. 17-19. Platelet aggregation was assessed using PAT I. PRP is rotated in siliconized glass flasks (37°C, 20 rpm)

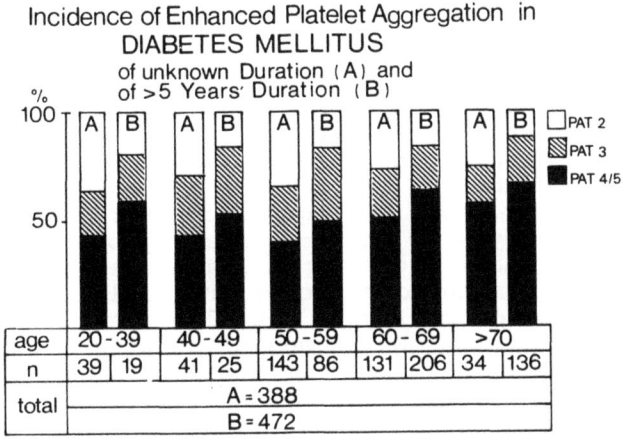

Fig. 18

Results

1. In an apparently "healthy" population an increasing number of individuals with enhanced platelet aggregation (PAT grades 4 and 5) were found with rising age (Figs. 15 and 17). In the age group 50-59 years, one third of the "normals" showed a markedly enhanced spontaneous aggregation tendency.

2. Three groups of diabetics were studied: 388 patients in whom the duration of the disease was unknown, 472 patients with diabetes of more than 5 years' duration, and 182 patients in whom diabetes was known for 10 years at most. Throughout the various age groups the incidence of enhanced platelet aggregation was markedly higher in diabetics than in the control group, which was most evident below 40 years of age. If diabetes had lasted for more than 5 years, there was another 10-15% increase of PAT

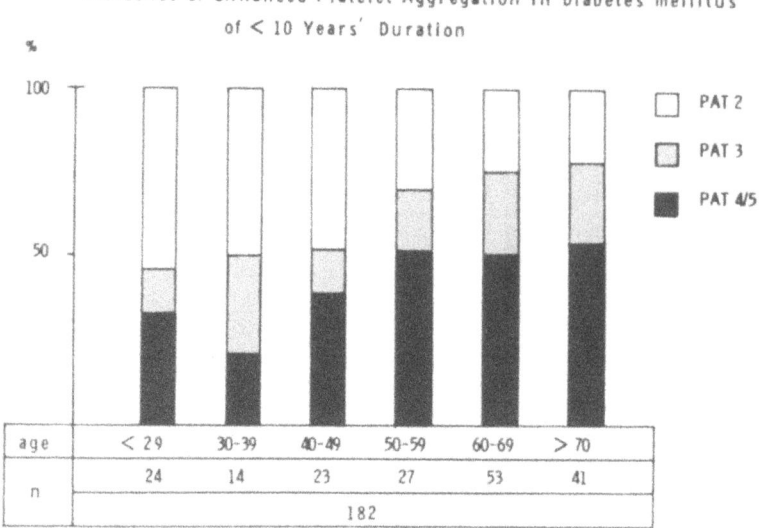

Fig. 19

grades 4/5 as compared with the unselected group of diabetics (Fig. 18). Similar results were obtained in diabetes of less than 10 years' duration, except that there was a lesser percentage of enhanced aggregation under 40 years (Fig. 19). Using the photometric PAT (III), aggregation curves were obtained from 124 healthy volunteers and 88 diabetics in whom the disease was known for less than 10 years (Fig. 20). Two age groups were compared: under 40 years, and 40-59 years. In the control group no spontaneous aggregation with a small angle alpha 2 is observed in 65% of the younger subjects. Enhanced aggregation (angle alpha 2 greater than 60°) is found chiefly in the older age group. The percentage of diabetics below the age of 40 with a normal aggregation test is roughly half that of the control group. The number of diabetics with strongly enhanced aggregation is close to 50% in both age groups.

3. In a follow-up study of 401 patients in whom a PAT I was performed between 5 days and 3 years after they had suffered a myocardial infarction, more than half of them had enhanced platelet aggregation in all age groups over 40 years (Fig. 21).

4. The incidence of PAT grades 4 and 5 in 273 patients with peripheral arterial occlusive disease was between 45 and 60% (Fig. 22).

5. In 148 patients a PAT had been performed within 6 months *prior* to a thromboembolic episode. These data may be called prospective because they were gathered before the accident which led to hospitalization. The incidence of normal platelet aggregation was strikingly low in all age groups, and enhanced aggregation was found in 54-78% of these patients (Fig. 23).

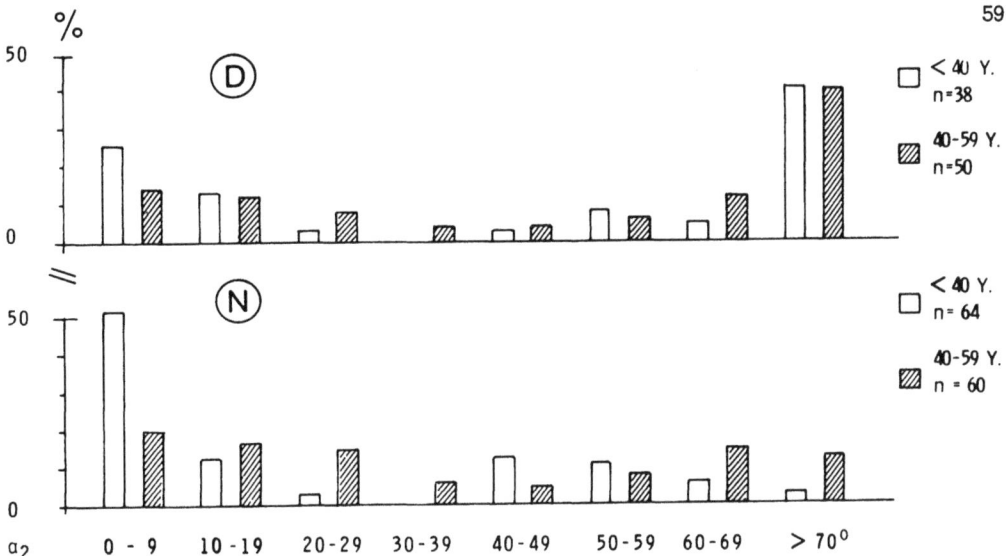

Fig. 20. PAT grade 2: no aggregation, single platelets; grade 3: mild aggregation, some reversible aggregate; grade 4: enhanced aggregation, partly irreversible; grade 5: strongly enhanced aggregation, irreversible aggregates. Platelet aggregation was assessed using PAT III: Angle alpha 2 was measured in 88 patients with diabetes mellitus (< 10 years duration) and 124 healthy controls. Two age groups are compared: < 40 years, 40-59 years

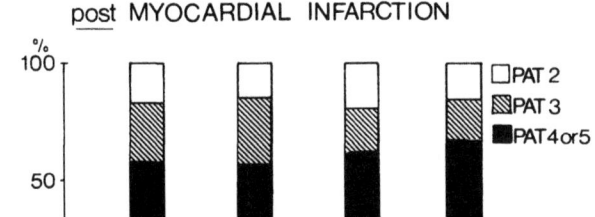

Fig. 21

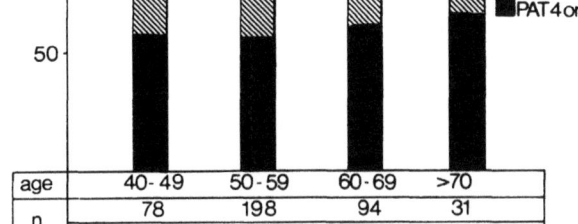

Figs. 21-24. Rotated PRP is then poured into plastic slides, which are rinsed and stained and platelet aggregation is studied microscopically

6. Two patient groups were selected for further evidence that enhanced platelet aggregation may be of importance in predicting thrombotic complications of vascular disease. Out of 41 patients who were seen before they suffered myocardial infarction, only one had a normal aggregation test, and none out of 24 who developed venous or arterial thrombosis. But by far more than half of these patients had strongly enhanced platelet aggregation (Fig. 24).

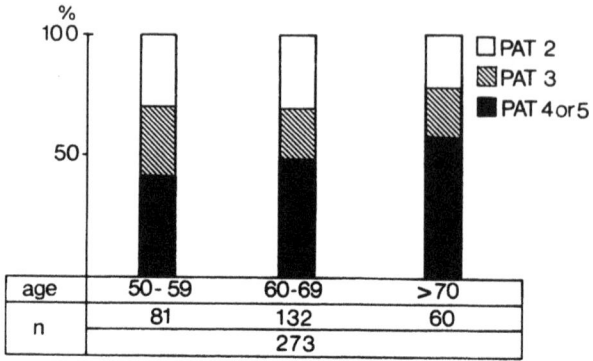

Fig. 22

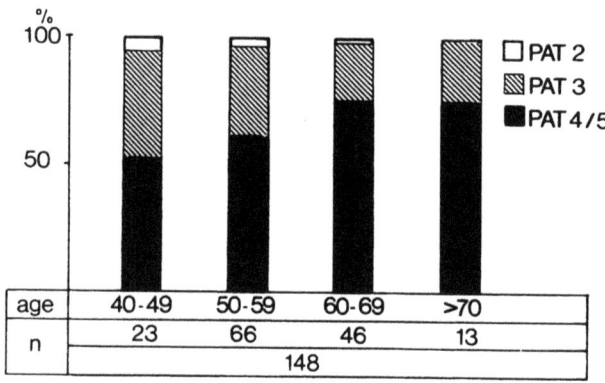

Fig. 23

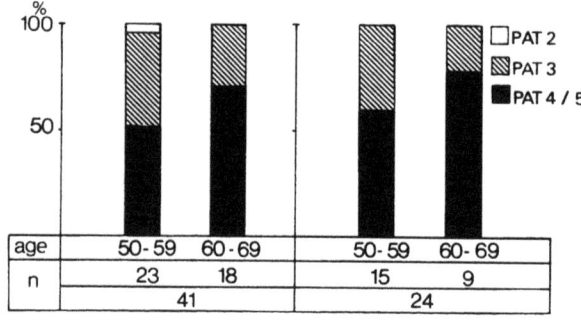

Fig. 24

Discussion

The number of "healthy" persons with enhanced platelet aggregation is growing with age. In our opinion this is correlated with the progress of atherosclerosis with advancing years.

In patients with complications of atherosclerosis, like recent myocardial infarction, peripheral occlusive disease, but also in diabetics, the incidence of enhanced platelet aggregation was significantly higher than in "healthy" subjects.

The highest incidence of enhanced aggregation was observed *prior* to thromboembolic attacks in atherosclerotic patients and in patients with diabetes of more than 5 years' duration and recent myocardial infarction. These groups were not different in statistical analysis, but altogether varied from age-matched "healthy" controls.

Following these findings it may be concluded that enhanced spontaneous platelet aggregation, if repeatedly evident in one of the aggregation tests (PAT I or III), by itself represents an independent risk factor for thromboembolic complications, especially in patients with advanced atherosclerosis.

Our recent results lead to the following hypotheses:

1. The first step in hemostasis or thrombus formation is the development of irritated platelets. This is an extremely rapid process. We do not yet know which factors are responsible for it or accelerate and slow it.

2. From in vitro investigations we only can expect indirect insight in these processes. Using citrated or heparinized blood we generally work with a population of irritated platelets.

3. In clinical investigations the time-dependent changes of the different test systems must be considered. Contradictory results obtained by different investigators using the same test in the same group of patients can now be explained if the tests were performed at varying times after blood sampling.

4. The PAT tests I and III show enhanced aggregation in a high percentage of patients with vascular disease and prospective data obtained so far make it likely that enhanced aggregation is a risk factor for thrombosis and probably also for the progress of atherosclerotic lesions.

Further investigations are necessary to elucidate the biochemical reactions which are responsible for the changes observed on platelet sensitivity and morphology.

References

Born, G.V.R.: Quantitative investigations into the aggregation of blood platelets. J. Physiol. (Lond.) 162, 67 (1962)

Breddin, K.: Die Thrombozytenfunktion bei hämorrhagischen Diathesen, Thrombosen und Gefäßkrankheiten. Thromb. Diath. haemorrh. Suppl. 27 (1968)

Breddin, K., Grun, H., Krzywanek, H.J., Schremmer, W.P.: On the measurement of spontaneous platelet aggregation. The platelet aggregation test III. Methods and first results. In preparation

Breddin, K., Ludwig, H., Ziemen, M., Bauer, M., Schaudinn, I.: Time dependent changes of ADP-collagen and epinephrine-induced in "spontaneous" platelet aggregation and morphologic platelet changes. In preparation

Fyfe, T., Glasg, M.B., Hamilton, E.: Effect of variation of the interval between venipuncture and measurement of platelet adhesiveness by the Payling Wright method. The Lancet 542, 543 (1967)

Krzywanek, H.J., Breddin, K.: Der Plättchenaggregationstest (PAT). Med. Lab. (Stuttg.) 15, 103 (1972)

Leuschner, M., Breddin, K.: Electronmicroscopic investigations. In preparation

O'Brien, J.R.: Some possible enzyme systems involved in platelets aggregation. J. Atheroscl. Res. 3, 262 (1961)

Reuter, H., Podolsak, W., Hagen, T., Linker, H., Stroder, J., Gross, R.: Investigations on the time dependance of platelet functions. J. gen. meth. Invest. Thromb. Res. 3, 307 (1973)

Warlow, Ch., Corina, A., Ogston, D., Douglas, A.S.: The relationship between platelet aggregation and time interval after venipuncture. Thromb. Diath. haemorrh. 31, 133 (1974)

Zucker, M.B., Borelli, J.: Reversible alterations in platelet morphology produced by anticoagulants and by cold. Blood 9, 602 (1954)

Pathogenesis

Chairmen: M. Anthony, T. Abe, E. J. Acheson,
J. Sterling Meyer, J. Marshall, K. J. Zülch

Thrombosis and Embolism as a Cause of Ischemic Cerebrovascular Disturbances. Analysis from a Serie of 1000 Patients*

K.J. Zülch and H. von Einsiedel-Lechtape

In this section we would like to describe one specific area where agents for the prevention of thrombosis and embolism may be important: cerebrovascular disease. This can be the result of a variety of disturbances and though there is rarely only one factor responsible (Yates and Hutchinson, 1961; Zülch, 1967), it is most frequently caused by local factors such as arteriosclerotic changes of the vessel wall, leading to stenosis, thrombotic occlusion of the lumen, or embolism. The latter usually occurs as macroembolism, mainly of cardiac origin, or as microembolism originating probably from ulcerated atherosclerotic neck arteries. Microemboli are at present considered as a frequent cause of Transient Ischemic Attacks, however, they play only a minor role in the pathogenesis of longer duration ischemic lesions (Fisher, 1961). The reported frequencies of embolism vary widely between 5% and 46% (Glynn, 1956; Jörgensen and Torvik, 1966). Kurtzke (1969), in reviewing several clinical, postmortem, and epidemiologic studies, suggested that about 10% of thromboembolic occlusive disease was due to embolism.

An important disadvantage and source of error in most clinical studies concerned with the relative frequencies and the natural history of thrombosis and embolism is the lack of an arteriographically established diagnosis.

The present investigation was undertaken to evaluate the extent in which thrombosis and embolism are causative in the pathogenesis of ischemic CVD and the factors that precede or accompany the permanent or temporary occlusion of a cerebral artery.

Methods. The case material was collected from 1000 charts of patients with the diagnosis of ischemic CVD who were admitted to the Neurologische Klinik Köln-Merheim from 1959 to 1973. The medical records were abstracted for information regarding age, sex, predisposing and contributing factors, clinical and laboratory findings, blood pressure readings, diagnostic procedures, and also therapy, outcome, pathogenesis, and, in a number of cases, pathologic anatomy.

---

* Supported by a grant of Verband der Lebensversicherungs-Unternehmen e.V., Bonn

The diagnosis of an ischemic stroke was primarily based on the neurologic examination; in 54.2% of the cases arteriography had been performed, 91% had at least one EEG examination, 49% a brain scan. In 18% the diagnosis could be confirmed by autopsy.

The differentiation into the etiologic subdivisions, thrombosis and embolism, was based on:

1. The criteria of the ad hoc Committee on CVD (1958) and the Joint Committee for Stroke Facilities (1972) and

2. The morphologic changes as verified by arteriography or autopsy.

Diverging from the International Classification of Diseases (1967) thrombosis in this material comprised patients with verified complete vessel occlusion only.

The criteria used to differentiate between thrombosis and embolism were:

1. For thrombosis, an observation of arteriosclerosis and complete obstruction of an arterial cerebral vessel as demonstrated by arteriography and/or autopsy;

2. For embolism, the abrupt onset of the clinical syndrome, when one of the following factors were present: predisposing cardiac conditions, single or multiple branch occlusions with or without dissolution in reangiography or embolic material within the arterial branches of the cerebral circulation at necropsy.

Thus, of the 1000 patients, in 297 the diagnosis of thrombotic occlusion or embolism could be assessed with relative certainty. Of these 297 patients, 204 (68.7%) had arteriography and 99 (33.3%) had autopsy. Electroencephalographic studies were carried out in 267 (28.9%) and brain scanning in 151 (50.8%).

The difficulties in determining the causative factors are best illustrated in the following case of a young man in whom an arterial thrombosis occurred without any arteriosclerosis: This 20-year-old healthy policeman arrived at a Spanish seaside hotel after a 2 h flight. On the same night he became intoxicated with alcohol and the next morning he awoke hemiplegic. After his return, he was found to have an occlusion of the internal carotid artery. In this case we did not find a convincing pathogenetic explanation. However, we had a very similar observation in a young student of gymnastics, who had the same unfortunate experience of waking up with a hemiplegia after a heavy drinking session during a skiing holiday in Austria.

When we consider that young people in many countries consume a great deal of alcohol without any similar consequences, we may be inclined to consider that alcoholic intoxication alone is not the causative factor. However, in the above cases there were no other visible factors.

f Causative Entities. The pathogenetic mechanism mainly ~~respons~~ible for the ischemic insult leading to hospital admission was considered to be thrombosis in 236 patients (23.6%) and ~~emboli~~sm in 61 (6.1%). It is, of course, not possible to calculate the true clinical incidence of cerebral emboli from these ~~cas~~es. In about half (40.9%) of the entire material the pre~~cise~~ etiology was never found. However, the percentage of emboli~~sm~~ was quite probably underestimated.

~~In~~ summary, thrombotic occlusion was four times as common as ~~emb~~olism.

Sex and Age. Males predominated in the total series and in the thrombotic group, in the latter with a ratio of 2:1. This is in agreement with the known male excess for arteriosclerotic lesions, especially in the extracranial carotid arteries (Fields et al., 1968; Dorndorf, 1969). In embolism, females outnumbered males slightly and thus differed significantly from the thrombotic group (Table 1).

The *age* distribution of thrombosis and embolism showed differences only in the younger age groups, i.e., below 41 years. Here there was a significant increase in the frequency of embolism, which was a reflection of the high proportion of women. The role of oral contraceptives is not yet fully established but several recent reports stress the influence of the estrogen content on the relative frequency of stroke in young women. The age-sex distribution of patients with a thrombotic occlusion shows significant differences which were not present in embolism: in thrombosis, the peak of onset was one decade later in women. This may be the consequence of the delay in the onset of arteriosclerosis in the female sex. Men constituted 64% (M/F ratio 1.8) of the thrombotic group and this differed significantly from the embolic in which 47.5% were males (M/F ratio 0.9).

Table 1. Age and sex distribution

Age	Thrombotic occlusion			Embolism		
	Male	Female	Total	Male	Female	Total
< 41	2.0	9.6	4.7	7.0	21.9	14.8
41-50	15.1	13.1	14.4	6.9	15.7	11.5
51-60	22.4	22.6	22.5	24.1	18.7	21.3
61-70	39.5	17.8	31.8	31.0	18.8	24.6
70	21.1	36.9	26.7	31.0	25.0	27.9
Total (%)	100.1	100.0	100.1	100.0	100.1	100.1
N	152	84	236	29	32	61
M/F ratio		1.81			0.91	

In summary, there was a general male preponderance of 2:1 in thrombosis and there was a relatively higher general frequency of embolism in the younger age groups, particularly in females. In thrombosis men prevailed and, remarkably, the peak of onset was one decade earlier in men.

Predisposing and Contributing Factors

1. Hypertension. It seems evident from prospective studies (Kannel et al., 1965; Veterans Administration Cooperative Study, 1967, 1970) that hypertension is one of the most important treatable risk factors for stroke. In the present study, a history of hypertension was recorded as well as the mean of hourly blood pressure readings while in the hospital. Several of the patients had, in spite of a previous diagnosis of hypertension, normal blood pressure values while in the hospital and vice versa. Despite this, a surprisingly good agreement between the corresponding data was found in both groups. Hypertension was present in 55% of the total series, in 57% of patients with thrombotic occlusions and in 39% of those with embolic occlusions. The significantly less elevated blood pressure values in embolism probably represent those of a normal population, while, on the other hand, hypertension may be one of the causative factors in thrombosis (on the basis of atherosclerosis).

2. Cerebrovascular Disease. A history of transient ischemic attacks (TIAs) is one of the factors which, with regard to stroke prevention, should be identified and treated successfully (Joint Committee for Stroke Facility, 1972). In our material, TIAs preceeded a stroke in 16 (7.2%) of the patients with thrombotic occlusion, but only in one case (1.7%) of the embolism patients. On the other hand, a history of one or more strokes occurred in roughly equal proportions in both groups (25.7% thrombotic versus 27.1% embolic).

3. Cardiovascular and Peripheral Vascular Disease. Prior evidence of cardiovascular disease was significantly higher in the embolic group (35 out of 63 cases = 66%), when compared with the thrombotic group (67 of 198 patients = 33.8%). However, as would be expected, peripheral vascular diseases were significantly lower in the embolic group (3.8%) in comparison with the thrombotic group (12.6%).

4. Diabetes was about equally distributed (14.8% in thrombosis, 14.0% in embolism). Although elevated *serum cholesterol or lipid levels* were found less often in embolism (28.1%) compared to thrombosis (40.5%), this difference is not statistically significant.

In summary, the higher incidence of TIAs in the history of patients with thrombosis was important, however, it can be explained on the basis of our pathogenetic concept of TIAs, where stenosis (which is a common precursor of thrombosis in the aged) is usually regarded as the main causative factor. Furthermore, hypertension is more frequent in the thrombotic group and this is in good correlation to the general concept, since this is one of the causative factors in the stenosing type of arteriosclerosis.

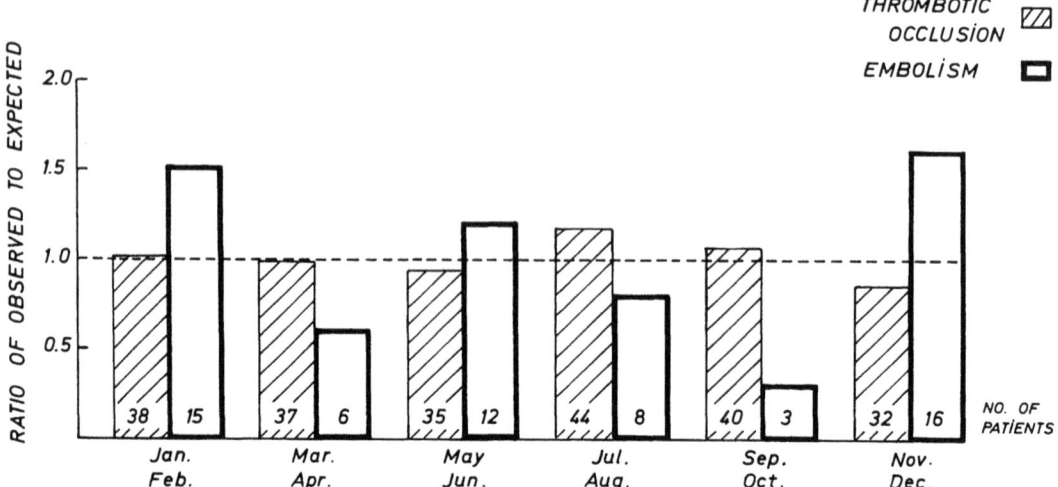

Fig. 1. Distribution by 2-month periods of onset of the ischemic stroke. The ratio of the observed to the expected number of cases is 1 (broken line), if figures are equal. No marked fluctuations in thrombosis, but in embolism a tendency for more attacks to occur in winter months

Seasonal, Weekly, and Daily Variations

Seasonal variations in the onset of the whole spectrum of cerebrovascular diseases have been established from death statistics (Wylie, 1962), however, the monthly trend in onset of ischemic cerebrovascular accidents has never been thoroughly investigated. For the total series we found an even distribution throughout the year and this has also been reported for Israel (Adler, 1969). In our series the exact date and hour of occurrence of the ischemic insult could be determined for the majority of patients. The distribution over the months of the year did show fluctuations with slight increases in July/August for thrombosis and in May/June and during the winter months (November-February) for embolism (Fig. 1). Yet for both groups the observed distribution did not differ markedly from the one expected.

Weekly Variations. The rhythm of the week did not markedly influence the onset of thrombosis (Fig. 2), whereas in embolism a certain - but statistically insignificant - prevalence for the weekend is clear. A similar trend was also seen in a study of carotid insufficiency in 50 cases of this series (Zülch, 1973).

In summary, there may be some weekly frequency variations of thrombosis and embolism which need to be clarified by investigations on a much larger scale.

Daily Variations. There were significant differences in the incidence of thrombosis and embolism according to the time of day. Patients with thrombosis showed a marked peak of onset between 4 and 8 a.m.; patients with embolism, a moderate peak between 8 a.m. and 1 p.m. (Fig. 3).

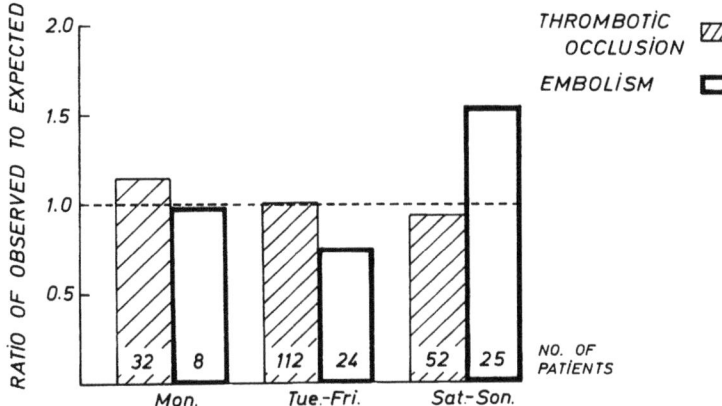

Fig. 2. Distribution of onset by days of the week in thrombotic occlusion and embolism. Almost even distribution in thrombosis, increased frequency during the weekend in embolism

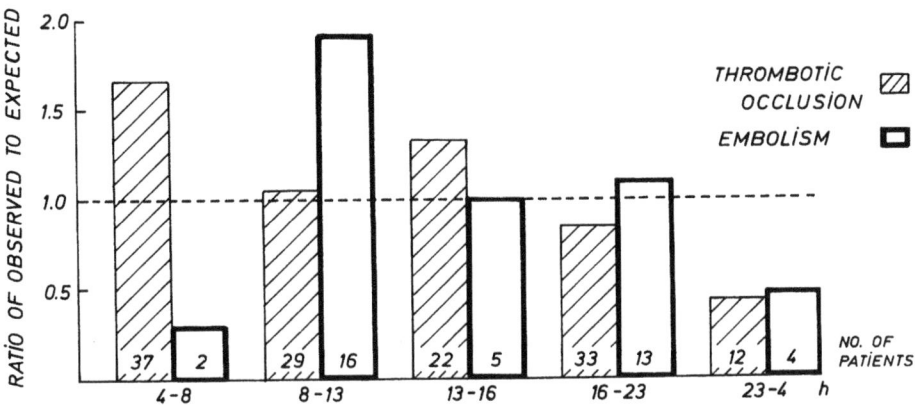

Fig. 3. Distribution of onset by hours of the day in thrombotic occlusion and embolism. Prevalence for thrombosis to occur between 4 and 8 a.m. and for embolism between 8 a.m. and 1 p.m. Rare incidence between 11 p.m. and 4 a.m. in both groups

Our studies concerning the rhythmic nycthemeric fluctuations of blood pressure indicated that these could well be the final factor in the induction of cerebrovascular insufficiency.

In summary, the hourly frequency index showed interesting trends: a marked peak between 4 and 8 a.m. in thrombosis and a moderate peak between 8 a.m. and 1 p.m. in embolism.

Table 2. Early prognosis

Outcome	Thrombosis		Embolism	
	No.	%	No.	%
No deficits	3	1.3	8	13.1
Improvement	90	38.1	32	52.5
No change or deterioration	46	19.5	5	8.2
Exitus	97	41.1	16	26.2
Total	236	100.0	61	100.0

Local Distribution of Thrombosis and Embolism

The topographic distribution of the various cerebral arteries involved was interesting. Although embolism, when found in the carotid system, was almost exclusively restricted to the middle cerebral artery, there was no statistically significant difference between thrombosis and embolism in the involvement of the carotid and the vertebrobasilar tree. Angiography demonstrated emboli to be lodged mainly in the MCA (in 65% of the cases), and less often in the PCA (13.1%). Emboli were rarely found in the carotid artery (4.3%) and in some cases they had migrated from the supraclinoid portion into the MCA (8.6%). None entered the ACA. In both thrombosis and embolism, the left hemisphere was involved more frequently, the L/R ratio being approximately 1.3:1.

On the other hand, as would be expected from the predilections of arteriosclerosis, thrombotic occlusions in the internal carotid were by far the most frequent (41.2%) followed by MCA occlusions (21.2%), PCA (9.7%), and vertebral artery occlusions (8.5%).

It may be noted that the traditional view of the left side predominance is seen in the total series as well as in the subdivisions, with almost the same ratio of approximately 1.3:1.

In summary: there was a prevalence of the MCA and a deficiency in the ACA in macroembolism and a prevalence of the carotid and MCA in thrombosis.

Early Prognosis

At the time of discharge from the hospital the outcome of patients was classified as "no neurological deficits," "improvement," "no change or deterioration," and finally "exitus." These showed highly significant differences between the two groups investigated (P < 0.001). In embolism the early prognosis was much better (Table 2): here 13.1% had no remaining neurological deficits and only 1/4 died. Complete recovery was observed in only 3 patients (1.3%) in the thrombotic group and nearly half of the patients died.

Autopsy Findings

The morphologic investigation of the autopsied patients revealed cerebral infarctions in all but one case of thrombosis. In embolism, infarctions were localized exclusively in the area of the MCA, including one "borderline" infarction. In thrombosis, the area of the MCA (44.3%) was the most often involved but involvement of the adjacent structures (27.2%) or borderline cases (2.3%) were quite frequent. Vertebrobasilar infarctions were present in 23.9% of the cases, whereas ischemic lesions in the area of the PCA (4.6%) or the ACA (3.4%) were rare. Only one patient had an isolated infarction in the vascular irrigation area of the anterior chorioidal artery.

Discussion

In this series of 1000 patients with obvious ischemic cerebrovascular disease, the relative frequency of thrombotic occlusion and also to a lesser extent of embolism should be interpreted with respect to the proportion having sufficient arteriographical examination (542 cases) or autopsy (additional 127 cases). The complete obstruction of a vessel which was thought to be occluded was thus found in 35.3% of those patients with an adequately visualized vascular system. This incidence is in agreement with that found in arteriographically studied ischemic stroke patients (Hass et al., 1968). Autopsy data, for understandable reasons, showed higher values; Berry (1959) found occluded vessels in 75% of cases with a single and large cerebral infarct which was not believed to be embolic. Moossy (1966), in 55% of brains having a recent infarction observed a thrombus in an intracranial artery. The incidence of embolism in our material seems somewhat lower than expected by Kurtzke (1969) who, in reviewing several clinical, postmortem, and epidemiologic studies, suggested that about 10% of thromboembolic occlusive disease is caused by embolism.

In our study, and this applies to both groups, approximately 45% of the patients were over 65 years of age. This is a higher proportion than that reported for a series of over 4700 cases with ischemic cerebrovascular disease where the patients of this age group comprised only 33% (Fields et al., 1968).

The role of oral contraceptives in the relatively high proportion of young women, especially in embolism, is not yet unequivocally established. Recent reports stress the increasing risk with increasing doses of estrogen (Inman et al., 1970). Likewise, the mechanisms of arterial occlusion - whether by thrombi, emboli, or dissecting aneurysms - are at present not clearly understood Gautier et al., 1974).

Males predominated in thrombotic occlusions, females in embolism. Comparable studies have shown similarly different M/F ratios (Wells, 1959; Fisher, 1961; Carter, 1964; Kannel et al., 1965). On the other hand, as pointed out by Kurtzke (1969), the sex ratios for the subdivisions of thromboembolism can not be accurately calculated from hospital series because they are too small.

Of patients with a thrombotic occlusion, 57% had arterial *hypertension*. This is a rather high incidence and though in agreement with some studies (Heyden et al., 1970), it deviates considerably from other series with angiographically proven vessel obstruction where distinctly lower incidences were found (Lindgren, 1958; Krayenbühl and Yasargil, 1964; Dorndorf, 1969).

As might be expected, the frequency of *cardiac and cardiovascular* disturbances in our patients with embolism was considerably higher than in the thrombotic group. This indicates that the heart was the most common source of emboli and this is in broad agreement with earlier investigations (Jörgensen and Torvik, 1966).

Seasonal variations in the occurrence of cerebrovascular and cardiovascular accidents are best established from death statistics and these indicate an increasing death rate during the winter months and a declining rate during the summer months (Mullins, 1936; Wylie, 1962). Reports on monthly variations in the occurrence of ischemic CVD are rare. McDowell et al. (1970), in a study involving more than 1000 patients with an established date of onset of nonembolic cerebral infarction, found the lowest incidence during June, July, and August and the highest during the months from November to May. In our material, no marked seasonal swings could be observed. The same equal distribution over the months was reported by Adler (1969) for Israel and by Bokonjić and Zec (1968) for Yugoslavia.

Daily variations in the onset of ischemic cerebrovascular disease have received little attention. Previous studies of the rhythmic nycthemeric fluctuations of blood pressure indicated that these could well be the final factor in inducing cerebrovascular insufficiency. In thrombotic occlusion, the accumulated onset between 4 and 8 a.m. is interpreted as being due to the "down" period of blood pressure shortly after midnight. Patients with these "midnight" strokes wake up in the morning with their symptomatology (Zülch and Hossmann, 1967; Hossmann, 1969). In our patients with embolism an increase was observed between 8 a.m. and 1 p.m., although this was not significant. Wells (1959), in his study on cerebral embolism, reported the same observation with 61% of the episodes occurring between 6 a.m. and noon. He suggested that the transition from the resting metabolic level of sleep to the active level of waking could be significant in precipitating embolism.

Embolism has been said to have a poor prognosis, even worse than that for thrombotic occlusion (Fisher, 1961). Though this does not account for our series, the mortality in the embolic group, i.e., 26%, is rather high especially if contrasted to the favorable outcome of Meyer's (1971) series. This consisted of 42 patients treated with hyperosmolar agents and the mortality was 9.5%. On the other hand, 56.7% of the patients from seven centers in the United States, with or without anticoagulant therapy, died within 2 months after randomization (Fisher, 1961).

In thrombotic occlusion, outcome and early mortality is strongly influenced by the age of the patient and *concomitant diseases*, including hypertension (Louis and McDowell, 1970). In this study, older age groups are well represented and the incidence of hypertension is also pronounced, and this might account for the unfavourable outcome in these patients.

Summary

From a total of 1000 unselected hospital patients with ischemic disease, 236 (23.6%) were diagnosed as having thrombotic occlusion of a cerebral artery and 61 (6.1%) cerebral embolization. Certain characteristics of these subdivisions were compared. Differences were seen in the male/female ratio and in the sex-age distribution of patients below 41 years of age. A high incidence of previous cardiac and cardiovascular disease and a rare occurrence of hypertension, transient ischemic attacks, and peripheral vascular disease was characteristic of patients with embolism. The incidence of diabetes and hypercholesterinemia was the same in both groups. Vascular accidents of both types showed an even distribution over the months of the year and the days of a week but marked variations during the hours of the day. The early prognosis was better among patients with embolism.

References

Ad Hoc Committee: A classification and outline of cerebrovascular diseases. Neurology (Minneap.) 8, 395-434 (1958)
Adler, E.: Stroke in Israel, 1957-1961. Epidemiological, Clinical, Rehabilitation and Psychological Aspects. Jerusalem/Israel: Polypress Ltd. 1969
Berry, R.G.: The vascular lesion in cerebral softenings. Trans. Amer. neurol. Ass. 84, 49-53 (1959)
Bokonjić, R., Zec, N.: Strokes and the weather. A quantitative statistical study. J. neurol. Sci. 6, 483-491 (1968)
Carter, A.B.: Cerebral Infarction. Oxford-London-Edingburgh-New York-Paris-Frankfurt: Pergamon Press 1964
Dorndorf, W.: Verlauf und Prognose bei spontanen zerebralen Arterienverschlüssen. Heidelberg: Dr. Alfred Hüthig Verlag 1969
Fisher, C.M.: Anticoagulant therapy in cerebral thrombosis and cerebral embolism. A national cooperative study, interim report. Neurology (Minneap.) 11 (part 2), 119-131 (1961)
Fields, W.S., North, R.R., Hass, W.K., Galbraith, J.G., Wylie, E.J., Ratinow, G., Burns, M.H., MacDonald, M.C., Meyer, J.S.: Joint study of extracranial arterial occlusion as a cause of stroke. I. Organization of study and survey of patient population. J.A.M.A. 203, 955-960 (1968)
Gautier, J.C., Rosa, A., Lhermitte, F.: Accidents vasculaires cérébraux et contraceptifs oraux. Rev. neurol. 130, 217-236 (1974)
Glynn, A.A.: Vascular diseases of the nervous system. A series of 315 cases. Brit. med. J. 1, 1216-1219 (1956)
Hass, W.K., Fields, W.S., North, R.R., Kricheff, I.I., Chase, N.E., Bauer, R.B.: Joint study of extracranial arterial occlusion. II: Arteriography, techniques, sites and complications. J.A.M.A. 203, 961-968 (1968)
Heyden, S., Heyman, A., Goree, J.A.: Non embolic occlusion of the middle cerebral and carotid arteries - a comparison of predisposing factors. Stroke 1, 363-369 (1970)

Hossmann, V.: Der Hirninfarkt und seine Abhängigkeit von exogenen und endogenen Faktoren des Blutdrucks. Inaugural Dissertation, Universität Köln 1969

Inman, W.H.W., Vessey, M.P., Westerholm, B., Engelund, A.: Thromboembolic disease and the steroidal content of oral contraceptives. A report to the committee on safety of drugs. Brit. med. J. 2, 203 (1970)

Jörgensen, L., Torvik, A.: Ischemic cerebrovascular diseases in an autopsy series. Part I: Prevalence, location and pre-disposing factors in verified thrombo-embolic occlusions, and their significance in the pathogenesis of cerebral infarction. J. neurol. Sci. 3, 490-509 (1966)

Joint Committee for Stroke Facilities: V. Clinical prevention of stroke. Stroke 3, 803-825 (1972)

Kannel, W.B., Dawber, T.R., Cohen, M.E., McNamara, P.M.: Vascular disease of the brain-epidemiologic aspects: the Framingham-study. Amer. J. publ. Hlth 55, 1355-1366 (1965)

Krayenbühl, H., Yasargil, M.G.: Verschluss der A. cerebralis media: Ergebnisse der klinischen und katamnestischen Untersuchungen. Schweiz. Arch. Neurol. Psychiat. (Chicago) 94, 287-304 (1964)

Kurtzke, J.F.: Epidemiology of Cerebrovascular Disease. Berlin-Heidelberg-New York: Springer-Verlag 1969

Lindgren, S.O.: Course and prognosis in spontaneous occlusions of cerebral arteries. Acta psychiat. scand. 33, 343-358 (1958)

Louis, S., McDowell, F.: Age: its significance in nonembolic cerebral infarction. Stroke 1, 339-453 (1970)

McDowell, F.H., Louis, S., Monahan, K.: Seasonal variation of non-embolic cerebral infarction. J. chron. Dis. 23, 29-32 (1970)

Meyer, J.S., Charney, J.Z., Rivera, V.M., Mathew, N.T.: Cerebral embolization: prospective clinical analysis of 42 cases. Stroke 2, 541-554 (1971)

Moossy, J.: Cerebral infarction and intracranial arterial thrombosis. Necropsy studies and clinical implications. Arch. Neurol. (Chicago) 14, 119-123 (1966)

Mullins, W.L.: Age, incidence and mortality in coronary occlusion: a review of 400 cases. Penn. med. J. 39, 322 (1936)

Veterans Administration Cooperative Study Group on Anti-Hypertensive Agents: Effects of treatment on morbidity in hypertension: Results in patients with diastolic blood pressures averaging 115 through 129 mmHg. J.A.M.A. 202, 1028-1034 (1967).

Veterans Administration Cooperative Study Group on Anti-Hypertensive Agents: Effects of treatment on morbidity in hypertension: II. Results in patients with diastolic blood pressures averaging 90 through 114 mmHg. J.A.M.A. 213, 1143-1152 (1970)

Wells, C.E.: Cerebral embolism. The natural history, prognostic signs, and effects of anticoagulation. Arch. Neurol. Psychiat. (Chicago) 81, 667-677 (1959)

Wylie, C.M.: Cerebrovascular accident deaths in the United States and in England and Wales. J. chron. Dis. 15, 85-90 (1962)

Yates, P.O., Hutchinson, E.C.: Cerebral infarction: The Role of Stenosis of the Extracranial Cerebral Arteries. Spec. Rep. Ser. med. Res. Counc. no. 300, London: Her Majesty's Stationery Office 1961

Zülch, K.J.: Morphology and pathogenesis of cerebral infarction. In: Anales del XII Congreso Latinoamericano de Neurocirurgia. Symp. Internacional de Investigaciones Neurologicas Lima, 1967, pp. 255-265

Zülch, K.J.: Die Carotis-Insuffizienz. Folia angiol. (Milano) 21, 47-62 (1973)

Zülch, K.J., Hennemann, U.: Predilection of Regional Intracranial Macro- and Microembolism. To be published

Zülch, K.J., Hossmann, V.: Über die 24-Stunden-Rhythmik des menschlichen Blutdrucks. Dtsch. med. Wschr. 92, 567-572 (1967)

Platelet Adhesiveness and Cerebral Vascular Disease Revisited

E. J. Acheson

In the western world cerebral vascular disease remains a major cause of morbidity and mortality and the pathogenesis has been the subject of much study and discussion in recent years.

Although it has been shown that platelet adhesiveness is increased in patients with cerebral vascular disease (Millar and Dalby, 1965; Danta, 1970), no relationship was found between the level of platelet adhesiveness and certain clinical factors (Acheson et al., 1972). Further, oral dipyridamole did not appear to influence the natural history of the disease (Acheson et al., 1969).

In a study of the natural history of 500 patients with cerebral vascular disease, the findings indicated that there may be an "active" form of cerebral vascular disease (Acheson and Hutchinson, 1971; Acheson, 1971; Hutchinson and Acheson, in press).

It is accepted that certain factors may influence the natural history of the disease. For example, hypertension is important in patients with cerebral vascular disease, and it has more than one role to play. The presence of coronary artery disease, as shown by the electrocardiogram, is a common finding in patients presenting with stroke, and particularly in patients with multiple stroke episodes.

Although the observations indicate that the site of disease is not important in terms of prognosis, it may be that the incidence of ischaemia in the vertebrobasilar territory should be halved, thus equalising the contribution of the two carotid and two subclavian arteries.

Nelson et al. (1973) examined the effect of ischaemia on the endothelium of carotid artery of monkeys. Striking abnormalities were frequently found in the ischaemic arteries and rarely, if at all, in the normal arteries. Their findings would suggest that the vessel wall may also play a part in the pathogenesis of cerebral vascular disease.

Although it is accepted that platelets have a role in the pathogenesis of cerebral vascular disease, studies of the natural history indicate that other factors also play a part and the relationship between these various factors is not yet established.

References

Acheson, J.: Factors affecting the natural history of "Focal cerebral vascular disease". Quart. J. Med. 40, 25-46 (1971)

Acheson, J., Danta, G., Hutchinson, E.C.: Controlled trial of dipyridamole in cerebral vascular disease. Brit. med. J. 1, 614-615 (1969)

Acheson, J., Danta, G., Hutchinson, E.C.: Platelet adhesiveness in patients with cerebral vascular disease. Atherosclerosis 15, 123-127 (1972)

Acheson, J., Hutchinson, E.C.: The natural history of "Focal cerebral vascular disease". Quart. J. Med. 40, 15-23 (1971)

Danta, G.: Platelet adhesiveness in cerebral vascular disease. Atherosclerosis 11, 223-233 (1970)

Hutchinson, E.C., Acheson, J.: Natural history of cerebral ischaemia. London: Saunders (in press)

Millar, J.H.D., Dalby, A.M.: Platelet stickiness in cerebrovascular disease. In: Proceedings of the 8th Int. Congr. Neurol. Vienna, Vol. 4, 1:483-487 (1965)

Nelson, E., Sunaga, T., Shimamoto, T.: Microvasculature in focal cerebral ischaemia and infarction. In: Cerebral Vascular Diseases. 8th Princeton Conference. McDowell, F.H., Brennan, R.W. (eds.). New York: Grune and Stratton 1973

The Significance of Platelet Aggregation in Amaurosis Fugax

J. Marshall

During recent years there has been considerable advance in our knowledge of factors associated with platelet aggregation. This has led to a reconsideration of the role of platelets in transient ischemic attacks (TIAs) in the cerebral circulation. The purpose of the present paper is to review some of the problems posed by the application of experimental knowledge to the clinical situation. The phenomenon of amaurosis fugax has been chosen in illustration because the retinal vessels are the only part of the cerebral circulation open to inspection.

Historically, three hypotheses have been proposed to explain TIAs whether in the cerebral or retinal circulation. The first of these was vasospasm; this hypothesis rested, not on evidence in its favour, but on the difficulty, in the state of knowledge current at the time, in envisaging any other mechanism for a brief interruption in blood supply on a focal basis. Spasm of cerebral arteries does occur but only in well-defined circumstances; these are irritation of the adventitia by blood as in subarachnoid haemorrhage, surgical manipulation, during the malignant phase of hypertension (Byrom, 1954) and in the preheadache phase of migraine (Shinhøj and Paulson, 1969; O'Brien, 1967). Spasm does not occur spontaneously in the genesis of TIAs as was well argued by Pickering (1951) many years ago.

The second hypothesis invoked for TIAs was the haemodynamic crisis, Denny-Brown and Meyer (1957) having shown that stenosis or occlusion of a vessel may not produce clinical evidence of ischemia until there is a general fall in perfusion pressure. This is undoubtedly a cause of some TIAs but the attractive nature of the hypothesis, and its therapeutic implications with regard to the dangers of hypotensive therapy, led to it being invoked in explanation of many more TIAs than justified by the evidence (Kendell and Marshall, 1963).

The third hypothesis - which brings us nearer to our particular subject - was recurrent emboli. The possibility of emboli being responsible for TIAs had previously been rejected by many workers on the grounds that it was unlikely that emboli would so constantly find their way to the same site as was apparently the case in many patients with TIAs of a stereotyped pattern. Amaurosis fugax provides an excellent example of stereotypy, many patients repeatedly experiencing transient loss of the whole or a part of one visual field in exactly the same manner. Objection to the embolic hypothesis did not take sufficient account of the existence of laminar flow in arteries. Because of laminar flow there is a high probability that emboli arising at one site will follow the same pathway through the vascular tree. The more central the source of emboli, the greater the number of bifurcations to be traversed, and the more turbulence present because of arterial disease, the less likely is this to happen. But when, for example, emboli arise at the origin of the internal carotid artery, the chance of them being carried consistently to the first branch, namely the ophthalmic, and so to the retina, is high.

The embolic hypothesis for TIAs received confirmation by the observation by Fisher (1959) of a patient during an attack of amaurosis fugax in which bodies, after temporary arrest, passed through the retinal circulation. Evidence of emboli in the retinal vessels and of their frequent origin from stenosis of the internal carotid artery has since been accumulated (Gunning et al., 1964) and there can be no doubt that emboli account for the majority of, but not all, cases of TIAs, a haemodynamic crisis being responsible for some attacks.

This development naturally led to increased interest in the source, structure, and behaviour of emboli with a view to achieving a parallel therapeutic advance. Anticoagulant therapy with dicoumarol and inanedione derivatives was widely prescribed and was undoubtedly effective in reducing the frequency of TIAs in some cases (Millikan, 1971). Failure in other cases led to a closer consideration of the nature of the emboli occurring in TIAs. These are not of one type; platelet aggregations, thrombus, atheromatous debris - particularly cholesterol crystals - calcareous deposits from heart valves, mycotic emboli, and fat are variously responsible, as may be illustrated in series of patients with amaurosis fugax (Ross Russel, 1968). With so wide a spectrum it is hardly surprising that anticoagulants of the dicoumarol and inanedione type - which have no effect upon platelet aggregation but reduce the frequency of embolisation arising from thrombus - were effective in some cases but not in others.

Meantime work on the cause of inhibition of platelet aggregation was continuing. Realization that the common aspirin had a marked inhibitory effect on platelet aggregation in relation to connective tissue damage led to its use in TIAs - particularly amaurosis fugax - in some cases with striking effect. Our own experience is illustrated by the case of a woman of 67 years with a 10-week history of attacks of loss of vision occurring every 2 to 3 days. Vision was lost for 5 min either in the entire field or in the lower half only. Examination of the eye and nervous system was normal. There was a loud bruit over the right carotid artery; platelet count was 250,000/c.mm. The effect of aspirin, dipyridamole, and a placebo on the frequency of the attacks is shown in Figure 1.

Experience of this type caused some workers to advocate the use of aspirin in all cases of TIAs or even prophylactically in older age groups. Such advocacy took little account of the various types of emboli causing TIAs nor of the various modes of action of the

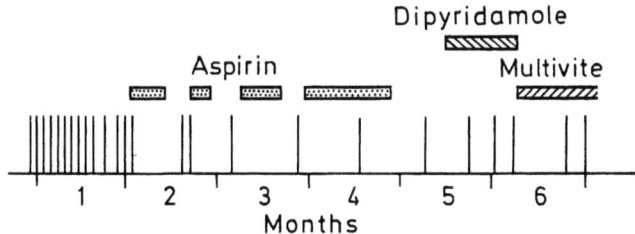

Fig. 1. Effect of aspirin 600 mg daily, dipyridamole 150 mg daily, and multivite on frequency of attacks of amaurosis fugax

available antiplatelet aggregation drugs. Whilst aspirin is effective in preventing platelet aggregation caused by connective tissue damage, it is not of great value in opposing the effect of ADP on platelets. Similarly, dipyridamole is particularly effective against platelet aggregations developing in relation to prostheses, such as mechanical heart valves, but how much of this effect is directly on the platelets and how much on the associated red cell damage which these valves cause is not known.

Just as dicoumarol was used without discrimination so it is with aspirin. This is not only harmful clinically, but may prove a handicap scientifically because failure to achieve expected results raises doubt about the basic scientific observations and assumptions. This can happen even when available evidence indicates that the phenomenon being treated does not have a unitary aetiology and that the drugs being used may act in a variety of ways.

The pathology of an embolus can often be determined in amaurosis fugax by the clinical features and the ophthalmoscopic appearances. Cholesterol emboli appear as yellowish refractile bodies, often multiple. Thrombus is seen as a solitary, large, white body, often giving rise to permanent obstruction of a vessel with corresponding visual field loss. Platelet aggregations are also white, but appear as a series of bodies which traverse the vessels, arrest briefly at bifurcations, and then pass on and disappear through the capillaries. It is platelet aggregates which give rise to the type of case in which a patient experiences many episodes of amaurosis fugax (or other TIAs) over a period of hours or days. Cholesterol emboli and embolisation of thrombus do not cause such rapidly repetitive attacks.

Clearly, clinical trials of drugs in amaurosis fugax (comparing a treated with a control group) which do not differentiate between the type of attack - hence the nature of the emboli - are likely to give inconclusive results. Amaurosis fugax due to platelet aggregates is by no means common; in the cases that do occur the attacks are brief and frequent, hence a trial in which the patient is his own control, with periods of treatment compared with a placebo, is more appropriate.

If progress in the treatment of TIAs is to be maintained, distinction must be made between those due to platelet aggregation, those due to thrombus, and those due to atherosclerotic debris, and therapy prescribed accordingly. However, this ignores one aspect of the subject which could be of importance, namely, that platelet aggregations, besides providing emboli, form the nidus of thrombus. The factors which determine whether a platelet aggregation on a vessel wall simply throws off platelet emboli, or whether fibrin is deposited, leukocytes involved and thrombus formed, giving rise to thrombotic emboli, are not understood. It could be argued that prevention of platelet aggregation should in turn prevent thrombus formation. On this hypothesis antiplatelet aggregation drugs should be effective not only against TIAs due to platelet emboli, but also against those due to thrombotic emboli. Clinical experience indicates that this is not the case. The reason why is not immediately clear and we must turn to those working with platelets for help with this and other problems.

In conclusion, though considerable advance has been made, many questions remain to be answered. There is already a sufficiency of knowledge to make the indiscriminate treatment of TIAs by one or other method no longer appropriate. The aetiology of TIAs can often be determined clinically, and sufficient knowledge is available to indicate appropriate therapy in many instances. Further work is required, however, particularly on the role of platelets.

References

Byrom, F.B.: The pathogenesis of hypertensive encephalopathy and its relation to the malignant phase of hypertension. Experimental evidence from the hypertensive rat. Lancet $\underline{2}$, 201-211 (1954)

Denny-Brown, D., Meyer, J.S.: The cerebral collateral circulation. 2. Production of cerebral infarction by ischemic anoxia and its reversibility in early stages. Neurology (Minneap.) $\underline{7}$, 567-579 (1957)

Fisher, C.M.: Observations of the fundus oculi in transient monocular blindness. Neurology (Minneap.) $\underline{9}$, 333-347 (1959)

Gunning, A.J., Pickering, G.W., Robb-Smith, A.H.T., Ross Russel, R.: Mural thrombosis of the internal carotid artery and subsequent embolism. Quart. J. Med. $\underline{33}$, 155-194 (1964)

Kendell, R.E., Marshall, J.: Role of hypotension in the genesis of transient focal cerebral ischaemic attacks. Brit. med. J. $\underline{2}$, 344-348 (1963)

Millikan, C.H.: Reassessment of anticoagulant therapy in various types of occlusive cerebrovascular disease. Stroke $\underline{2}$, 201-208 (1971)

O'Brien, M.D.: Cerebral-cortex-perfusion rates in migraine. Lancet $\underline{1}$, 1036 (1967)

Pickering, G.: Vascular spasm. Lancet $\underline{2}$, 845-850 (1951)

Ross Russel, R.W.: The source of retinal emboli. Lancet $\underline{2}$, 789-792 (1968)

Skinhøj, E., Paulson, O.B.: Regional blood flow in internal carotid distribution during migraine attack. Brit. med. J. $\underline{2}$, 569-570 (1969)

Mechanisms of Platelet 5-Hydroxytryptamine Release in Migraine

M. Anthony

The clinical relationship between migraine and cerebrovascular disease is tenuous, to say the least, except perhaps for the fact that some patients experience transient neurologic symptoms during a migraine attack and the recent report that the incidence of hypertension, a predisposing factor to atherosclerosis, is higher in migrainous subjects than in the population at large (Leviton et al., 1974).

However, significant biochemical changes take place during migraine, consisting mainly of serotonin release from platelets, a phenomenon which could have some bearing on the accelerated aggregation responses of platelets of migrainous subjects reported by Hilton and Cumings (1972). Apparently, platelets from migrainous patients, whether collected during headache or headache-freedom, aggregate more readily after 1 min preincubation with serotonin than those of normal controls. It is, therefore, possible that migrainous platelets have defective serotonin receptors, which accept serotonin less readily or have a deficient mechanism for retaining the amine within the platelet cell.

Work, so far, confirms that migraine is a low serotonin (5HT) syndrome, in that plasma levels of serotonin fall during the attack (Anthony et al., 1967), and more recently Hilton and Cumings (1972) reported that the serotonin content of platelets was reduced during the migrainous episode.

Patients who suffered from frequent and severe attacks of migraine were admitted to the hospital for investigation. Blood was collected before, during, and after an attack of migraine and platelet-rich plasma (PRP) or isolated platelets were used for estimating serotonin, as previously described (Anthony et al., 1967, 1969).

This study comprises 61 patients. In 21, serotonin was estimated as total plasma serotonin using PRP, whilst in the remaining 40 the serotonin content of platelets was assessed. Of the 61 patients, 53 showed a fall in serotonin (plasma or platelet) during migraine greater than 10% (range 10-72%). Mean values for the 24 h before the headache, for the duration of the migraine attack and the 24 h following the attack were 0.63, 0.38, and 0.58 $\mu g/10^9$ platelets, respectively (Table 1). Mean fall was 40% and this was statistically highly significant (analysis of variance, $P < 0.001$).

Table 1. Plasma serotonin levels ($\mu g/10^9$ platelets)

	Migraine	
		Statistical significance (Analysis of variance)
Preheadache	0.63	$P < 0.001$
Headache	0.38	
Postheadache	0.58	$P < 0.001$
Number of patients	61	
Number of patients showing fall	53	
Mean fall	40%	

Table 2. Incubation of platelets with excess serotonin ($\mu g/10^9$ platelets)

	Mean			
	Preheadache	Headache	Postheadache	Statistical significance
Platelet 5HT (incubated with excess 5HT)	1.70	1.76	1.97	NS
Plasma 5HT (non-incubated)	0.75	0.41	0.76	$P < 0.001$

The mechanism of serotonin loss from platelets during migraine was investigated in the following manner:

a) Incubation of Platelets with Excess Serotonin

Ten patients suffering from frequent attacks of migraine had blood collected before, during, and after an attack of headache, as previously described. PRP from each blood specimen was divided into two parts. One was used for the estimation of plasma serotonin, whilst a 2 ml portion was incubated at 37°C for 90 min with excess serotonin, giving a final concentration of 1.5 µg/ml. At the end of that period the platelets were isolated and their serotonin content estimated. For the nonincubated specimens the mean plasma serotonin levels before, during, and after an attack of migraine (in $\mu g/10^9$ platelets) were 0.75, 0.41, and 0.76 respectively, whilst for the incubated specimens the corresponding levels were 1.70, 1.76, and 1.97 (Anthony et al., 1969). Statistical comparison between the preheadache and headache values showed a highly significant difference for the nonincubated specimens, but none for the incubated specimens (Table 2). The inference from these results is that the ability of platelets to take-up serotonin is not impaired by the headache process.

Table 3. Adenine nucleotides in platelets during migraine (μmole/10^9 platelets)

	24 h Preheadache	Headache	24 h Postheadache	Statistical significance
ATP	101	79	81	NS[a]
ADP	29	32	22	NS
AMP	12	9	10	NS
5HT	4.93	2.89	4.70	$P < 0.001$
ATP/5HT	20.5	27.3	17.2	$P < 0.001$

[a] NS - not significant

b) Adenine Nucleotide Content of Platelets

Serotonin uptake by platelets is an active process and most probably involves ATP, as the amount of serotonin taken up is proportional to the ATP content of platelets. The platelet content of all adenine nucleotides (ATP, ADP, and AMP) was estimated during the various phases of the migraine attack in nine migrainous patients. No significantly statistical difference was found in ATP, ADP, and AMP content of platelets before, during, and after the headache (Anthony et al., 1969). Results are summarized in Table 3. The increase in ATP/5HT ratios during headache reflects only the serotonin loss during the attacks. Thus, in spite of the loss of serotonin from the platelet, the platelet energy stores appear to be unaffected by the biochemical process of the headache, which may account for their readiness to absorb serotonin in vitro and for the quick restoration of normal serotonin levels once the migraine attack is over.

c) Cross-Incubation Experiments

The hypothesis that a factor with a serotonin-releasing effect could be responsible for serotonin loss was investigated as follows. Platelet-poor plasma (2 ml) was prepared from each specimen of blood collected until 2 days after the migraine attack. On the second day sufficient blood was collected and the platelet aliquots from duplicate 2 ml portions of PRP were mixed with 2 ml platelet-poor plasma from:

1. Two specimens from the headache-free period,

2. Two specimens from the blood that produced the platelet aliquots,

3. And two specimens from the migraine attack.

The specimens were incubated at 37°C in a metabolic water bath for 90 min. At the end of that period the platelets were isolated and used for serotonin estimation. Results of the study are shown in Table 4. Of the 30 patients investigated, 24 showed a significant fall in plasma serotonin during the migraine attack.

Table 4. Cross-incubation experiments

Headache-free platelets with PPP

	$\mu g/10^9$ platelets	Statistical significance
Preheadache	0.48	$p < 0.001$
Headache	0.38	
Postheadache	0.50	$p < 0.001$
Number of patients	30	
Number of patients showing fall	24	
Mean fall	21%	

Table 5. Cross-incubation experiments

Headache-free platelets with filtrate PPP

	$\mu g/10^9$ platelets	Statistical significance
Preheadache	0.41	$p < 0.01$
Headache	0.34	
Number of patients	16	
Number of patients showing fall	11	
Mean fall	17%	

The serotonin content of platelets when incubated with "migraine plasma" was significantly lower than when incubation took place with "headache-free plasma" in 18 cases.

In 16 patients, instead of using whole plasma a filtrate of it was used with a molecular weight of less than 50,000. Platelets contained less serotonin when incubated with the plasma filtrate from the migraine attack than when incubation took place with headache-free plasma filtrate in 11 out of 16 cases (Table 5). It would appear, therefore, that in the majority of patients there appears in the plasma, during a migraine attack, a factor with serotonin releasing properties. Further, since serotonin release from platelets also occurs with a filtrate of plasma, it would appear that this serotonin releasing factor is of molecular weight of 50,000 or less. This would tend to exclude most of the plasma proteins and antigen-antibody complexes. It could well be a fatty acid, an amino acid, a polypeptide, or one of the other monoamines, e.g., tryptamine or tyramine.

Table 6. Plasma free fatty acid (FFA) changes in migraine

	Preheadache	Headache	Postheadache
Mean FFA levels µmole/ml	0.569	0.733	0.648
Standard error of mean	0.119	0.138	0.135

Number of patients	15
Number showing rise	11
Mean % of rise	36.6

d) Plasma Free Fatty Acids

Certain free fatty acids (FFA), particularly palmitic, stearic, and behenic are known to be potent releasers of platelet serotonin in vitro (Inouye et al., 1970). Their small molecular size suggests the possibility of their acting as the serotonin releasing factor described above. Hockaday et al. (1971) demonstrated a two- to three-fold increase in FFA during migraine in patients who related their attacks to fasting, in contrast to those who developed no such headache under similar circumstances. In a more recent study (Anthony, 1973) 11 of 15 patients showed a significant rise of FFA during migraine and 9 of these showed a simultaneous fall in platelet serotonin content. The values of FFA for the preheadache, headache, and postheadache periods were 0.569, 0.733, and 0.648 µmole/ml respectively, whilst mean rise for the group was 36.6% (Table 6).

Conclusions

Platelet serotonin loss during migraine is not associated with ATP depletion nor impairment of the ability of platelets to take up serotonin when incubated with excess of the amine in vitro. In that respect, the serotonin release reaction is similar to that produced by reserpine where there is little change in the ATP content of platelets and unlike that produced by tyramine where there is disruption of monoamine/ATP complexes (Pletscher et al., 1971). On the other hand, although certain FFA release platelet serotonin in vitro, they appear to do so through ADP release, since their aggregating effect can be inhibited by adenosine (Haslem, 1964). In view of this observation the elevation of plasma FFA during migraine reported in this study, which in the majority of cases is associated with platelet serotonin loss without adenine nucleotide depletion, is not easily explained.

All that can be concluded at present is that the serotonin releasing factor in migraine is an endogenous substance released into the circulation during the attack with a reserpine-like action on platelet serotonin. It has a low molecular weight and could well be an amine, a lipid, an amino acid, or some other tissue component. Its exact nature remains to be elucidated.

References

Anthony, M.: Proc. Aust. Ass. Neurol. 10, 87-89 (1973)
Anthony, M., Hinterberger, H., Lance, J.W.: Arch. Neurol. (Paris) 16, 544-552 (1967)
Anthony, M., Hinterberger, H., Lance, J.W.: Res. clin. Stud. Headache 2, 29-59 (1969)
Haslam, R.J.: Nature (Lond.) 202, 765-768 (1964)
Hilton, B.P., Cumings, J.N.: J. Neurol. Neurosurg. Psychiat. 35, 505-509 (1972)
Hockaday, J.M., Williamson, D.H., Whitty, C.M.W.: Lancet 2, 1153-1155 (1971)
Inouye, A., Shio, H., Sorimachi, M., Kataoka, K.: Experientia (Basel) 26, 308-309 (1970)
Leviton, A., Malvea, B., Graham, J.R.: Neurology (Minneap.) 24, 669-672 (1974)
Pletscher, A., Da Prada, M., Berneis, K.H., Tanzer, J.P.: Experientia (Basel) 27, 993-1120 (1971)

Experimental Observations on Platelet Emboli in Focal Brain Ischemia

C. Fieschi, F. Volante, N. Battistini, and E. Zanette

Models of experimental pathology must show analogies with the spontaneous human pathology.

Premises to this model of transient focal cerebral ischemia are: (1) that the transient ischemic attacks (TIA) in man can be caused by platelet emboli; (2) that the intracarotid ADP or arachidonic acid (AA) infusion in the experimental animal will be able to form similar platelet emboli in vivo in the cerebral circulation, and would not provoke important systemic effects (Fisher, 1952, 1959; Russel, 1961).

This condition of absence of systemic effects was not entirely realized until now. The ADP infused (8 mg/ml, 1 ml/min times 5 min) into the internal carotid artery of rabbits in a concentration of 10^{-2} M, after being diluted 1:10 in the arterial blood, and 1:150 in the systemic blood is equally able to transiently provoke bradycardia and severe arterial hypotension (Fieschi et al., 1975). As a consequence, the action which we observe on the cerebral circulation is a result of the combined effect of both the local embolization and the systemic hypotension.

However, the systemic hypotension produced by the ADP is entirely reversible during the experiment; and such effect is not observed when doses of intracarotid AA are employed (Fieschi et al., in preparation). With the exception of the pH electrode experiments, adrenaline (1×10^{-3} mg) was added to the ADP perfusate.

Arachidonic acid is a potent physiologic platelet aggregating agent in vitro (Silver et al., 1973; Vargaftig and Zirinis, 1973) and in vivo (Silver et al., 1974; Furlow and Bass, 1975). In our experiments sodium arachidonate[1] ($6,6 \times 10^{-3}$ M) was infused for 15 s in the internal carotid artery of rabbits in a dose of 0.30 mg/kg b.w.

The effects on systemic blood pressure consist of a decrease by less then 30% for 60 s, while local effects are very pronounced. A massive embolization of the pial and intracerebral vessels takes place, with a complete ischemia of the cerebral cortex for

---

[1] Arachidonic acid (99% pure), Sigma Chemical Co, St. Louis, Missouri

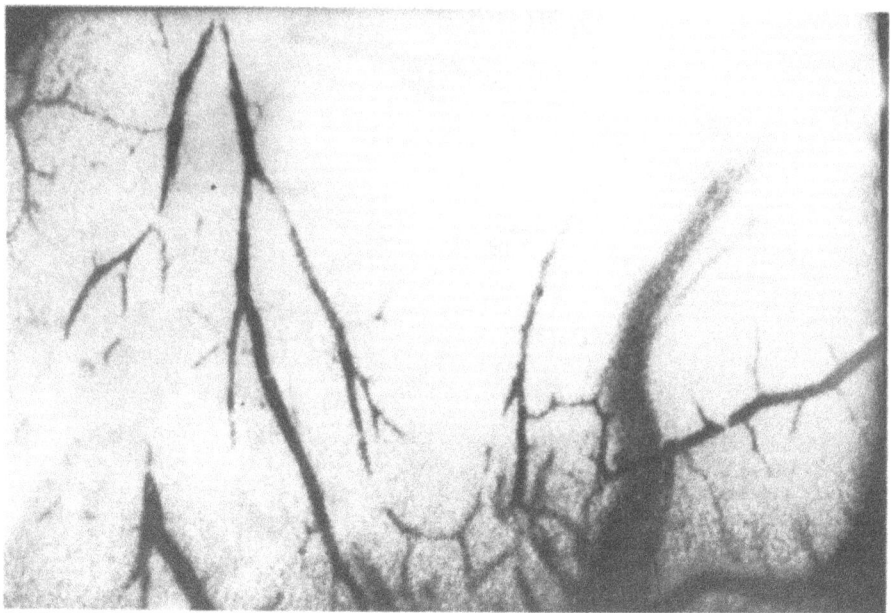

Fig. 1. Pial circulation 3 min after internal intracarotid infusion of sodium arachidonate 0.30 mg/kg b.w. in 15 s. Massive embolization with partial resumption of circulation

at least 3 min, followed by a gradual recovery of the circulation (Figs. 1 and 2). Further emboli and persistent ischemic areas are subsequently observable for at least 1 h.

The present section deals with results obtained with the ADP model: by this method we have studied 60 adult male rabbits, weighing approximately 2.5 kg.

The method (Fieschi et al., 1975; Zanette et al., 1972) is as follows: isolation and retrograde cannulation of the right lingual artery with the ligature of all other collateral branches of the external carotid, so that all the ADP is injected exclusively in the right internal carotid artery. Recordings are of the arterial pressure, of the end-tidal CO_2, and EEG, with periodic measurement of the arterial PO_2 and PCO_2. The animals anesthetized with pentothal (30 mg/kg b.w.) were paralyzed and artificially ventilated, and their body temperature controlled.

The direct observation of the pial vessels was effected through a cranial window with a Zeiss surgical microscope.

Samples of mixed cerebral venous blood were taken from the internal jugular vein to count platelets. A diminution in the platelets count in the cerebral venous blood coincides with the observation of embolies in the cortical arteries, and is an expression of the entity and timing of the platelet aggregation in cerebral vessels.

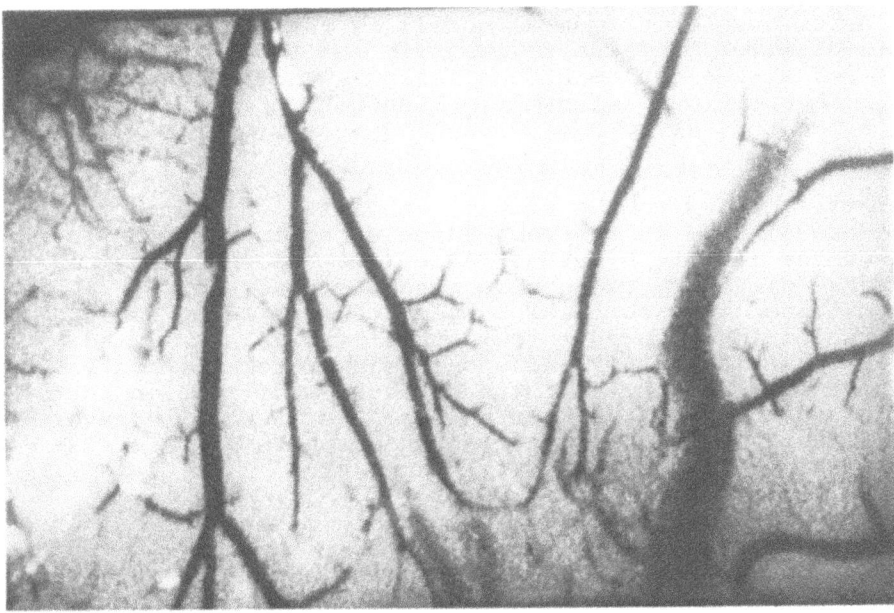

Fig. 2. Same, 10 min after infusion. Many arterioles are still blocked by platelet emboli

Platelet embolies were observed through the cranial window in such a number and size as to block simultaneously many arterial branches, determining a visible cortical ischemia and EEG alterations.

These events were constant at the concentrations of ADP used, but they were also entirely transient. The embolies became gradually fragmented and removed, with a restitution of the internal jugular vein platelet count, of the pial circulation, and recovery of functional activity as monitored by the EEG. This was previously described in detail (Fieschi et al., 1974, 1975; Zanette et al., 1972).

More recent observations regard the size of tissue ischemic areas produced by such platelet emboli. This was initially studied with a qualitative (in terms of blood flow) autoradiographic method with radioactive antipyrine (I^{131}) (Reivich et al., 1969). The antipyrine is infused intravenously for 40 s, starting at the 4th min of intracarotid ADP infusion, when emboli are still at their maximum. At this time, the ischemic areas are numerous both at the surface and in the depth of the brain, extending to 18.7% of the territory irrigated by the internal carotid artery. This we consider a direct experimental demonstration of a phenomenon which is well known clinically, that pure platelet embolies are able to produce significant and rather large areas of focal cerebral ischemia. In our case, this is not followed by any sign of biochemical or histopathologic permanent tissue damage (Fieschi et al., 1974). An antiaggregant therapy blocks this mechanism and the consequent cerebral ischemia: we will discuss these therapeutic effects in the last section of our paper.

An advantage of such experimental models is that it is possible not only to observe, but also to quantify the effects of platelet emboli and tissue ischemia. The following data have been obtained in a research made with the collaboration of Dr. T. Duffy at the Department of Neurology, Cornell University of New York (to be published).

In the parietal cortex on the side of embolization, compared to the contralateral side in the same animals, an increase of lactate, and of the lactate/pyruvate ratio (not reported here), and a decrease of energy-rich phosphates was observed. These data were obtained from cortical samples taken after freezing in vivo with liquid nitrogen through the intact skull, immediately after cessation of ADP infusion. Values from the contralateral, control side also indicate a slight modification that might be due to the arterial hypotension; the changes, however, were significantly higher in the right hemisphere, site of ischemia produced by platelet emboli.

Another method allows a continuous recording of biochemical changes in the cortex (Heuser et al., 1974) based on the registration of H^+ and K^+ concentration by surface microelectrodes. This part of the study has been done in collaboration with Dr. D. Heuser (Heuser et al., 1976). The infusion of ADP in the carotid artery produced an acidotic shift and an increase of the extracellular K^+.

Both effects are transient, being recovered (K^+ first, pH later within a short time of cessation of ischemia due to ADP infusion). These changes are strictly unilateral in spite of the systemic arterial hypotension produced by ADP. Measurements of the local CBF using ^{14}C antipyrine method (Reivich et al., 1969) during the ADP infusion have further demonstrated that the reduction in cerebral blood flow is confirmed to the right hemisphere, while autoregulation protects the contralateral one against hypotension (Fieschi et al., 1974).

When ADP was infused into the femoral artery of the same rabbits, the same hypotension but no cortical pH or K^+ changes were observed. Therefore, the biochemical changes are a specific consequence of the cerebral ischemia caused by the platelet emboli. Finally, in collaboration with Prof. G. Weber, it has been shown that hypercholesterolemia induced by high cholesterol diet, does not modify the pattern of platelet aggregation, transient emboli, and cerebral ischemia in the rabbit.

This rabbit model is now being used to study the effects of antiaggregating agents with clinical application in cerebral TIAs.

Acetylsalicylic acid (ASA) was injected in rabbits, 20 min prior to the ADP infusion, in a single dose of 20 mg/kg i.v. (as lysinacetylsalicylat = 36 mg). Such a dose corresponds to 1 g ASA in a patient weighing 50 kg. In this therapeutic range - but not with a single dose of 15 mg/kg b.w. - ASA showed a complete protective effect lasting up to 18 h (Table 1). After 24 h, however, the protective effect had almost completely disappeared. One may

Table 1. Protective effect of ASA (20 mg/kg i.v.) against ADP + adrenaline in vivo

	20 min	2 h	18 h	24 h
Protected	10	10	3/4	1/4
Aggregated	0	0	1/4	3/4

Table 2. Protective effect of dipyridamole against ADP + adrenaline in vivo

	30 min	2 h	24 h after ASA
15 mg/kg	- -	- -	+ +
20 mg/kg	- +	- +	
40 mg/kg	+ +	+ +	

presume that at this time, a single dose much lower than 20 mg/kg would be sufficient to restore it. In fact, in those animals which were protected longer after 24 h, an injection of another antiaggregant such as dipyridamole (15 mg/kg b.w.) completely reblocked the cerebral platelet emboli (Table 2). Note that the use of dipyridamole alone does not protect from ADP-induced cerebral emboli, it is efficacious only at a very high dose such as 40 mg/kg b.w. Therefore, the protection given by 15 mg/kg of dipyridamole 24 h after the administration of ASA, confirms: a) that ASA has a cumulative effect; b) that the clinical use of association between nonsteroid antiinflammatory agents and other antiaggregant drugs such as dipyridamole, is based on rational experimental evidence. Spontaneous platelet aggregation in patients may be less dramatic than that induced by massive doses of ADP or arachidonic acid: it therefore may not be significant clinically, given the fact that therapeutic doses of dipyridamole up to 3 mg/kg b.w. are not effective, even in association.

In conclusion: selective ADP infusion into the internal carotid artery of rabbit constantly produces cerebral platelet emboli and transient focal ischemia of the nervous tissue of a sufficient size to cause clinical effects. These observations confirm the pathogenetic importance of the platelet emboli in focal cerebral human ischemia.

The biochemical and EEG modifications are completely transient and therefore, no permanent morphologic alterations are produced in the nervous tissue. These effects are more pronounced and longlasting with arachidonic acid. The animal models described reproduce the mechanism of human transient ischemic attacks due to the platelet emboli.

The efficacy of drug treatment and prevention can thus be suitably studied, allowing for the differences between spontaneous and experimental pathology.

References

Fieschi, C. et al.: In preparation
Fieschi, C., Volante, F., Battistini, N. et al.: Pathology of Cerebral Microcirculation. Cérvos-Navarro, J. (ed.). Berlin: Walter de Gruyter 1974, Chap. 11, p. 251
Fieschi, C., Battistini, N., Volante, F. et al.: Stroke $\underline{6}$, 6 (1975)
Fisher, C.M.: Arch. Opthal. (Chicago) $\underline{47}$, 167 (1952)
Fisher, C.M.: Neurology $\underline{9}$, 333 (1959)
Furlow, T.W. Jr., Bass, N.H.: Science $\underline{187}$, 658 (1975)
Heuser, D., Astrup, J., Lassen, N.A. et al.: Acta physiol. scand. (In press, 1974)
Heuser, D., Fieschi, C., Volante, F.: Platelet emboli and Focal cerebral ischemia: an experimental study on the circulatory and metabolic effects of intracarotid infusion of ADP and arachidonic acid in rabbit. In: The Cerebral Vessel. Cérvos-Navarro et al. (eds.). New York: Raven Press 1976
Reivich, M., Jehle, J., Sokoloff, L. et al.: J. appl. Physiol. $\underline{27}$, 206 (1969)
Russel, R.W.R.: Lancet $\underline{2}$, 1422 (1961)
Silver, M.J., Smith, J.B., Ingarman, C. et al.: Prostaglandins $\underline{4}$, 863 (1973)
Silver, M.J., Hoch, W., Kocsis, J.J. et al.: Science $\underline{183}$, 1085 (1974)
Vargaftig, B.B., Zirinis, P.: Nature (New Biol.) $\underline{244}$, 114 (1973)
Zanette, E., Cavallini, C., Fieschi, C. et al.: Boll. Soc. ital. Biol. sper. $\underline{48}$, 770 (1972)

Observations on Platelet Aggregability in the Acute Phase of Untreated Strokes

G. L. Lenzi, F. Laghi Pasini, and A. Vittoria

It is known from previous reports (Guiraud et al., 1972) that in the ischemic cerebrovascular accidents there is a modification of platelet aggregability. Therefore we attempted to determine (1) if this modification was detectable in the acute phase of untreated stroke; (2) what was his time-course; (3) if there was a relationship with the clinical decourse.

In order to assess that, we compared the platelet aggregability of two groups of subjects, one group conventionally labeled "stroke", the other "nonstroke". There were 22 patients examined. The "ictus" or stroke group consisted of 11 patients affected by ischemic cerebrovascular episodes ranging from transient ischemic attacks to complete stroke without substantial recovery. Eight cases (72%) were examined during the first 12 h from the onset of symptomatology and before beginning any therapy. In six cases (55%) in which no antiaggregating drugs were administered, platelet aggregability was tested throughout the following 5-10 days.

The nonstroke group consisted of patients affected by different diseases not due to ischemic mechanisms, i.e., brain tumors, facial paralysis, depression, etc.

Platelet aggregability was determined in vitro by means of the Born aggregometer, utilizing ADP in scalar doses (Hardistry et al., 1972). Therefore we were able to classify the subjects under study on the basis of the minimal amount of ADP producing the typical biphasic aggregation curve. The study was performed in single-blind fashion, i.e., the determination of platelet aggregability was performed by one of the AA, unaware of patient's symptomatology, diagnostic judgment, and eventual therapy.

In general, the two groups showed similar hematologic values (sugar, nitrogen) and mABP. Blood lipids and cholesterol values were more elevated in the nonstroke group than in the stroke group, where they ranged within normal standards. We are inclined to refer this aspect to the large dyshomogeneity of the nonstroke group, since previous AA (Ladurner et al., 1973) had shown a general opposite behavior.

Figure 1 shows that the ADP doses suitable to obtain the classical biphasic curve are significantly smaller in the stroke group than in the nonstroke one (t-test was less than 0.01). It may be

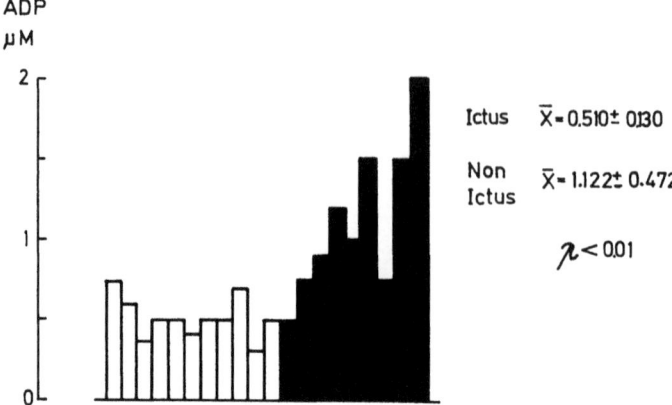

Fig. 1. Minimal ADP doses (µM) producing the appearance of the typical biphasic aggregation curve. "Ictus" patients: white columns. "Nonictus" patients: dark columns. The difference between the means of the two groups resulted statistically significant ($P < 0.01$)

relevant to note that these calculations were performed after discarding two nonstroke cases where the aggregability was very poor, that is, requiring a dose of 4 µM ADP.

Time-courses of platelet aggregability in these patients where general conditions allowed a restrain from antiaggregating-drug treatment, are shown in Figure 2. We must note that the drugs eventually received by these patients do not interfere with platelet aggregability. In all these cases, with one exception, the platelet aggregability shows a slight, but still significant (t-test was less than 0.05) decrease, in respect to the value observed at the onset of the stroke. In contrast, we did not find any correlation between the aggregability as judged by the amount of ADP required to obtain the biphasic curve with the Born aggregometer, and the interval between the onset of symptoms and the test. In fact, some patients showed 4-5 days after the stroke a greater "absolute" aggregability in comparison to other patients tested after a few hours from the appearance of the symptomatology. Meanwhile, all stroke aggregability determinations performed after a 5-day interval revealed an increased aggregability over the nonstroke group.

The correlation between clinical decourse and platelet's aggregability deserve some comment. In fact, no clear correlation was found in the present study. Some patients showed a poor improvement of the clinical picture in contrast to a rapid decrease of platelet hyperaggregability. Other patients had a fast, complete clinical recovery but platelet aggregability was still reported as highly increased. This negative result may be related to the small number of patients that completed the study. In fact, the two main conditions for admitting a patient into the stroke group, i.e., the absence of any therapy in the acute phase and the possibility of avoiding antiaggregating drugs through a

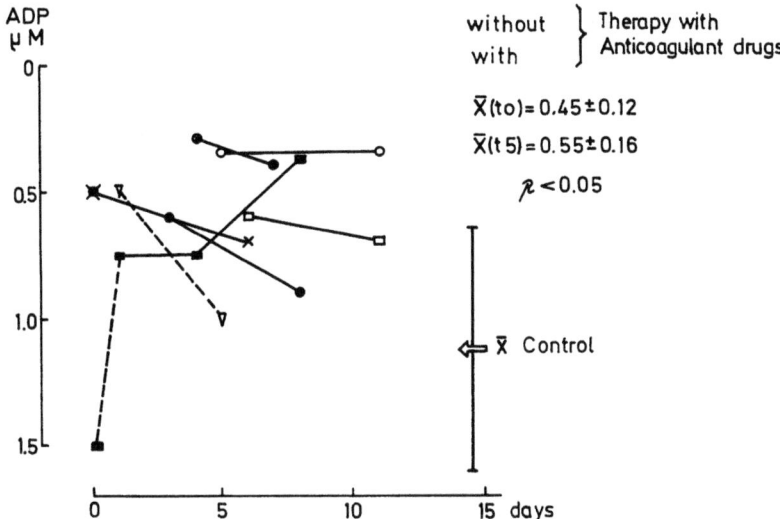

Fig. 2. Time-behavior of the in vitro platelet's aggregability, in patients presenting with ischemic episodes, as judged by the minimal ADP dose that produces the classical biphasic curve. Continuous line: patients not receiving antiaggregating treatment. Interrupted line: patients receiving antiaggregating treatment. The difference between the amount of ADP at the first determination ($\bar{x}(t_o)$) and the amount of ADP in the same patients after five days ($\bar{x}(t_5)$) resulted statistically significant ($P < 0.05$). Either mean are significantly different from the mean of the control group ($\bar{x}$ controli)

10-day period, resulted in us having to discard a large number of ischemic cerebrovascular patients.

In spite of that, and on the basis of the strict conditions observed, we are inclined to confirm the presence of platelet hyperaggregability in the acute as well as in the later phases of a stroke. This platelet hyperaggregability does not seem to require an altered lipemic background, it tends to decrease through the following days and actually does not show any evident correlation with the initial neurologic picture and one's future evolution.

References

Guiraud, B., Boneu, B., Boneu, A., Geraud, G., David, J.: Abnormal platelet behavior in atheromatous disease of cerebral vessels. Management by agents inhibiting platelet aggregation. 6th Int. Conf. Cerebral Vascular Disease, Salzburg. Meyer, J.S., Lechner, H., Reivich, M., Eichhorn, O. (eds.). 1972, pp. 31-40

Hardistry, R.M., Hutton, R.A., Montgomery, D., Rickard, S., Trebilcock, H.: Secondary platelet aggregation: a quantitative study. Brit. J. Haemat. 19, 307 (1970)

Ladurner, G., Ott, E., Holasek, A., Lechner, H.: Serum lipids and lipoproteins in cerebrovascular disease. 6th Int. Conf. Cerebral Vascular Disease, Salzburg. Meyer, J.S., Lechner, H., Reivich, M., Eichhorn, O. (eds.). p. 24-30 (1972)

Microembolism in the Nervous System

J. M. Williams, S. Hohmann, N. C. R. Merrillees, B. L. Oppermann, and P. M. Robinson

Abstract

We found microthrombotic emboli commonly in patients who developed complications after open-heart surgery, but rarely in patients who recovered uneventfully. Our experiments in dogs suggested that all patients undergoing open-heart surgery have emboli of simethicone and fat.

Subsequent experiments in rats showed that microthrombotic emboli consistently produced changes in fine structure in autonomic ganglia whereas total continuous or intermittent obstruction of the blood supply failed to produce changes regularly.

Introduction

In a clinical study of 35 patients undergoing cardiopulmonary bypass surgery, white plugs were observed in the retinal arteries of 11. Of these 11 patients, 10 developed neurologic complications in the postoperative period or died. In 2 patients at necropsy, the actual plugs during surgery were removed by careful dissection and examined by light or electron microscopy. The plugs were identified as platelet microthrombi (Swank et al., 1954; Williams, 1974).

In a parallel study in dogs, we surveyed the problem of microembolism by examining the retinal vessels after prolonged cardiopulmonary bypass perfusion without filtration (Williams et al., 1974). We then subjected dogs to cardiopulmonary bypass perfusion identical to that used in patients (i.e., with filtration) and, as expected, retinal microthrombotic emboli were greatly reduced (Williams and Davis, 1973). However, from our examination of trypsin digest preparations of retinal vessels of both patients and dogs after open-heart surgery, we concluded that emboli of simethicone (the defoaming agent used during the oxygenation of the blood) and of fat, are present in all patients undergoing cardiopulmonary bypass surgery, whether dacron-wool filtration is used or not (Williams and Davis, 1973).

Therefore we postulated that microthrombotic emboli were more damaging to central nervous and other tissues than were emboli of simethicone or fat.

The present study is the first part of an investigation into the effects of microthrombotic emboli on nervous tissue at a cellular level.

We chose automatic ganglia because they have a relatively simple neural organisation which has been described in detail in the literature (Elvin, 1963a,b; Williams and Palay, 1969). We decided to examine first the effects of microthrombotic emboli on the fine structure of nerve cells, accessory cells, and blood vessels.

Materials and Methods

Albino rats, each weighing between 100 and 450 g were intubated under general anaesthesia (ether, followed by subcutaneous sodium pentobarbital 0.1 mg/g). An endotracheal tube was passed through an incision in the trachea, which was exposed by separating the strap muscles of the neck, and respiration was maintained by a ventilator. In all experiments, the fine structure of the superior cervical ganglion on the same side as the surgical procedure (left), was compared with the fine structure of the superior cervical ganglion on the opposite (right) side, which served as a control.

Surgical Procedures

a) Induction of a Thrombus in the Left Common Carotid Artery

By placing cotton threads in the left common carotid artery, we produced a thrombus in the major artery supplying the nutrient vessels of the left superior cervical ganglion. Eighteen experiments were done. In each of five experiments, the threads were in situ for 30 min, in seven for 60 min, in one for 90 min, in three for 120 min, in one for 165 min and in one for 180 min.

b) Total Obstruction of the Blood Supply of the Left Superior Cervical Ganglion

By clamping the left common carotid artery, the left internal carotid artery, and the left external carotid artery above the origin of the ascending pharyngeal artery, we totally obstructed the blood supply of the left superior cervical ganglion. In each experiment the arteries were freed from the vagus and sympathetic nerve chains before being clamped.

Nineteen experiments were done in which the blood supply of the left superior cervical ganglion was totally obstructed. In each of three experiments the arteries were clamped for 10 min, in four for 30 min, in six for 60 min, and in six for 240 min.

That the blood supply of the left superior cervical ganglion is totally obstructed during the above clamping procedure had been demonstrated by carbon-black injection in a previous set of experiments (Fig. 1). The carbon-black injections were done in the following way. The animals were anaesthised and the clamps placed as above. The vascular system was perfused through the left ventricle with phosphate-buffered formol-saline and the perfusion continued for 20-30 min. After this time, the fixative was replaced by a 10% gelatin solution containing enough carbon-black in suspension to make it opaque. The specimen was then left overnight, the superior cervical ganglion from each side was removed and photographed (Fig. 1).

c) Intermittent Partial Obstruction of the Blood Supply of the Left Superior Cervical Ganglion

We clamped and unclamped the left common carotid artery at 5 min intervals. Seven experiments were done. In each of three experiments the left common carotid artery was clamped intermittently for 30 min, in one for 45 min, in two for 240 min and in one for 360 min.

At the end of all of the experiments described above we immersed the tissues in fixative: firstly the left superior cervical ganglion, then the right superior cervical ganglion, and lastly, in those experiments in which a thrombus was produced, the segment of the left common carotid artery where the threads had been placed. The tissue remained in fixative (1% osmium tetroxide in 0.50 M cacodylate buffer, pH 7.2-7.6) for 60 min, was then block stained in 2.5% magnesium uranyl acetate for 60-90 min, dehydrated in ascending concentrations of acetone and embedded in Spurr's low viscosity embedding medium.[1] Each ganglion and each portion of the left common carotid artery were embedded separately. Sections were stained with lead citrate (approximately 1%) and examined with a Siemen's microscope. At the completion of the assessment of all the sections, each block was sectioned again and the new sections reexamined by two of us independently.

Results

a) Induction of a Thrombus in the Left Common Carotid Artery

Each segment of the left common carotid artery, where the threads had been placed, clearly demonstrated loss of endothelium of the vessel wall with the formation of a thrombus in the lumen (Fig.5).

---

[1] Spurr's low-viscosity embedding medium, Polysciences, Inc., Warrington, Penna.

Fig.1A-D. (A) Injected vessels from untreated rat. Common carotid artery CC, ▶ external carotid artery EC, internal carotid artery IC, ascending pharangeal artery AP, and superior cervical ganglion SCG. Although not shown in figure, the major arterial supply to superior cervical ganglion is derived from the ascending pharangeal artery. Magnification × 13. (B) Higher power view of a specimen from another untreated rat. The plexus of vessels within the ganglion can be seen. Magnification × 24. (C) Region of the carotid bifurcation from the right (control) side of experimental animal. Magnification × 11. (D) Region of the carotid bifurcation from the left (clamped) side of same animal as (C). Clamps placed on common carotid, internal carotid, and external carotid arteries under anaesthesia prevented subsequent entry of carbon-black. This demonstrated occlusion of the vessels. Magnification × 11

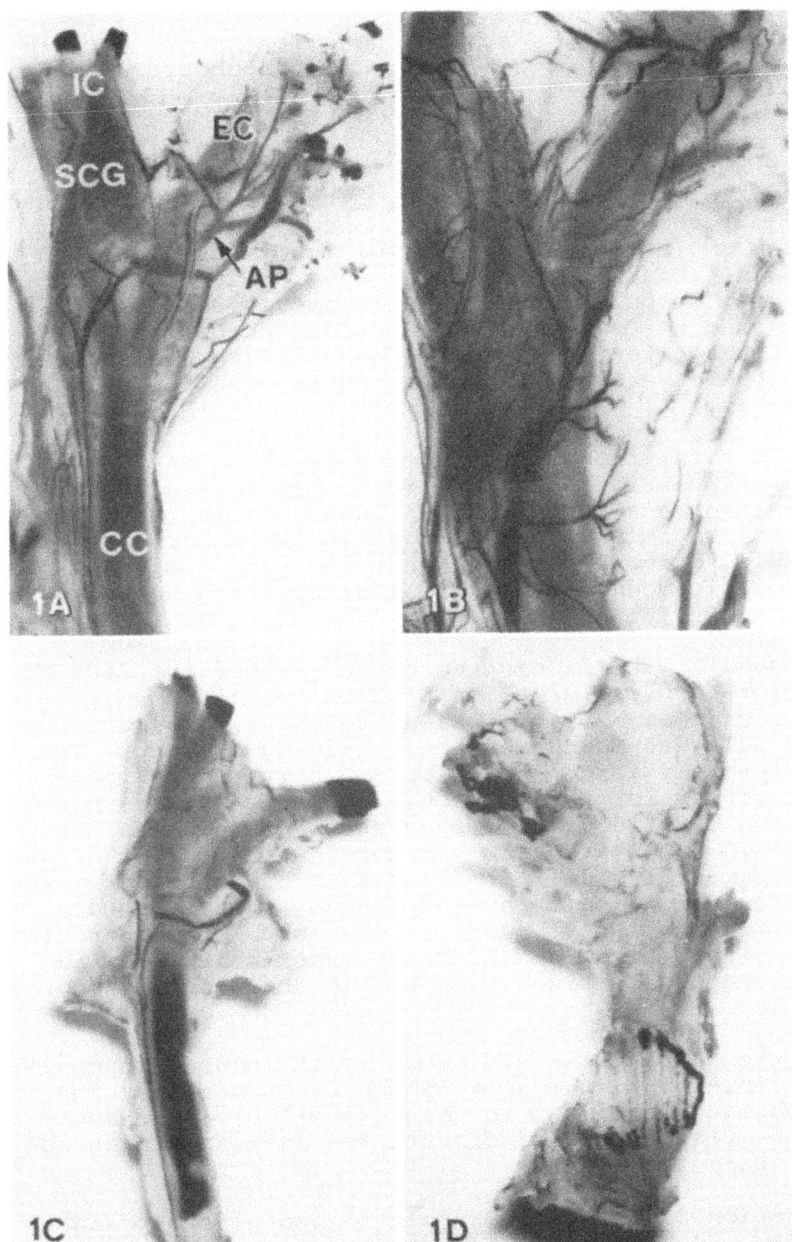

Fig. 1

This observation supported our suspicion that the changes in fine structure of the superior cervical ganglion on the experimental side were due to the presence of microthrombotic emboli in its nutrient arteries.

In 15 of the 18 rats we found substantial changes in fine structure in the left superior cervical ganglion when threads were placed in the left common carotid artery. The prominent features were: disrupted mitochondria and swollen endoplasmic reticulum in ganglion cells and satellite cells, swollen axons, intercellular debris (Figs. 2B, 3B), and swollen endothelium in blood vessels (Fig. 4B). Commonly, intercellular debris (sometimes with macrophages and extravasated blood cells) accumulated near blood vessels with swollen or disrupted endothelium. In severely damaged ganglia there were occasional shrunken ganglion cells and satellite cells.

In 11 of these 15 rats, the fine structure of the contralateral (control) ganglia appeared to be identical with that of ganglia taken from normal animals (Figs. 2A, 3A, and 4A). In the other 4 of these 15 rats we found minor changes in the control ganglion - for example, a few axons and dendrites were swollen, and in some areas, small amounts of intercellular debris had accumulated.

No change was found in the fine structure of either the left or the right superior cervical ganglion of 2 of the 18 rats with induced thrombi. Of these 2, 1 rat bled profusely throughout the experiment. No cause was found for this bleeding tendency. The other rat, a relatively large animal of 350 g (the first of a series of 8 relatively large rats), only two threads were placed in the left common carotid artery. Because two threads in the left common carotid artery regularly induced changes in the ganglion on the experimental side in smaller rats and because we suspected a slow blood flow would augment tissue injury due to microaggregation, we inserted six threads in the left common carotid artery of rats of this size in subsequent experiments. We then regularly found changes in fine structure in the ganglion on the experimental side.

In the last of the 18 rats in this group (with induced thrombi), six threads were placed in the left common carotid artery for 180 min (the longest experiment in this group). Severe changes were found in the fine structure of both the left and right superior cervical ganglia.

b) Total Obstruction of the Blood Supply of the Left Superior Cervical Ganglion

In 19 experiments in this group, we were unable to produce changes regularly in either the left or right superior cervical ganglion, and in fact, damage was found more frequently in the right (control) ganglion in this group than was found in the control ganglia in the group with induced thrombi.

In 11 of the 19 rats in this group, the changes in the ganglion on the operated side were the same as those described above for

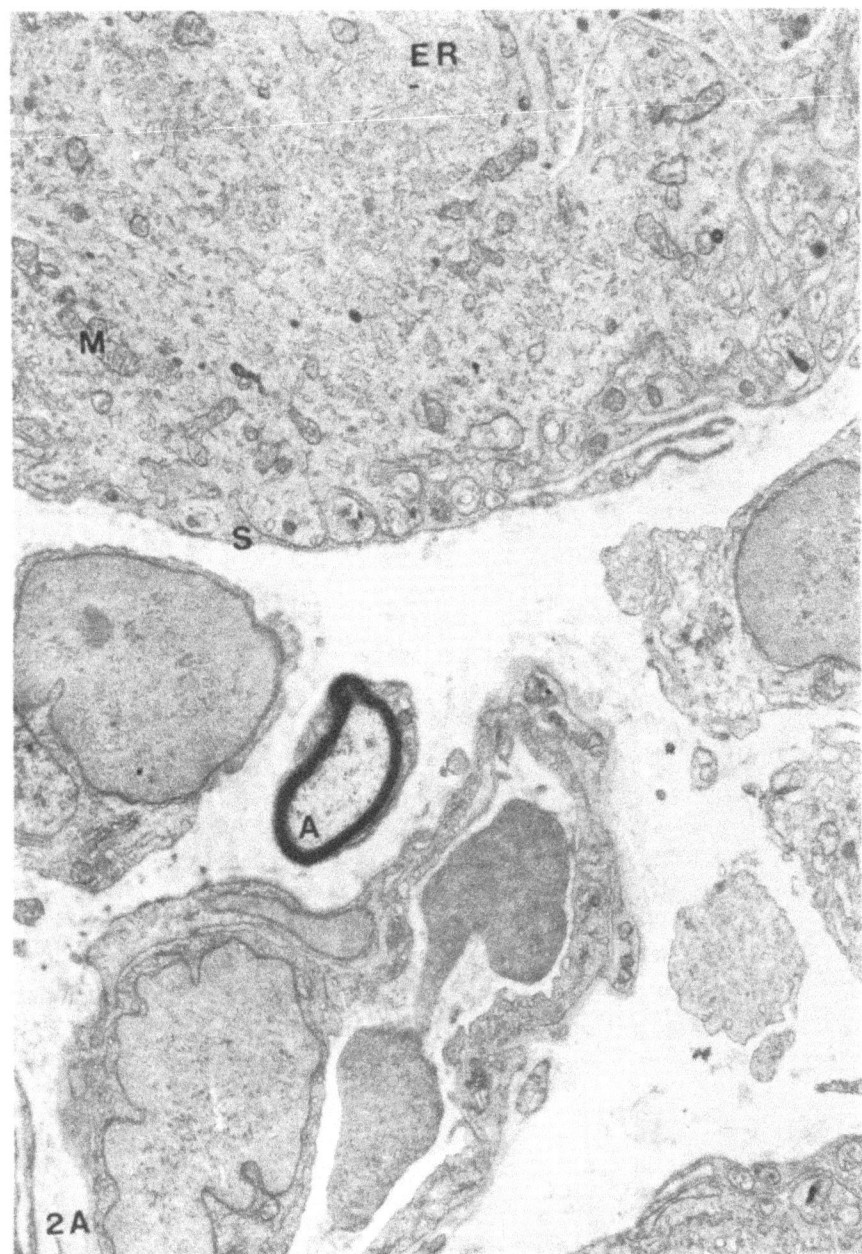

Fig. 2A

Figs. 2, 3, and 4. These show changes in fine structure in the right and left superior cervical ganglia of a rat 90 min after formation of a thrombus (Fig.5) was initiated in left common carotid artery of that rat

Fig. 2A and B. (A) Right control ganglion. (B) Left ganglion. In 2B, compared with control (2A), the mitochondria M, endoplasmic reticulum ER, and axons A are swollen. Furthermore, in experimental ganglion (2B), there is much intercellular debris ID and a satellite cell S shows swelling of organelles. Magnification × 7000

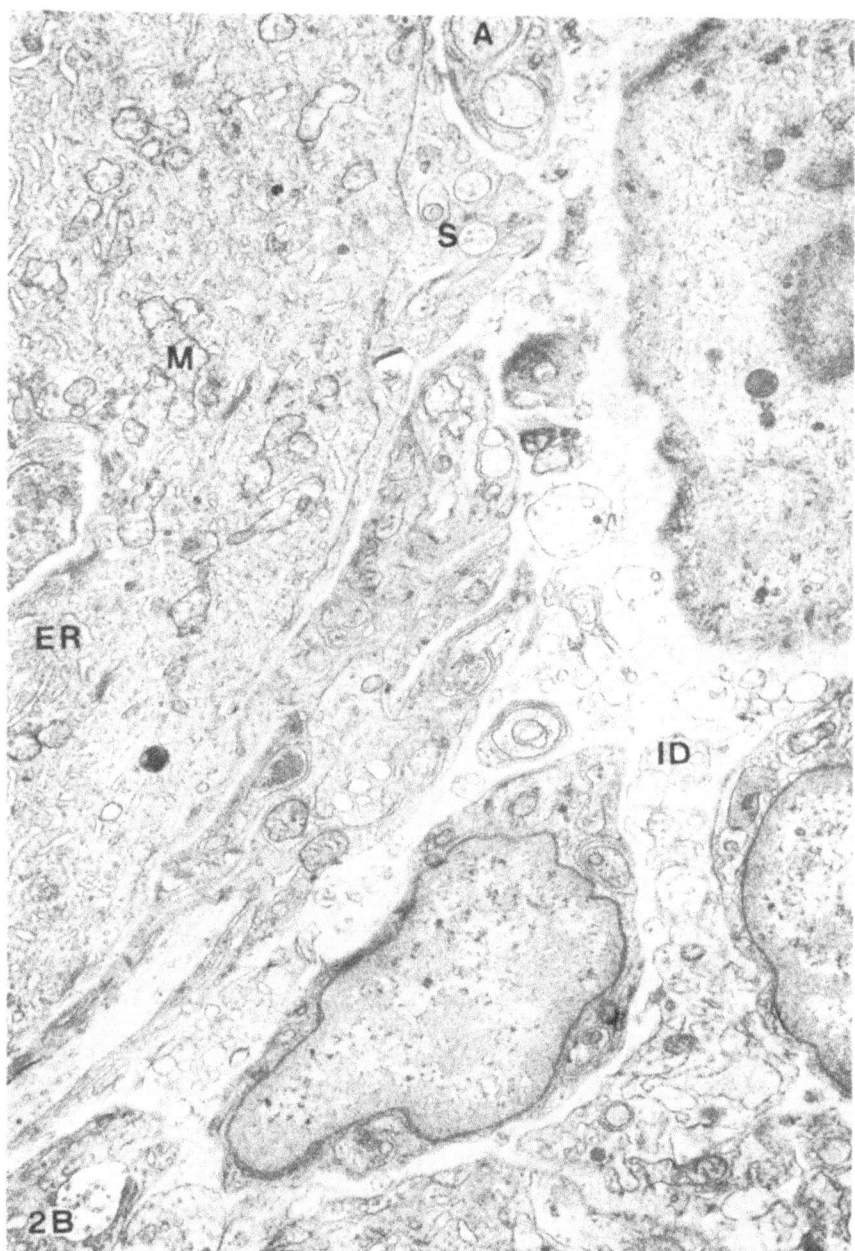

Fig. 2B

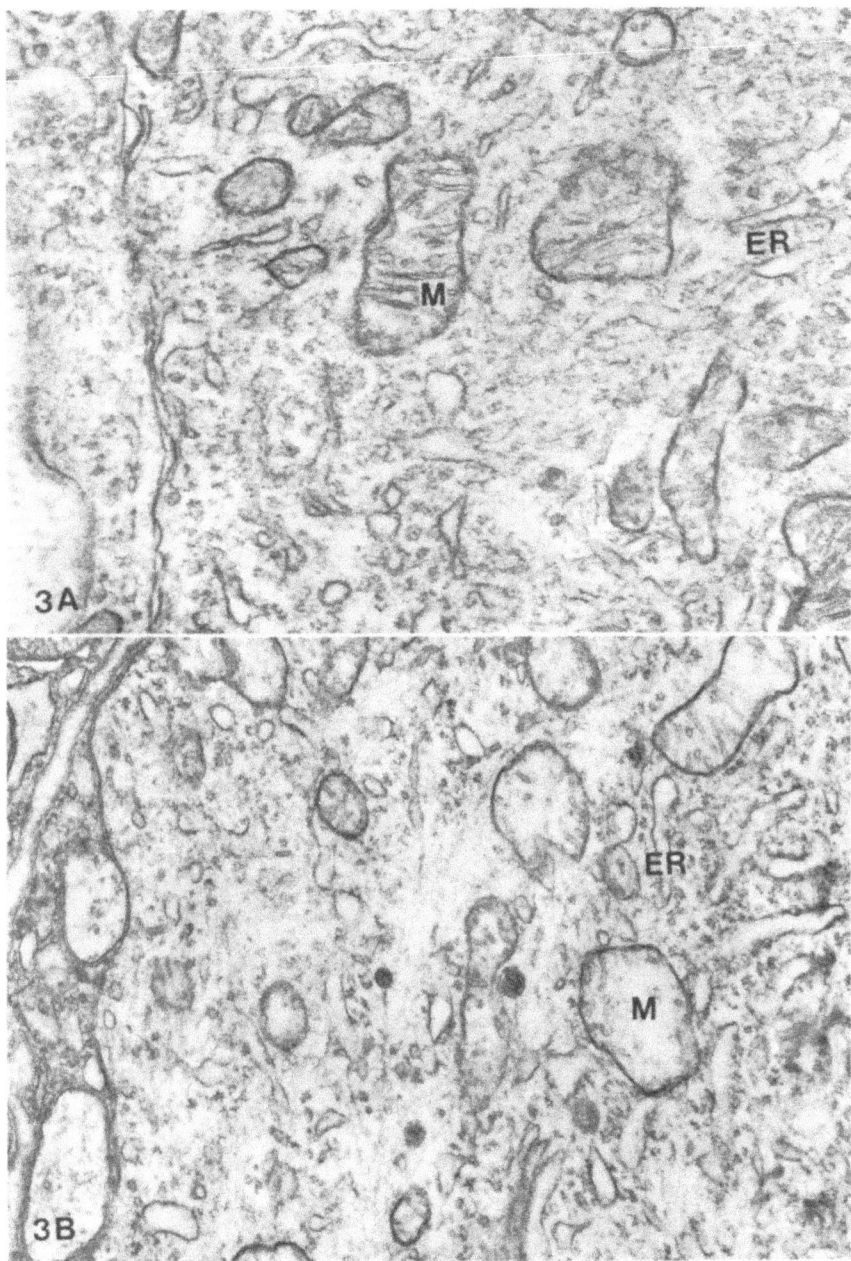

Fig. 3A and B. (A) Right control ganglion shown in 2A. (B) Left ganglion shown in 2B. Compared with those in control (3A), the mitochondria M and endoplasmic reticulum ER in 3B are swollen. Magnification × 15,000

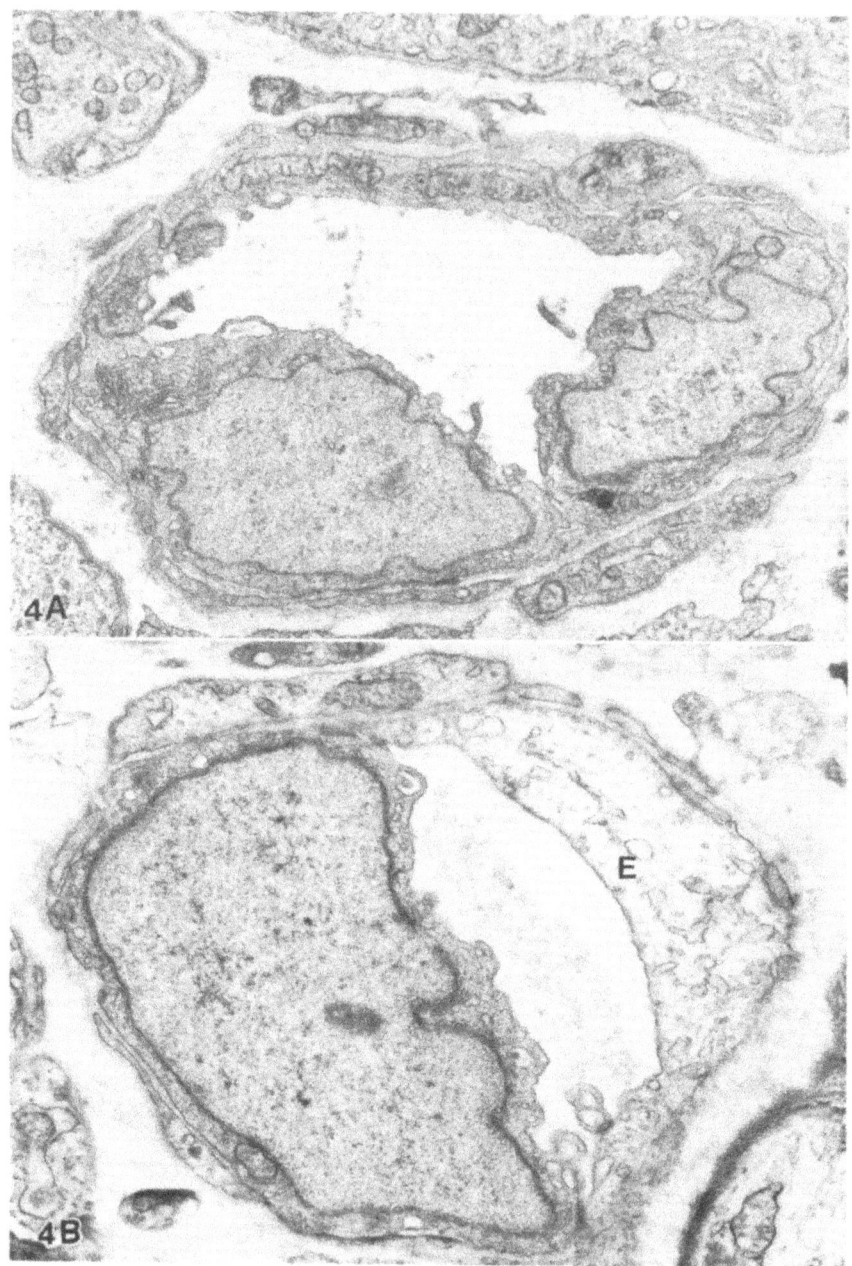

Fig. 4A and B. (A) Blood vessel in right (control) ganglion shown in 2A. (B) Blood vessel in left ganglion shown in 2B. Note swollen endothelial (E) cell in 4B. Magnification × 12,000

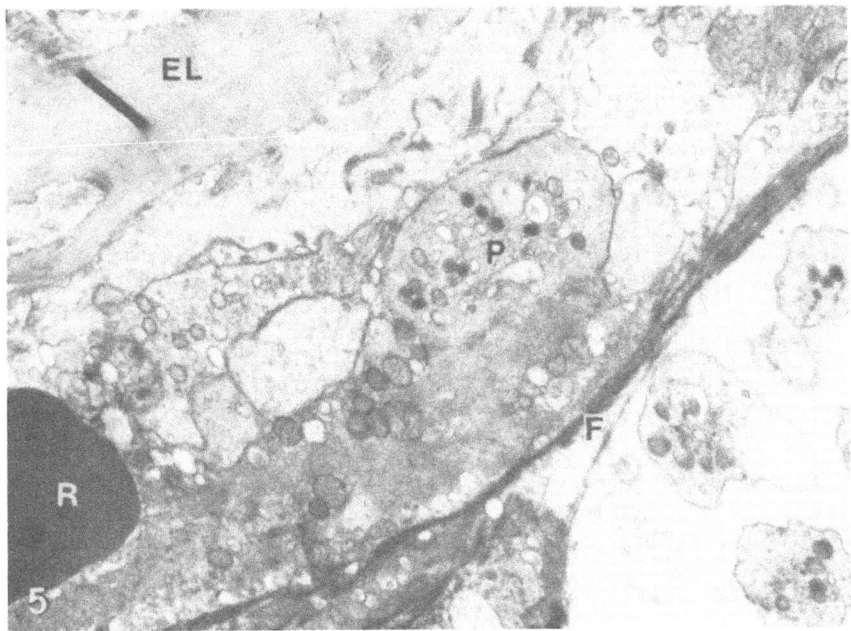

Fig. 5. Electron micrograph of a thrombus induced in the left common carotid artery. The endothelium has been destroyed. Against the intact innermost elastic lamina E, are platelets P, red blood cell R, fibrin F, and other debris. Magnification × 9000

the experimental side in the "thrombotic" rats. In 7 of these 11 rats, minor changes in fine structure were also found in the control ganglion; no change was found in the control ganglion in the other 4 animals.

In another 4 of the 19 rats in this group, we found severe damage in the *control* ganglion, but only minor changes in the ganglion on the experimental side of 3 of the 4. There were no changes at all in the experimental ganglion of 1 animal.

In the remaining 4 of the 19 rats in this group, we found minor changes in fine structure in both the left and right ganglia.

c) Intermittent Partial Obstruction of the Blood Supply of the Left Superior Cervical Ganglion

In 5 of the 7 rats in this group, we found no change in fine structure in either the left or right ganglia, with the exception of a small amount of intercellular debris in the experimental ganglia of 1 rat, and in both the left and right ganglia of another rat.

In two experiments in which the left common carotid artery was clamped intermittently for 240 min and 360 min respectively, we found minor changes in fine structure in the left ganglia, and an accumulation of a small amount of intercellular debris in the right ganglion.

Discussion

Since we consistently found changes in fine structure in the left superior cervical ganglion when the formation of a thrombus (i.e. with platelets) was induced in the left common carotid artery, we inferred that the changes in fine structure were due to microthrombotic emboli. This hypothesis is consistent with the known effect of microthrombotic emboli in other organs.

Kulvin and David (1967) used fluorescein angiography in monkeys to demonstrate leakage from retinal arteries through which microthrombotic emboli had passed. Mustard (1973) considered that platelets, during aggregation, contribute to an inflammatory response by liberating some constituents which increase the permeability of vessels. Nachman (1973) suggested that cationic proteins released from aggregated platelets may initiate the early vascular abnormalities associated with microthrombotic emboli. From a study of myocardium kidney and lung, Jorgensen et al. (1970) concluded that platelet aggregation in the microcirculation, even when transient, may be an important factor in causing vascular lesions and mural thrombosis. Furthermore, Mustard (1973) suggested that enzymes, such as elastase which are released from platelets during aggregation, may contribute to tissue injury.

Connell et al. (1973) examined the fine structure of the lungs of patients after 1 h of extracorporeal circulation without filters. They reported extensive occlusion of capillary beds by aggregates of leukocytes in various stages of disintegration, and widespread degenerative changes in the cellular and noncellular components of the interalveolar septum. Sometimes the capillary endothelium was focally swollen and had ruptured, thus facilitating the free dispersion of lysozymal granules in the interstitial tissue. Furthermore, they reported protein exudates in the interstitial and alveolar air spaces. The removal of microaggregates by Dacron-wool filtration during bypass reduced the extent of these degenerative lesions.

a) Induction of a Thrombus in the Left Common Carotid Artery with Threads

We found that the changes induced in fine structure in the superior cervical ganglion, when thrombus partially occluded the left common carotid artery, were disruption of organelles of ganglion cells and satellite cells, swelling of axons, swelling of the endothelium of blood vessels, and an accumulation of intercellular debris. Occasional macrophages and blood cells were found outside capillaries, and, rarely, we found disrupted capillaries. Since few platelet aggregates were found in the sections of nervous tissue, we concluded that the changes in fine structure were due primarily to products of thrombosis, such as platelet enzymes, and not to impaired nutrition consequent upon the blockage of small vessels by microaggregates.

Our findings are consistent with those of Meyer et al. (1962) who showed that cerebral embolism with small particles of pumice, oil, or air produce catastrophic cerebral damage much more severe than that seen after occlusion of the carotid or middle cerebral arteries. Furthermore, fibrin and platelets began to propagate at sites of endothelial damage caused by the passage of irregular pumice particles.

In one large rat two threads were ineffective. Because we believed that two threads and the thrombi induced by them failed in reducing the flow of blood in the experimental ganglia of large rats sufficiently to allow microthrombi enough time to pause and liberate "toxic" constituents, we placed six threads in the left common carotid artery of large rats in subsequent experiments. In fact, six threads regularly produced changes in fine structure in the experimental ganglion of relatively large rats. This explanation needs to be proved in a larger series.

Microaggregates may pass through the capillary circulation and be distributed at random. We frequently found such microaggregates in the retinal capillaries of dogs which had undergone total cardiopulmonary bypass perfusion without filtration (Williams et al., 1974). We suggest that widely distributed microaggregates induced the changes in fine structure in both the left and right superior cervical ganglion of the rat in which six threads were placed for 180 min.

The structure of the thrombi induced in the left common carotid arteries in the present study, closely resembled the structure of deposits which Hovig et al. (1970) induced on the inner lining of a flow chamber and on collagen fragments within the chamber after 5-20 min of blood flow. The deposits were composed of aggregated platelets and of a peripheral fibrin layer resembling that shown in Figure 5.

b) Total Obstruction of the Blood Supply of the Left Superior Cervical Ganglion by Clamping

The surgical procedure described in this group required much dissection. Inadvertent damage to blood vessel walls may have induced microthrombotic emboli in rats in this group. It is probable the changes in fine structure in both the control and experimental ganglia in this group were due to microthrombotic emboli induced during dissection.

c) Intermittent Partial Obstruction of the Blood Supply of the Left Superior Cervical Ganglion by Clamping

It is probable that microthrombi, forming at the site of intermittent obstruction of the left common carotid artery, induced changes in fine structure in the superior cervical ganglion on the experimental side, and that those microthrombi that pass through the capillary circulation, and, therefore, were distributed at random, induced changes in the control ganglion. We suggest that the changes in fine structure in both sides were minor because there was no reduction in blood flow between the periods of obstruction of the left common carotid artery.

In experiments in dogs undergoing prolonged cardiopulmonary bypass perfusion with filtration, we found that dipyridamole, added to the perfusate, greatly reduced the number of microthrombotic plugs appearing in the retina (Williams et al., 1974).

This was consistent with the report of Rittenhouse et al. (1972) that dipyridamole, added to the blood prime prior to circulation in a pump oxygenator, lessened the rise in screen filtration pressure (a measure of the degree of microaggregate formation during cardiopulmonary bypass perfusion (Swank et al., 1954). They concluded that dipyridamole can reduce the degree of microaggregate formation in blood circulated for prolonged periods in the pump oxygenator.

Emmons et al. (1964) showed that both the intravenous administration of dipyridamole (0.43-0.86 mg/kg b.w.) and oral dipyridamole (200-600 mg/day) regularly reduced the spontaneous platelet aggregation that occurs when platelet rich plasma of patients is stirred. During cardiopulmonary bypass, dipyridamole reduces microaggregation formation in the peripheral circulation of patients (Mielke et al., 1973).

Agents such as dipyridamole alter the platelet surface interaction. It is hoped that tissue damage caused by substances "toxic" to nervous tissue, which are released when platelets aggregate, may be limited by such agents. We are investigating this possibility in our laboratory.

References

Cornell, R.S., Page, U.S., Bartley, T.D., Bigelow, J.C., Webb, M.C.: The effect on pulmonary ultrastructure of Dacron-wool filtration during cardiopulmonary bypass. Ann. thorac. Surg. 15, 217-229 (1973)

Elvin, L.G.: The ultrastructure of the superior cervical sympathetic ganglion of the cat. 1. The structure of the ganglion cell processes as studied by serial sections. J. Ultrastruct. Res. 8, 403-440 (1963a)

Elvin, L.G.: The ultrastructure of the superior cervical sympathetic ganglion of the cat. 2. The structure of the preganglionic end fibers and the synapses as studied by serial sections. J. Ultrastruct. Res. 8, 441-476 (1963b)

Emmons, P.R., Harrison, M.J.G., Honour, A.J., Mitchell, J.R.A.: Effect of dipyridamole on human platelet behaviour. Lancet 2, 603-606 (1965)

Hovig, T., Jorgensen, L., Rowsell, H.C., Mustard, J.F.: The structure of thrombus-like deposits formed in extracorporeal shunts. Amer. J. Path. 59, 75-99 (1970)

Jorgensen, L., Hovig, T., Rowsell, H.C., Mustard, J.F.: Adenosine Diphosphate-induced platelet aggregation and vascular injury in swine and rabbits. Amer. J. Path. 61, 161-176 (1970)

Kulvin, S.M., David, N.J.: Experimental retinal embolism. Studies with high speed fluorescein cinematography. Arch. Opthal. (Chicago) 78, 774-788 (1967)

Meyer, J.S., Gotoh, F., Tazaki, Y.: Circulation and Metabolism following experimental cerebral embolism. J. Neuropath. exp. Neurol. 21, 4-24 (1962)

Mielke, C.H., de Leval, M., Hill, J.D., Macur, M.F., Gerbode, F.: Drug influence on platelet loss during extracorporeal circulation. J. thorac. cardiovasc. Surg. 66, 845-854 (1973)

Mustard, J.F.: The platelet as an inflammatory cell. Part 17 in Platelets, Drugs and Thrombosis. Canad. med. Ass. J. 108, 453 (1973)

Nachman, R.L.: The platelet as an inflammatory cell. In: Transactions of the Eighth Conference on Cerebral Vascular Diseases. McDowell, F.H., Brennan, R.W. (eds.). New York and London: Grune and Stratton 1973, pp. 281-285

Rittenhouse, E.A., Hessel, E.A., Ito, C.S., Merendino, K.A.: Effect of dipyridamole on microaggregate formation in the pump oxygenator. Ann. Surg. 175, 1-9 (1972)

Swank, R.L., Osborn, J.: Blood filter for extracorporeal circulation. In: Microcirculatory Approaches to Current Therapeutic Problems. Proceedings of the 6th European Conference on Microcirculation, Aalborg, 1970. Ditzel, J., Lewis, D. (eds.). Basel: Karger Publ. Co. 1971, pp. 59-64

Swank, R.L., Roth, J.G., Jensen, J.: Screen filtration pressure method and adhesiveness and aggregation of blood cells. J. appl. Physiol. 19, 340-346 (1954)

Williams, I.M.: Central nervous system dysfunction with open-heart operations. Proc. Aust. Ass. Neurol. 10, 1-6 (1973)

Williams, I.M., Davis, B.B.: Retinal emboli in total cardiopulmonary bypass surgery. Proc. Aust. Soc. med. Res. 3, 124 (1973)

Williams, I.M., Farmer, S., Dixon, J.: Effect of dipyridamole on retinal embolism associated with cardiopulmonary bypass surgery in the dog. Graefes Arch. klin. exp. Opthal. 189, 251-263 (1974)

Williams, I.M.: Retinal vascular occlusions in open-heart surgery. Brit. J. Opthal. (in press, 1974)

Williams, T.M., Palay, S.L.: Ultrastructure of the small neurons in the superior cervical ganglion. Brain Res. 15, 17-34 (1969)

Influence of Plasma Components in the Development of Conditions of Increased Platelet Aggregation Found in a Number of Vascular Diseases

R. Breda, B. Bizzi, and G. Leone

Our studies on changes in platelet aggregation found in subjects with vascular diseases began a few years ago when, using the Born aggregometer, a high incidence of abnormalities was observed in diabetics (Leone, 1970; Leone and Bizzi, 1971; Leone et al., 1974).

Our work demonstrated that, when decreasing doses of ADP are used to bring about aggregation of platelet-rich plasma (PRP), in normal subjects irreversible aggregation is possible up to 1-2 µM of ADP (as final concentration in the mixture), whereas irreversible aggregation with lower doses of ADP (0.25-0.5 µM) is possible in a large number of diabetics. Our results are similar to those obtained by Kwaan et al. (1972) and Passa et al. (1974). Collagen and thrombofax-induced platelet aggregation instead was found to be consistently normal in diabetics.

Abnormalities in the curve of ADP-induced platelet aggregation were actually observed in diabetics with overt signs of vascular disease. Extensive studies have shown that changes in platelet aggregation occurred frequently in patients with vascular disease, irrespective of the presence of pancreatic diabetes. This finding is in agreement with studies carried out by other authors who have demonstrated an increased ADP-induced platelet aggregation in arteriosclerotic patients, when such an increase was assessed on the basis of minimum dose required to induce irreversible aggregation.

This was found to be true both when the cerebral region was prevailingly involved (Danta, 1970) and the coronary (Zahavi and Dreyfuss, 1969) as well as the lower limbs (Dawis, 1973).

ADP-induced platelet aggregation, which is routinely performed on platelet-rich plasma, is known to be affected not only by platelets themselves, but also by plasma components to a considerable extent.

Experiments performed in vitro on washed platelets demonstrated the important role played by fibrinogen, prothrombin complex factors and contact factors. Ardlie and Han (1974) in their late work postulate that contact factors, thrombin, and fibrinogen are involved in ADP-induced platelet aggregation: the release reaction fails to occur in the latter if the washed platelets are resuspended in plasma adsorbed with alumina or treated with celite. As a matter of fact, we believe that with a high ADP con-

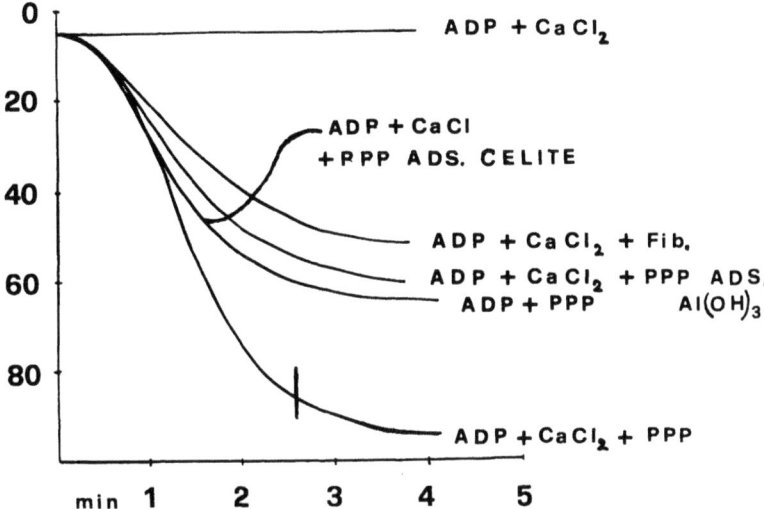

Fig. 1. ADP-induced platelet aggregation (8 µM) of platelets washed and resuspended in different solutions

centration, irreversible aggregation can be obtained even with plasma adsorbed with alumina (Fig. 1).

The above plasma factors appear to have a peculiar influence on ADP-induced platelet aggregation. Collagen-induced platelet aggregation as pointed out earlier does not appear to be changed in patients with vascular disease nor is it affected by fibrinogen, contact factors, or the prothrombin complex.

Subjects with vascular diseases are known to commonly exhibit a high level of fibrinogen and the presence of circulating activated contact factors: hence, after discarding the direct responsibility of platelets in inducing enhancement of platelet aggregation with low doses of ADP, we felt it was worthwhile to consider the part played by the above components.

Thus, from plasma of subjects with vascular diseases exhibiting enhanced ADP-induced aggregation, platelets were separated according to Walsh (1972). This method is based on a density gradient made with albumin, and allows platelets to be freely obtained from plasma factors commonly adsorbed to them, which preserve an almost normal response to ADP.

In most cases considered, platelets of subjects with vascular diseases, which exhibited such an anomaly, after resuspension in normal plasma gave normal curves for ADP-induced aggregation.

By contrast, platelet-poor plasma (PPP) of subjects with vascular diseases, with enhanced response to ADP, added to platelets of a normal subject (PRP), could enhance ADP response (Fig. 2).

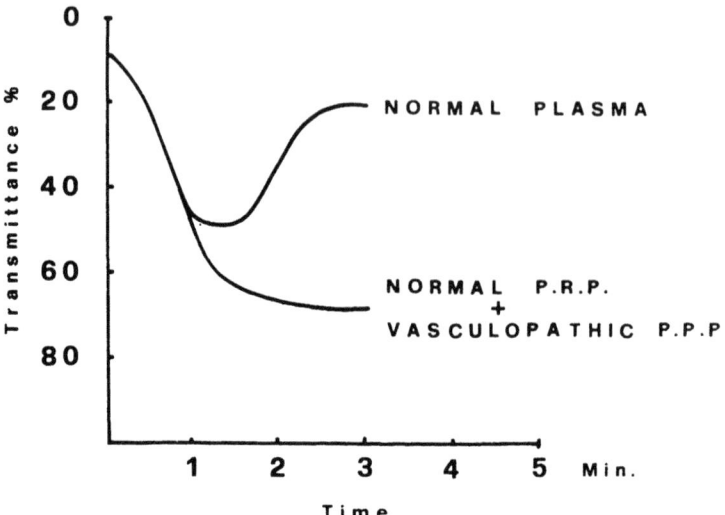

Fig. 2. ADP-induced platelet aggregation (0.5 µM) of normal platelet-rich plasma (PRP) and a mixture of normal PRP and platelet-poor plasma (PPP) of a subject with vascular disease

Such a phenomenon is not related to fibrinogenemia, as the latter and the minimum dose of ADP required to induce irreversible aggregation are unrelated and the effect favoring platelet aggregation is found to occur in serum, too.

Figure 3 summarizes our experiments on the effect of platelet-poor plasma (PPP) - treated according to different methods - of patients with vascular disease exhibiting the anomaly, on ADP-induced aggregation of normal platelet-rich plasma (PRP). Since our results were not always in agreement, we have reported the event occurring most frequently.

Plasma adsorbtion with alumina, which is known to remove factors II, VII, IX, did not affect the ability of plasma from vascular patients to enhance ADP-induced platelet aggregation in six out of seven cases.

Treatment of plasma with celite, aimed at removing contact factors, in most cases (five out of seven) has removed the effect promoting ADP-induced platelet aggregation in plasma of subjects with vascular disease.

Conversely, it was possible to observe that in subjects with vascular disease who exhibited such an alteration (Fig. 4), the contact product eluate with celite and resuspended in saline can enhance ADP-induced platelet aggregation similarly to plasma.

In conclusion it can be postulated that a large part of subjects with arteriosclerotic vasculopathies exhibit enhanced ADP-induced platelet aggregation, based on the minimum dose likely to bring about irreversible aggregation.

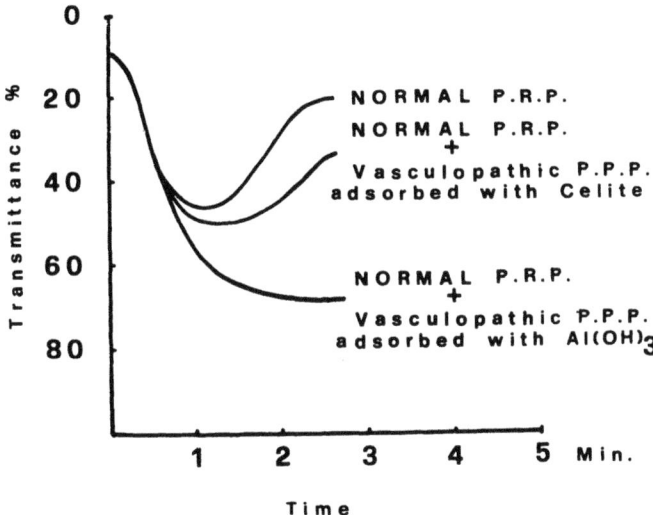

Fig. 3. ADP-induced platelet aggregation (0.5 µM) of normal PRP, a mixture of normal PRP and PPP of subject with vascular disease adsorbed with AL(OH)$_3$, and a mixture of normal PRP and PPP of subject with vascular disease adsorbed with celite

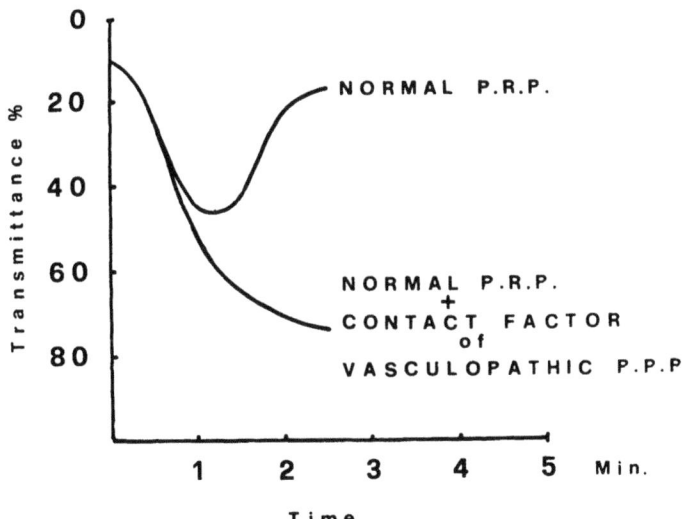

Fig. 4. ADP-induced platelet aggregation (0.5 µM) of normal PRP and a mixture of normal PRP with contact factor, eluted from plasma of a subject with vascular disease

Among the causes for an increased ADP-induced platelet aggregation, plasma factors are known to play a major role.

As to plasma factors in our experience, which is still limited and not final, the presence of activated contact factors is likely to play a decisive role.

References

Ardlie, N.G., Han, P.: Enzymatic basis for platelet aggregation and release: the significance of the platelet atmosphere and the relationship between platelet function and blood coagulation. Brit. J. Haemat. 26, 331 (1974)

Danta, G.: Second phase platelet aggregation induced by adenosine diphosphate in patients with cerebral vascular disease and in control subjects. Thromb. Diath. haemorrh. (Stuttg.) 23, 159 (1970)

Dawis, J.W.: Defective platelet disaggregation associated with occlusive arterial disease. Angiology 24, 391 (1973)

Kwaan, H.C., Colwell, J.A., Cruz, S., Suwanwela, M., Dobbie, J.C.: Increased platelet aggregation in diabetes mellitus. J. Lab. clin. med. 80, 236 (1972)

Leone, G.: Intervento in "Round Table Conference on Normal and Modified Platelet Aggregation". Acta med. scand., Suppl. 525, 277 (1970)

Leone, G., Bizzi, B.: Il comportamento della adesività e della aggregabilità piastrinica nei diabetici. Gazz. int. Med. Chir. 76, 368 (1971)

Leone, G., Bizzi, B., Accorrà, F., Boni, P.: Platelet Function in diabetes mellitus. In: Platelet Aggregation and Drugs. London: Academic Press 1974

Passa, P., Benoussan, D., Levy Toledano, S., Caen, J., Canivet, J.: Etude de l'agregation plaquettaire au cours de la retinopathie diabetique. Atherosclerosis 19, 277 (1974)

Walsh, P.N.: Albumin density gradient separation and washing of platelets and the study of platelet coagulant activities. Brit. J. Haemat. 22, 205 (1972)

Zahavi, J., Dreyfuss, F.: An abnormal pattern of adenosine diphosphate induced platelet aggregation in acute miocardial infarction. Thromb. Diath. haemorrh. (Stuttg.) 21, 76 (1969)

Mechanism Associated with Platelet Adhesiveness in Cerebrovascular Disease*

R. L. Swank

Physiologic and pathologic studies indicate that platelet aggregates (with or without leukocytes) can occlude conduits of the microvasculature causing increased vascular resistance, blood gas and pH changes, and EEG and focal neurologic findings (Hissen and Swank, 1965; Alweis et al., 1967; Swank and Edward, 1968; McNamara et al., 1972; Singh et al., 1967). In the lungs, the first structural changes seen with the electron microscope are occlusion then digestion of the endothelia of vascular channels. The interstitial tissues, and finally the epithelia of the alveolar air sacs are destroyed and the vascular contents enter the air spaces (Connell and Swank, 1973). Many of these microemboli quickly pass through the lungs and secondarily embolize other organs of the body (Connell et al., 1972). In the brain (Figs. 1, 2) the neuropile surrounding the microcirculation is fragmented and is destroyed leaving wide pericapillary, periarteriolar, and perivenular spaces. The lung lesions spontaneously recover in 2-4 weeks if not too extensive. The brain lesions can no longer be identified at the end of 1 week. The microemboli negotiate the brain capillaries in a few minutes and are infrequently seen in the brain, although they do accumulate in the sinusoids of the liver.

It would appear that the tight junctions between the endothelial cells of the cerebral microcirculation separate sufficiently to allow fluids and an occasional red cell to pass. These accumulate perivascularly attended by tissue destruction.

Platelet adhesiveness and aggregation occur in vitro in blood stored for transfusion (Swank, 1961). They occur also in vivo under a wide variety of circumstances, including trauma (Swank et al., 1964), hypotension (Swank, 1962), and in patients with intraarterial blood cell aggregation (Swank and Davis, 1966). The venous blood from organs rendered ischemic contains abnormally adhesive and aggregated platelets (Hirsch et al., 1964), and similar changes have been demonstrated after high fat meals (Cullen and Swank, 1954; Swank and Bartsch, 1971). The biological agents ADP, serotonin, adrenalin, and nonadrenalin all present in the body, and released to the circulation during periods of

---

*Supported by funds held by the Medical Research Foundation of Oregon and by the Advancement Fund of the University of Oregon Medical School

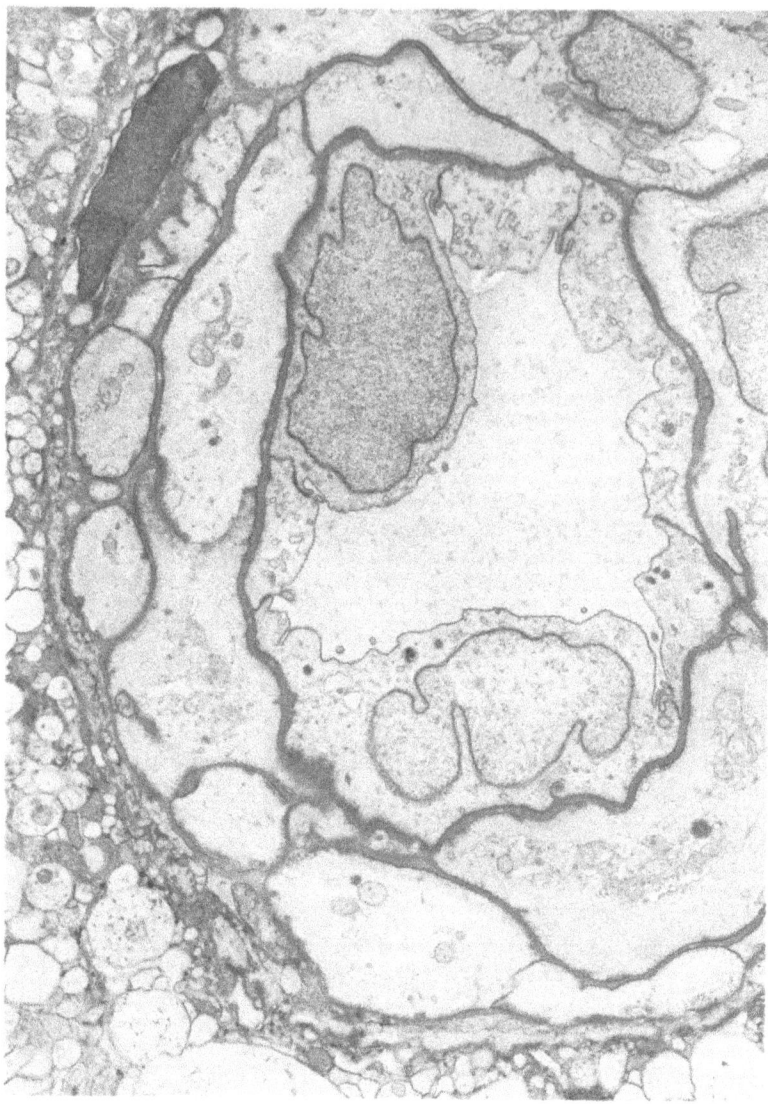

Fig. 1. Normal cerebral arteriole from dog brain. Note how closely attached the nerve and glial elements are to outer boundaries of the arteriole

stress, also cause platelet adhesiveness and aggregation accompanied by aggregation of all cells (Born, 1962; Swank et al., 1963; Swank and Bartsch, 1971).

The numerous causes of platelet aggregation and the ease with which these microemboli traverse the microcirculation of the lung suggests that general health, aside from microemboli from ulcerated atheromatous plaques in the carotid artery, may be an important cause of cerebral vascular incidents including transient as well as more permanent neurologic deficits.

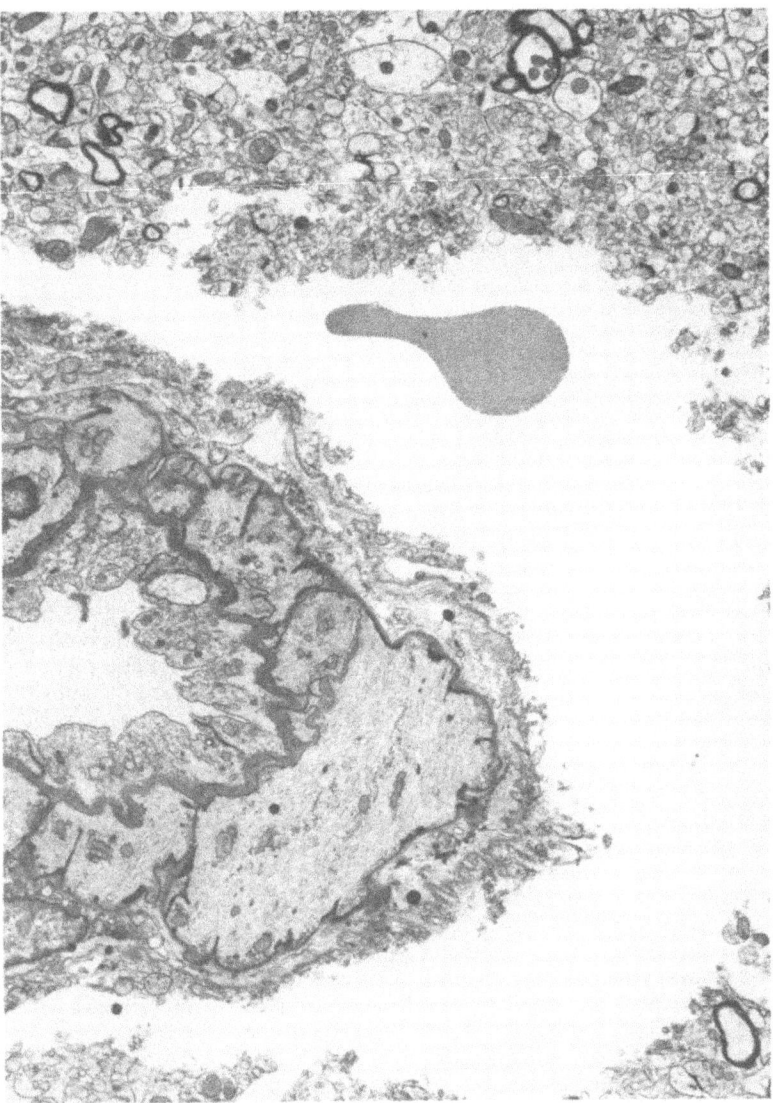

Fig. 2. Cerebral arteriole from dog following transfusion with 1-week-old blood. Emboli have passed through the lung vascular channels and secondarily embolized the cerebral circulation. Note that nervous and glial elements have become separated from arteriole leaving perivascular spaces containing debris and single red cell

In vascular disease of the extremities, kidneys, or heart, for example, platelets are probably continuously damaged during transit. These go directly to the lungs. If they are not destroyed or phagocytized, and manage to negotiate the microcirculation of the lung, they secondarily embolize the brain (and other organs). If the lesions are widely spaced, the functional and structural damage to the brain will be insignificant. However, saturation of an area with microemboli would be expected to cause

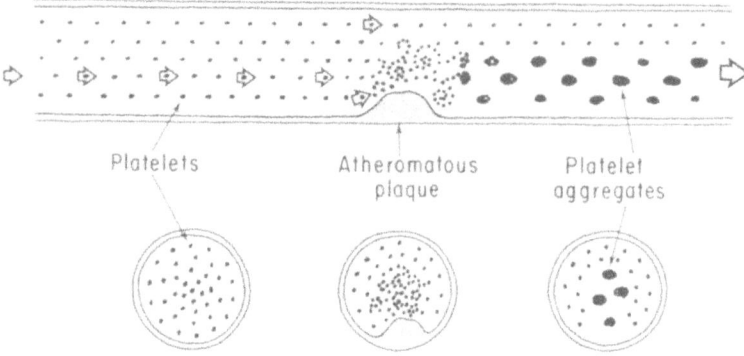

Fig. 3. Schematically illustrates effect of turbulence on adhesive platelets caused by atheroma protruding into lumen of artery. As blood approaches atheromatous plaque, platelets and other cells maintain their relative positions and have minimal contact. As blood stream is distorted by atheroma mixing occurs, platelets come in contact, and aggregate. They also contact and adhere to the rough endothelium over the atheroma

recognizable transient or more permanent neurologic deficits. Microemboli of the same size have produced large demyelinated lesions, some with areas of central softening in experimental animals (Swank and Hain, 1952). A cascading or self-perpetuating mechanism for further microemboli information could result.

Several other factors which might contribute to aggregate formation of adhesive platelets and/or to increase vascular resistance should be mentioned at this time. The first to be considered is turbidity produced in the normal laminar flow of blood in an artery by a partial occlusion of its lumen. Atheromatous plaques, large enough to significantly interfere with the blood flow through an internal carotid artery, are associated with cerebral ischemic attacks. Lesser contributions which do not significantly decrease the blood flow, however, may have an important influence because of the turbulence which they cause. Adhesive platelets during laminar flow would have minimum opportunity to come in contact and form aggregates, or even to come in contact with the endothelia of the artery. However, during turbulence the platelets would come into contact frequently, and consequently form aggregates (Fig. 3). They would also come into contact with the atheromata where they could adhere to form aggregates. These could later break free and enter the brain.

Another factor infrequently considered is the daily change in blood viscosity (Seaman et al., 1965). Decreases in viscosity accompanied by a decrease in hematocrit and the total protein content of the circulating blood occur during sleep. During the day the viscosity, hematocrit, and total protein of the blood increase. The decreased viscosity during inactivity at night is probably due to sequestration of red cells and other blood components in the capillary beds. If true this would tend to increase cerebral and other vascular resistance at night, which in turn could contribute to the relatively high frequency of strokes at this time.

It has been shown that a high fat intake causes increased viscosity of blood (Swank, 1954), slowing of the circulation, and a reduction in oxygen availability in cerebral tissue (Swank and Nahamura, 1960). On the other hand, whole blood and plasma viscosities decreased in patients with cerebral vascular disease when they follow a low fat diet for 3-6 months (Swank, 1959). It seems most likely that a lowering of the blood viscosity would be followed by an increased blood flow, and also by a tendency of platelets to be less adhesive (Swank and Bartsch, 1971).

References

Alweis, C., Abeles, M., Magnus, J.: Perfusion of cat brain with simplified blood after filtration through glass wool. Amer. J. Physiol. 213, 83 (1967)

Born, G.V.R.: Aggregation of blood platelets by adenosine diphosphate and its reversal. Nature (Lond.) 194, 927 (1962)

Connell, R.S., Swank, R.L.: Pulmonary microembolism after blood transfusions: An electron microscopic study. Ann. Surg. 177, 40 (1973)

Connell, R.S., Swank, R.L., Webb, M.C.: Secondary microembolism after transfusions: An electron microscopic study. Presented at the 1st Int. Congr. Biorheology, Lyon, France, September 1972

Cullen, C.G., Swank, R.L.: Intravascular aggregation and adhesiveness of the blood elements associated with alimentary lipemia and injections of large molecular substances: Effect on blood-brain barrier. Circulation 9, 335 (1954)

Hirsch, H., Gaehtgens, P., Sobbe, A.: Änderungen des Siebungsdrucks nach Ischämia von Gehirn, Extremität und Niere. Pflügers Arch. ges. Physiol. 281, 191 (1964)

Hissen, W., Swank, R.L.: Screen filtration pressure and pulmonary hypertension. Amer. J. Physiol. 209, 715 (1965)

McNamara, J.J., Burran, E.L., Larson, E., Omiya, G., Sushiro, G., Yamase, H.: Effect of debris in stored blood on pulmonary microvasculature. Ann. thorac. Surg. 14, 133 (1972)

Seaman, G.V.F., Engel, R., Hissen, W., Swank, R.L.: Circadian periodicity of viscosity and of some protein components in circulating blood. Nature (Lond.) 207, 833 (1965)

Singh, S.N., Carter, C.C., Swank, R.L., Blachly, P.H.: Relationship between post-cardiotomy delirium, clinical neurological changes and EEG abnormalities. J. thorac. cardiovasc. Surg. 54, 557 (1967)

Swank, R.L.: Effects of high feedings on viscosity of blood. Science 120, 427 (1954)

Swank, R.L.: Blood viscosity in cerebrovascular disease: Effect of low fat diet and heparin. Neurology 9, 553 (1959)

Swank, R.L.: Alteration of blood on storage: Measurement of adhesiveness of "ageing" platelets and leucocytes and their removal by filtration. New Engl. J. Med. 265, 728 (1961)

Swank, R.L.: Adhesiveness of platelets and leucocytes during acute exsanguination. Amer. J. Physiol. 202, 261 (1962)

Swank, R.L., Bartsch, W.: Influence of aggregated blood cells on cerebrovascular disease. K.J. Zülch: Cerebral circulation and stroke. Berlin-Heidelberg-New York: Springer-Verlag 1971

Swank, R.L., Davis, E.: Blood cell aggregation and screen filtration pressure. Circulation 33, 617 (1966)

Swank, R.L., Edward, M.J.: Microvascular occlusion by platelet emboli after transfusion and shock. Microvasc. Res. 1, 15 (1968)

Swank, R.L., Hain, R.F.: The effect of different sized emboli on the vascular system and parenchyma of the brain. J. Neuropath. exp. Neurol. 11, 280 (1952)

Swank, R.L., Nahamura, H.: Oxygen availability in brain tissue after lipid meals. Amer. J. Physiol. 198, 217 (1960)

Swank, R.L., Fellman, J.H., Hissen, W.W.: Aggregation of blood cells by 5-hydroxytryptamine (serotonin). Circulat. Res. 13, 392 (1963)

Swank, R.L., Hissen, W., Bergentz, S.E.: 5-hydroxytryptamine and aggregation of blood elements after trauma. Surg. Gynec. Obstet. 119, 779 (1964)

Regional Intravascular Coagulation and Microthrombosis in Traumatic Brain Lacerations in Man

S. Coccheri and C. Testa

The morphologic evolution of traumatic brain lacerations (TBL) in man has been the object of electron microscopic studies by Testa and Giovanelli (1967) and Testa et al. (1970, 1974). Acute TBL and chronic cerebral traumatic scars were examined on a series of microbiopsies performed during surgical operations. The interest of the authors was mainly focused on the role of the cerebral capillaries in the evolution of TBL.

It was soon evident that the center of the lesion had little if any morphologic interest, as this was the site of total acute cellular necrosis in acute cases. In chronic cerebral traumatic scars, the same central region formed the well-known glial-connective scar (Penfield, 1954). The most interesting data came, as a consequence, from the border zone, where the surgeon thinks he is on a "normal" brain tissue. A series of ultrastructural pathologic alterations were found in acute TBL (operated 15-36 hrs after the trauma) and some of these findings will be summarized and briefly discussed here.

In the acute stages the endothelium of the capillaries bordering the TBL shows scattered necrosis (Fig. 1), the basement membranes being still intact 36 h after injury. Throughout the lumen of the affected vessels the occurrence of microthrombi is a frequent finding (Fig. 2). The high incidence of this finding can be confirmed by optic microscopy on semithin sections (Fig. 3a and b). Microthrombi apparently contain platelet materials. The "normal brain tissue" of the zone bordering the TBL is, as a consequence, involved in a pathologic process, the extension of which is difficult to state, as microbiopsies were not performed - for evident reasons - remote from surgical lesions.

During the first few hours of evolution of a TBL the astrocytes are affected by severe intracellular edema. This is possibly due to ischemic factors, depending upon the large number of obstructed capillaries. Extracellular exudation is minimal in the acute stage of TBL and will take the main role only later (months), when the basement membranes rupture.

These data led us to undertake an investigation on the relations between head injury and blood coagulation. Our attention was focused on regional microthrombosis of the cerebral capillaries surrounding the injured area rather than on disseminated microthrombosis. The working hypothesis was that a localized brain

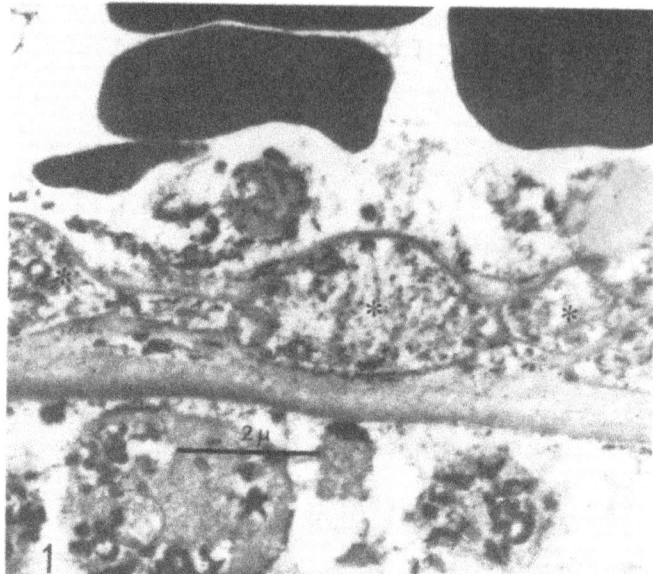

Fig. 1. Electron-microscopic photograph. Cerebral capillary at the border of a surgical resection over a cerebral temporal laceration. Asterisks mark endothelial necrosis. Basement membrane is intact. (From J. Neurosurg. Sci., 1974)

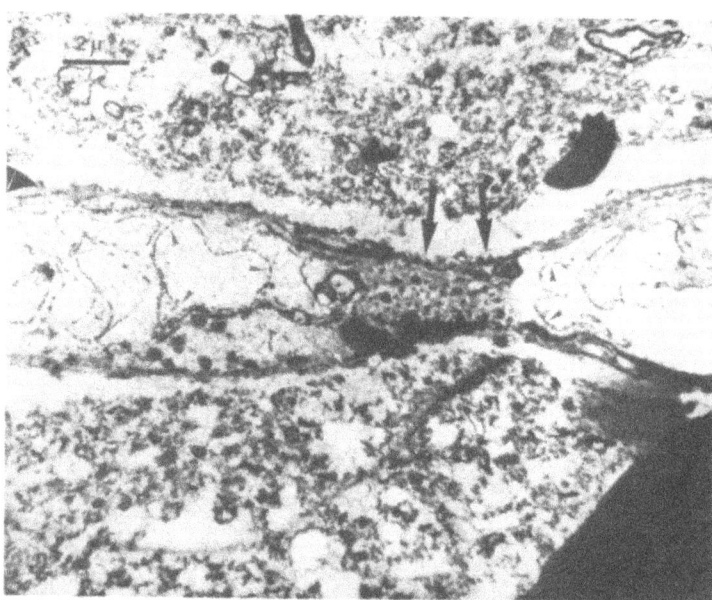

Fig. 2. Electron-microscopic photograph. Longitudinal section of cerebral capillary from cortical biopsy over frontal acute traumatic edema. Arrows point to thrombus. Figures indicated with triangles can be interpreted as "erythrocytic phantoms". (From J. Neurosurg. Sci., 1974)

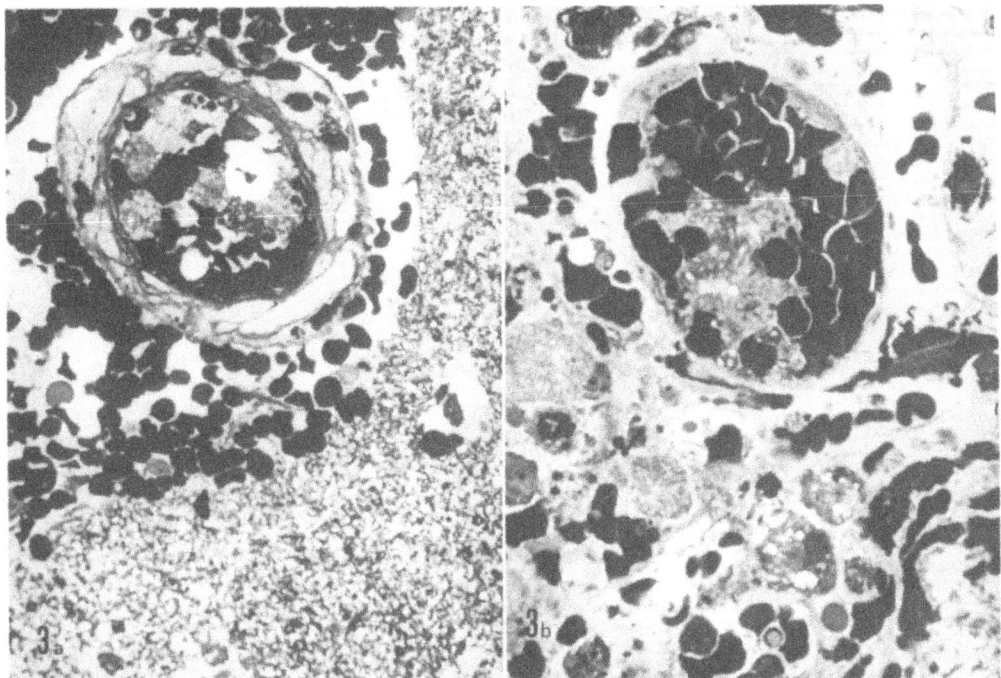

Fig. 3a,b. Optic photomicrograph (from a semithin section, araldite inclusion). Transversal section of two capillaries in border zone of a cerebral traumatic temporal laceration. Lumen is occluded by a microthrombus

injury like TBL could elicit regional intravascular coagulation, which not necessarily influenced the blood clotting values in the peripheral blood. Therefore, our blood clotting investigations were performed on samples of carotid and jugular blood from the same patients.

Materials and Methods

Blood clotting investigations were performed on four patients with TBL. Blood samples were withdrawn 14-16 hrs after injury, and always before craniotomy, from both the common or internal carotid artery and jugular vein omolateral to the injury. In this preliminary series blood clotting investigations were limited to prothrombin time, PTT, fibrinogen, paracoagulation tests (ethanol gelation and protamin sulphate tests), FDP, platelet count, and platelet aggregation test (PAT) according to Breddin. Heparin-thrombin time was also measured in platelet-poor plasma. (O'Brien et al., 1974).

Results

The results of blood clotting tests are summarized in Table 1. In the jugular blood flowing back from the injured side of the

Table 1. Traumatic brain lacerations (four cases)

		CAR	JUG
PTT (s)	M.V.	38	29
Fibrinogen (mg %)	M.V.	340	328
Ethanol gelation	Positive	0	2
Protamine sulfate	Positive	2	4
FDP (µg/ml)	M.V.	22	36
Platelet count (1000/mm^3)	M.V.	168	111
PAT	Stage	Unchanged	
Platelet factor 4 (s)	M.V.	24	18

CAR = Blood from carotid artery

JUG = Blood from jugular vein

M.V. = mean value

head, as compared with the arterial blood, we recorded the following changes:

a) A shortening of PTT

b) The occurrence of a positive ethanol gelation test in two cases vs. no positive result in the arterial blood

c) The occurrence of a positive protamine sulfate test in all cases, vs. two positive results in the arterial blood

d) An increase or further increase in serum FDP

e) A decrease or further decrease in the platelet count

f) A shortening of the heparin-thrombin time in platelet-poor plasma.

No differences were recorded for the fibrinogen level, nor for platelet aggregation.

Discussion

Microthrombosis of extracerebral capillaries and especially of lung capillaries is a known pathologic finding in patients who die after head trauma (Eeles and Sevitt, 1967). However, there are only a few reports in the literature concerning the influence of head trauma on blood coagulation. Changes in the blood clotting profile suggesting an intravascular coagulation syndrome have been described in single cases after self-inflicted gunshot wounds of the head (Mc Gehee and Rapaport, 1969; Keimowitz and Annis, 1973). In the patient described by Keimowitz and Annis massive hemorrhage occurred during craniotomy. Bleeding apparently subsided after a single i.v. administration of heparin. Laboratory features compatible with disseminated intravascular coagulation were recorded by this author, such as fibrinogen and platelet fall, a nearly normal euglobulin lysis time, and a high level of circulating fibrinogen degradation products (FDP) in serum. At necropsy, massive cerebral edema and "haemorrhagic gastrointestinal ulcerations with fibrin thrombi" were also found.

Clotting parameters have been investigated by Goodnight (1973) in a series of 12 patients with head injury. Although there was no evidence of shock, infection, hemolysis or renal failure, laboratory features suggesting a low-grade disseminated intravascular coagulation (DIC) were recorded. According to Goodnight, low-grade DIC is an early, even if transient, complication of head trauma, which should be carefully considered. Since DIC is a composite condition it is not surprising to notice that some authors who described similar cases (Milza and Goldoni, 1971; Druskin and Drianski, 1972; Tremoulet et al., 1974) stressed in turn hyper- or hypocoagulability as the main feature of the syndrome.

In a recent paper, Drayer and Poser (1975) also described two cases of severe craniocerebral injury which resulted in a comatose or stuporous state, with no hypotension, hypoxia, or acidosis, but complicated by profuse bleeding at craniotomy. Abnormal clotting values suggesting DIC were found, but no evidence of disseminated microthrombosis was shown at necropsy. Marked edema of the brain was a prominent feature of both cases.

Changes in the clotting values of the peripheral blood, compatible with DIC, can therefore occur in head trauma (as also stressed by Doni, 1974) but they are not always associated with disseminated microthrombosis, nor does hemorrhage due to consumption coagulopathy always represent a crucial problem in the management of these patients.

The scattered endothelial necrosis found by Testa et al. (1974) in the area bordering the TBL seems to be a selective phenomenon. In fact, neurons and glia of the same zone are not affected by acute necrotic alterations. This could be attributed to a peculiar sensitivity of the endothelium to mechanical forces. There is morphologic evidence that intracapillary thrombosis of the necrotic endothelial lining is mainly due to platelet aggregation probably followed by fibrin formation of minor degree.

The results of our blood clotting studies performed on the venous blood flowing back from the injured side, as compared with arterial blood, suggest that a regional process of intravascular platelet clumping and blood coagulation is taking place in the first hours after injury. The fall in platelet count in the venous blood indicates platelet retention in the cranial area, which can probably be compensated in the general circulation by release of blood cells and especially platelets from storage pools.

Antiheparin activity in platelet-poor plasma is probably related to platelet factor 4 (PF 4), a meaningful indicator of the platelet "release reaction". Increase of antiheparin activity in the platelet-poor plasma (PPP) obtained from jugular blood probably indicates that platelets retained in the cerebral microcirculation underwent aggregation and release, and is therefore consistent with the morphologic finding of altered platelets or platelet materials in the capillaries surrounding the central area of TBL.

Soluble fibrin monomer complexes (SFMC) are present in the returning blood, as demonstrated by ethanol gelation und protamine sulphate tests. The latter test seems to be more sensitive to these conditions as this was positive also in the arterial blood of two cases. The presence of SFMC usually indicates a thrombin effect, but might also be related to the paracoagulating action of PF 4, the platelet antiheparin factor, which seems to be increased in the blood of the jugular vein.

The finding of an increase of serum FDP, mainly in the blood flowing back from the injured side, is also consistent with the hypothesis of a regional intravascular coagulation with secondary local low-grade activation of fibrinolysis.

Several speculations can be made about the pathogenesis of this regional intravascular coagulation:

1. Endothelial injury and necrosis inducing massive platelet adhesion, release of platelet constituents and amplified aggregation, with subsequent minor fibrin formation favoured by the availability of platelet factor 4.

2. Liberation from the injured nervous tissue of phospholipids, capable of activating blood coagulation and platelet clumping.

3. Release from the damaged area of serotonin and catecholamines. The relations between serotonin, catecholamines, platelet aggregation and disorders of cerebral blood flow have been discussed elsewhere in this volume by Meyer and Welch.

4. Thrombogenic effect of involvement of hypotalamic structures as suggested by Brisman et al. (1973).

Finally it is noteworthy to mention, in relation to our findings, the case report by Maurice-Williams (1974) in which fatal progressive cortical thrombophlebitis followed a relatively minor and localized head injury.

It is still to be demonstrated that the observed microthrombi can be the starting point for microembolism and consequently remote capillary occlusion (Bousser et al., p. 179). If this is true, one could extend to the acute TBL syndrome the definition of thromboembolic disorder. This was advocated (see page 172) for several cerebrovascular nontraumatic diseases to which the TBL syndrome might be more closely related than previously suspected.

Implications regarding early antithrombotic treatment in TBL, as possibly supporting surgery and other measures, are intriguing, but any recommendation in this sense would sound premature at the present stage of knowledge.

Conclusions

Extensive capillary microthrombosis probably due to platelet aggregation is observed in brain tissue surrounding areas of traumatic brain lacerations. Blood coagulation and platelet tests performed on venous blood from the injured side show evidence of platelet retention, release reaction and aggregation in the cranial area, with a minor degree of fibrin formation and removal.

Localized traumatic brain lacerations are therefore responsible for a process of regional or local intravascular coagulation, different from systemic DIC as observed after massive brain injury. The implications of this concept for the course and evolution of TBL are briefly outlined.

References

Anderson, J.M., Braun, J.K.: Brain ischemia and DIC. Lancet I, 373 (1972)
Astrup, T.: Assay and content of tissue thromboplastin in different organs. Thromb. Diath. Haemorrh. 14, 401 (1965)
Bousser, M.G., Bara, L., Prost, R.J., Samama, M.: This volume
Brisman, M.D., Mendell, J.: Thromboembolism and brain tumors. J. Neurosurg. 38, 337 (1973)
Coccheri, S., Alessandri, M., Fregni, L., Fiorentini, F.: Significance and comparison of two methods for platelet aggregation. In: Platelet Function and Thrombosis, a Review of Methods. Adv. Exp. Med. Biol. 34, New York-London: Plenum Press 1972, p. 97
De Gaetano, G., Vermylen, J., Verstraete, M.: Platelet aggregation by Thrombofax. Studies on the mechanism of action. Experientia 29, 1136 (1973)
Doni, A.: L'emergenza emorragica nei craniolesi. Atti I Congr. Soc. It. Medico Chir. di Pronto Soccorso, Bologna, 1974. Pisa: Ediz. Pacini 1974, p. 56
Drayer, B.P., Poser, Ch.M.: Disseminated intravascular coagulation and head trauma. Two case studies. J. amer. med. ass. 231, 134 (1975)
Druskin, M.S., Drijanski, R.: Afibrinogenemia with severe head trauma. J. amer. med. ass. 219, 755 (1972)
Editorial: Defibrination with head injury. Lancet, June 15, 1974, p. 1206
Eeles, G.H., Sevitt, S.: Microthrombosis in injured and burned patients. J. Pathol. Bact. 93, 275 (1967)
Goodnight, S.H., Kenoyer, G., Rapaport, S.I., Patch, M.J., Lee, J.A., Kurze, T.: Defibrination after brain-tissue destruction. A serious complication of head injury. New Engl. J. Med. 290, 1043 (1974)
Keimowitz, R.M., Annis, B.L.: Disseminated intravascular coagulation associated with massive brain injury. J. Neurosurg. 39, 178 (1973)
Mc Gehee, W.G., Rapaport, S.I.: Systemic hemostatic failure in the severely injured patient. Surg. Clin. North Am. 48, 1247 (1968)
Maurice-Williams, R.S.: Post-traumatic progressive cortical thrombophlebitis. Brit. Med. J. 3, 24 (1974)

Meyer, J.S., Welch, K.M.A.: This volume
Milza, P.G., Goldoni, C.: Osservazioni tegrafiche sul processo coagulativo nei gravi traumatizzati cranio-encefalici. Boll. Soc. It. Biol. Sper. $\underline{47}$, 530 (1971)
Niewiarowski, S., Thomas, D.P.: Platelet factor 4 and adenosine diphosphate release during human platelet aggregation. Nature (Lond.) $\underline{222}$, 1269 (1969)
O'Brien, J.R., Etherington, M.D., Jamieson, S., Lawford, P.: Heparin thrombin clotting time and platelet factor 4. Lancet $\underline{II}$, 656 (1974)
Penfield, W., Jasper, N.: Epilepsy and the Functional Anatomy of the Human Brain. Boston: Little, Brown 1954, pp. 281-349
Testa, C., Giovanelli, M.: Studio ultrastrutturale di tre cicatrici cerebrali post-traumatiche operate. Min. Neurochir. $\underline{11}$, 224-229 (1967)
Testa, C., Bollini, C., Columella, F.: Cerebral capillaries in the evolution of traumatic lacerations in man: an ultrastructural study. Folia Angiol. $\underline{96}$, 84-89 (1970)
Testa, C., Calbucci, F., Columella, F.: The border-line of surgical resection in human acute cerebral traumatic lacerations. An ultrastructural study. J. Neurosurg. Sci. $\underline{18}$, 80-87 (1974)

Platelet Aggregation in Cerebrovascular Patients

M. Prencipe, F. M. Pissarri, V. Cecconi, and A. Agnoli

Introduction

The role of platelet embolism in cerebral ischemic lesions has recently been evaluated (Ross Russel, 1968; McBrien et al., 1963). How the platelet embolism is formed has yet to be clarified: is it set off by factors which cause a "physiologic" platelet aggregation, or is there an alteration in the platelets themselves or, more generally, in blood coagulation, which facilitates the occurrence of this phenomenon (Boneau et al., 1972; Danta, 1970). This study will try to show whether, using Born's method (Born and Cross, 1962), it is possible to discover if there is an enhanced platelet aggregation in patients who have had a cerebral ischemic lesion which could play a part in causing the arterial emboli.

Material and Methods

Healthy Volunteers. The examination included 15 healthy subjects, 10 men and 5 women, from 15 to 35 years old (average 29 years). During the last 3 months before study began they did not take any drugs which might effect platelet aggregation.

Cerebrovascular Patients. Altogether 47 cerebrovascular patients (CVP) were examined, 30 men and 17 women, from 57 to 73 years old (average 67 years). All patients were examined about a year after their stroke (minimum 6 months, maximum 20 months). Clinical and cerebral spinal fluid examinations and angiographic studies allowed us to exclude patients with hemorrhagic lesions. No patient had been submitted up to that point to an antiaggregant therapy.

Preparation of the Plasma. Plastic syringes containing 1 ml of calcium oxalate were used to take 9 ml venous blood samples, taking care to avoid venous stasis.

The platelet-rich plasma (PRP) was prepared by centrifuging the oxalated blood at 185 rpm/15 min. Platelet-poor plasma (PPP) was obtained by centrifuging the PRP at 4200 rpm/10 min. An initial count was made of the platelets in the PRP, after which it was diluted with PPP to give a platelet concentration of $300000/mm^3$. Platelet aggregation curves induced by adenosine diphosphate (ADP) were obtained by the turbidometric method of Born. ADP was

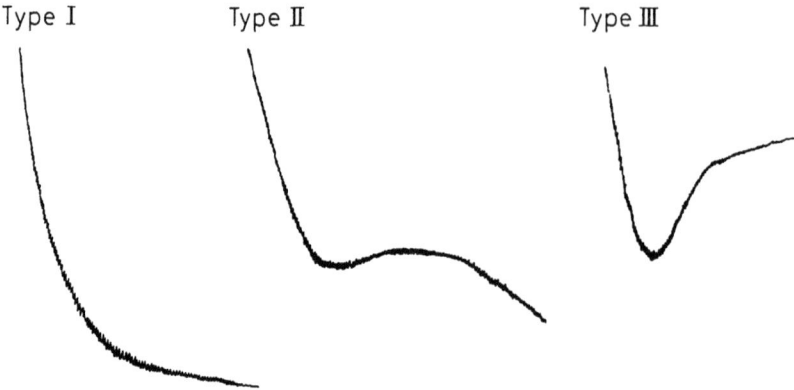

Fig. 1. Platelet aggregation curves: Type I (irreversible aggregation); Type II (double-wave aggregation); Type III (reversible aggregation)

added to stirred oxalated plasma in order to obtain a final concentration of 0.2 μM and 0.8 μM.

Platelet aggregation curves were evaluated using a qualitative method of subdivision into different types (Fig. 1):

Type 1: Irreversible aggregation

Type 2: Double-wave aggregation

Type 3: Reversible aggregation.

To this classification correspond three different aggregation degrees: high, medium, slight.

Results

Healthy Subjects. At the final concentration of 0.8 μM of ADP, there were 20% subjects of Type 1 aggregation, 50% of Type 2, and 30% of Type 3.

At the final concentration of 0.2 μM of ADP, there was no subject with a Type 1 aggregation, 20% of Type 2, 80% of Type 3 (Fig. 2).

It is thus clear that in healthy subjects, by reducing the final concentration from 0.8 to 0.2, the irreversible aggregation disappears.

Cerebrovascular Patients. At the final concentration of 0.8 μM of ADP, there were 41% patients of Type 1 aggregation, 53% of Type 2, and 6% of Type 3.

At the final concentration of 0.2 μM of ADP, there were 19% patients of Type 1 aggregation, 28% of Type 2, and 53% of Type 3.

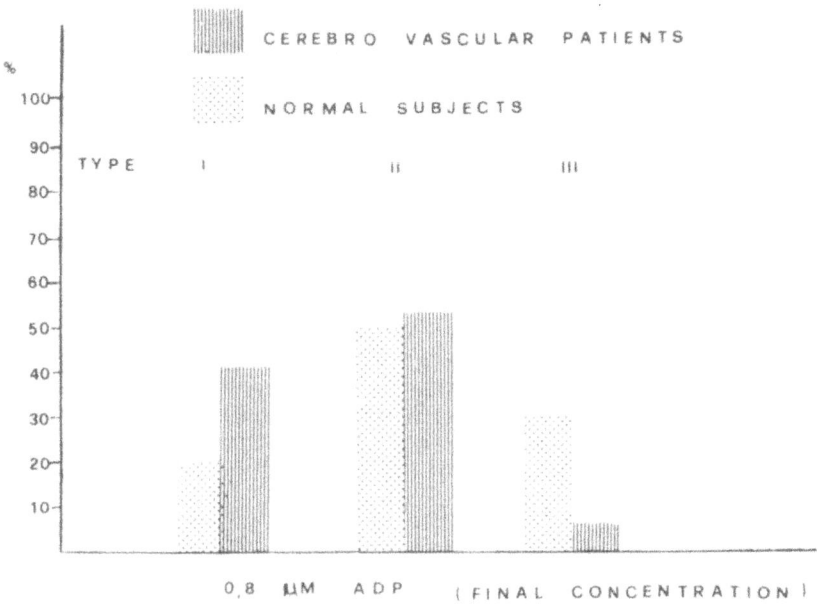

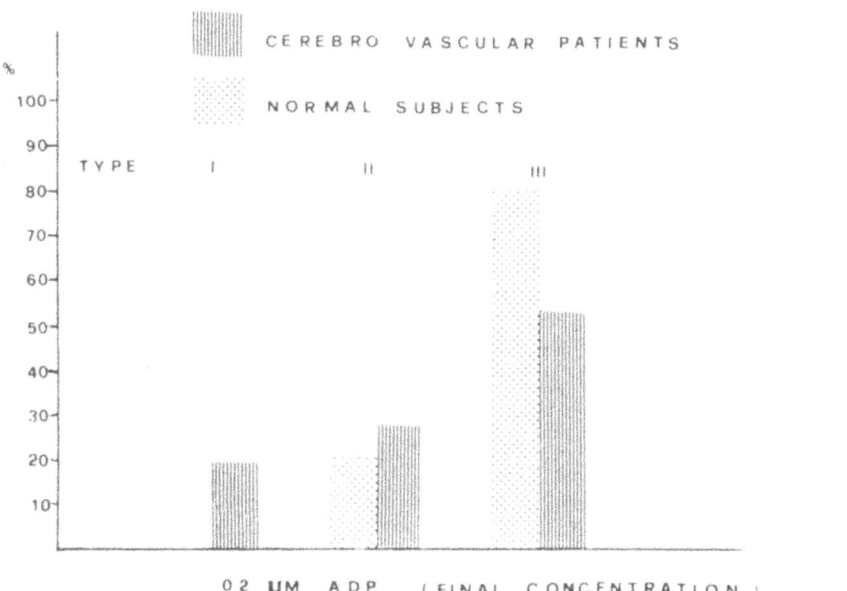

Fig. 2a,b. Type of platelet aggregation curves in normal subjects and cerebrovascular patients at 0.8 µM (top) and 0.2 µM (bottom) of ADP

A comparison between healthy subjects and CVP, showed that the 0.8 concentration determined a releasing reaction (Type 1 plus Type 2), in a higher percentage of CV patients (94%), than in the healthy subjects (70%).

This is more evident when dealing with a 0.2 concentration: 20% in the healthy subjects and 47% in the CV patients. At this concentration the Type 1 curve was absent in the healthy subjects, whereas 19% of the CV patients had it.

At the above percentage, it results that 9 CV patients out of 47 had an irreversible aggregation curve.

These patients with an enhanced platelet aggregation had been compared to 38 other cerebrovascular patients. More specifically we have taken into consideration some risk factors such as hypertension, diabetes, and dyslipidemia, which can effect the platelet aggregation (Ferguson et al., 1973; Kwaan et al., 1972).

These comparisons have shown that there were no significant differences between these nine CV patients and the others.

On the contrary, the electrocardiogram has shown a serious coronary heart disease in six out of the nine patients (66%); whereas among the other patients there was a percentage of 38%.

Moreover, in seven patients out of the nine (77%), the stroke was preceded by one or more transient ischemic attacks.

Of the other two patients, one had a stroke in evolution, and the other one had a carotid stenosis.

Conclusions

The platelet aggregation, studied with Born's method, shows remarkable individual variations even among the young, healthy subjects. It is not possible therefore to establish normal values according to which a single subject may be classified as normal or pathologic.

Moreover, the relatively high differences between the group of young, healthy subjects and the one of the pathologic patients, show that platelet aggregation variations do not play a decisive pathogenetic role in the majority of the cerebrovascular patients.

Nevertheless, a high percentage of ischemic heart diseases (66%) and transient ischemic attacks (77%), shows that this alteration may be a qualitatively important cofactor; just as diabetes and hypertension in the pathogenesis of central ischemic lesions.

Naturally these data are not sufficient for a thorough quantitative analysis of the blood-clotting alterations.

A comparison between normal and pathologic subjects of the same age, cannot bring exhaustive results either.

Only a prospective epidemiologic analysis could explain the relationships between ischemic cerebrovascular diseases and platelet aggregations.

In conclusion we can assume that in a small number of cerebrovascular patients only, the permanent enhanced platelet aggregation can promote arterial emboli.

In other patients the platelet embolus may be consequent to a sudden and transient enhanced platelet aggregation.

Finally, the cerebral ischemia can also be caused by other factors which cannot be related to this phenomenon.

These data can then be considered as a further stimulus to the prosecution of controlled trials on the effectiveness of anti-aggregant therapy. Apart from establishing the importance of these drugs in the prevention of cerebral emboli, these studies will provide further information on the pathogenetic mechanisms of ischemic cerebrovascular diseases.

References

Boneu, B., Guiraud, B., Fernet, P.: Traitement anti-agrégant plaquettaire par l'aspirine. Bases biologiques. Applications aux accidents vasculaires cérébraux. Nouv. Presse méd. 1, 863-868 (1972)
Born, G.V.R., Cross, J.: Platelet aggregation: photometric method. J. Physiol. (Lond.) 168, 178 (1962)
Danta, G.: Second phase platelet aggregation induced by adenosine di-phosphate in patients with cerebral diseases and in control subjects. Thromb. Diath. haemorh. (Stuttg.) 23, 159-169 (1970)
Ferguson, J.C., Dunnigan, M.G., Philiph, A.D.: Platelet adhesiveness in hyperlipidemic subjects. Atherosclerosis 18, 489 (1973)
Kwaan, H.C., Colwell, J.A., Cruz, S., Suwanwela, M., Dobbie, J.C.: Increased platelet aggregation in diabetes mellitus. J. Lab. clin. med. 80, 236 (1972)
McBrien, D.J., Bradley, R.D., Asheton, N.: The nature of retinal emboli in stenosis of the internal carotid artery. Lancet I, 697-699 (1963)
Ross Russel, R.W.: The source of retinal emboli. Lancet II, 789 (1968)

In Vivo Effect of Cyclic AMP and Related Drugs on Platelet Function*

F.J. Pareti and P.M. Mannucci

Several inhibitors of platelet aggregation have been shown to increase the intracellular level of cyclic AMP in other tissues either by stimulation of adenylate cyclase (PGE_1, adenosine, and isoprenaline) or by inhibition of cyclic AMP phosphodiesterase (Robinson et al., 1971). The existence of a hormonally responsive adenylate cyclase in platelets has also been reported in vitro. The inhibitory effect of PGE_1, adenosine, and isoprenaline has been shown to be associated with an increase of intracellular formation of cyclic AMP. This effect is greatly enhanced by the simultaneous addition of an inhibitor of cyclic AMP phosphodiesterase, like papaverine, theophylline, and pyrimidopyrimidine compounds (Mills and Smith, 1972).

These observations have led to the development of a unifying concept that these compounds are linked by the common effect on platelet cyclic AMP. A study has been carried out to evaluate whether the relationship existing between the inhibition of platelet aggregation and the increase of endogenous cyclic AMP induced by such drugs could also be demonstrated in vivo.

Intravenous infusion of drugs known to affect platelet cyclic AMP levels was performed in normal volunteers.

Platelet aggregation, platelet retention, PF3 availability, and PF4 release were carried out before 30 and 90 min after the infusion, together with the determination of endogenous cyclic AMP. Glucagon (70 mg/kg), aminophylline (15 mg/kg), and dipyridamole (0.8 mg/kg), given intravenously, had no pronounced effect either on platelet aggregation or on platelet cyclic AMP levels; higher doses could not be used in order to avoid pronounced side effects.

Isoprenaline was found to be a weak inhibitor of platelet aggregation and the effect was well correlated in time and magnitude with a slight increase of cyclic AMP.

---

*This article has been published in "L'aggregazione piastrinica nella patogenesi della vasculopatie cerebrali" abstract of the Round Table of Rome 30/31.10.1974. Edition: Boehringer Ingelheim, Firenze

Intravenous infusion of dibutyryl cyclic AMP and of cyclic AMP were followed by a definite inhibition of platelet aggregation induced by collagen, ADP, and adrenaline.

PF3 availability was also decreased within 90 min after the infusion. Platelet release reaction was also inhibited as shown by the decrease of PF4 release induced by the same aggregating agents.

Platelet retention was also reduced. Primary hemostasis did not seem to be affected by the infusion of these cyclic nucleotides, as shown by the lack of significant prolongation of the standardized bleeding time. The effect of intravenous infusion of dibutyryl cyclic AMP and of cyclic AMP seems to be related to the ability of such compounds to penetrate from plasma into platelets where they behave like endogenous cyclic AMP (Salzman and Levine, 1971), rather than to their enzymatic convertion to 5-AMP and adenosine, which are inhibitors of platelet aggregation in vitro.

The latter possibility seems to be excluded by the absence in plasma of 3'-5' C-AMP phosphodiesterase (Song and Cheung, 1973) and by the demonstration of the greater and more prolonged effect of dibutyryl cyclic AMP which penetrates more easily in the cells than exogenous cyclic AMP (Ryan and Durick, 1972).

Concomitant infusion of dibutyryl cyclic AMP (5 mg/kg) and of dipyridamole (0.8 mg/kg) produced an inhibitory effect on platelet aggregation induced by ADP, adrenaline, and collagen. When these drugs were given separately at the above mentioned doses, they did not induce any effect on platelet aggregation. These data suggest the existence, in vivo, of a relationship between inhibition of platelet aggregation induced by many drugs and their ability to enhance the intracellular cyclic AMP levels.

References

Mills, D.C.B., Smith, J.B.: The control of platelet responsiveness by agents that influence cyclic AMP metabolism. Ann. N.Y. Acad. Sci. 201, 291 (1972)

Robinson, G.A., Butcher, R.W., Sutherland, E.W.: Cyclic AMP. New York and London: Academic Press 1971

Ryan, W.L., Durick, M.A.: Adenosine 3'-5' monophosphate and N^6-2'-O-dibutyryl-adenosine 3'-5' monophosphate transport in cells. Science 177, 1002 (1972)

Salzman, E.W.: Cyclic AMP and platelet function. New Engl. J. Med. 286, 358 (1972)

Salzman, E.W., Levine, L.: Cyclic 3'-5'-adenosine monophosphate in human blood platelet II. Effects of N^6-2'-O-dibutyryl cyclic 3'-5'-adenosine monophosphate on platelet function. J. clin. Invest. 50, 131 (1971)

Song, Y., Cheung, W.Y.: Cyclic 3'-5' nucleotide phosphodiesterase properties of the enzyme of human blood platelets. Biochem. biophys. Acta 242, 593 (1973)

Platelet Hyperreactivity and Decreased Survival in Chronic Cerebrovascular Patients. Chronic Defibrination Syndrome?

G.G. Neri Serneri, E. Silvestrini, G.F. Gensini, and R. Abbate-Gensini

Summary

Some platelet functions have been investigated in a group of 35 subjects with chronic cerebrovascular disorders (CVP) and in a group of 30 age-matched controls. In CVP an increased average platelet adhesiveness to glass beads exists ($P < 0.01$), an increased platelet aggregation by ADP (0.15 µM), but not by collagen, in respect to age-matched controls. However, a wide overlapping has been observed for the data of CVP and of controls. Furthermore, these changes are not specific, as they have also been observed in a group of subjects with history of myocardial infarction. In CVP the residual aggregation at the 5th min by ADP (0.5 µM) is increased in respect to controls. The residual aggregation of control PRP is increased by addition of PPP of CVP. No differences have been observed in nucleotides and ^{14}C-serotonine release from the platelets of CVP and of controls. In CVP the average survival time of ^{51}Cr-platelets is decreased (7.6 ± 1.9 days instead of 9.1 ± 1.2; $P < 0.01$). The contemporary determination of platelet and fibrinogen survival time investigated by ^{75}Se-seleniomethionine resulted in an average decreased in CVP in respect to the controls. Fibrinogen survival time resulted 7.4 ± 0.9 days instead of 9.3 ± 1.8 for controls ($P < 0.01$) and platelet survival time 8.5 ± 2.1 days instead of 10.2 ± 1.1 for controls ($P < 0.01$). In the CVP a good correlation can be observed between platelet and fibrinogen survival time ($r = 0.783$, $P < 0.01$).

Heparin administration has provoked, in CVP, a lengthening of ^{131}I-fibrinogen half-time, of ^{51}Cr-platelet survival time and a decrease in FDP serum concentration.

These results suggest that in many CVP a slow chronic defibrination exists, probably due to activation of the thrombin system which could account for the increased platelet reactivity and the decreased platelet survival.

Introduction

Recent advances in the physiology of platelets and in the assessment of their role in the beginning of atherosclerosis and mainly in the production of its thrombo-occlusive complications promoted a lot of investigations on platelet aggregation (Murphy and Mustard, 1962; Pfleiderer and Rucker, 1964; McDonald and Edgill, 1959; Nestel, 1961; Neri Serneri et al., 1968; Stormorken, 1970; Goldenfarb et al., 1971). In the neurologic field, platelet aggregation has become particularly meaningful after the recognition of platelet aggregates in transitory ischemic accidents (Fisher, 1959; Ross Russel, 1961).

The availability of antiaggregating drugs has further increased the importance of the investigations on platelet aggregation in various arterial disorders. The investigations in this field have been numerous, but in most of these, platelet functions have been studied only in part, employing only one or two tests.

Moreover, for a rational therapy also, the finding of one increased platelet aggregation is insufficient, because platelet hyperreactivity, although of importance, cannot be disjointed from the study of coagulation, since platelet function and blood clotting are strictly interdependent.

Due to this reason, we believed it necessary to investigate in a single study platelet functions in correlation to fibrinogen turnover, because an increased turnover may suggest an activation of clotting processes.

We thus investigated whether in cerebral vascular patients a change of platelet aggregation and adhesiveness exists, whether it is a specific one, and whether the eventual increased platelet aggregation is primary, or rather a consequence of clotting disorders. In fact, changes of blood clotting have been reported in artherosclerotic subjects for quite sime time (McDonald, 1964; Naimi et al., 1963; Neri Serneri et al., 1970).

Materials and Methods

A. Subjects Examined

We examined as a whole, 92 subjects, divided in 3 groups: 30 controls, 35 subjects with chronic cerebrovascular disorders (CVP), and 27 subjects with a history of myocardial infarction (MIP). The subjects were matched by age (Fig. 1) and stage of illness for all groups. In fact, all patients were studied for at least 3 months after the last ischemic episode. We considered the subjects with a history of ischemic stroke and the subjects with repeated (at least three) transitory ischemic attacks with or without history of previous ischemic stroke affected by chronic cerebrovascular disorders. A group of 40 normal volunteers served as control subjects. These were asymptomatic persons who had normal findings on routine health examinations and who denied having taken medication recently.

B. Methods

1. Platelet Adhesiveness to Glass Beads. Platelet adhesiveness to glass has been studied on blood collected in 3.8% citrate in a ratio of 1:9 by a modification of the Hellem's method as previously reported (Neri Serneri et al., 1968). The filtering column

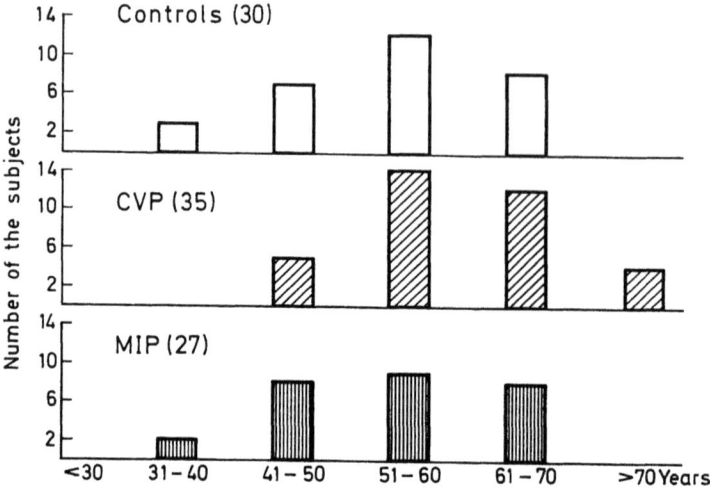

Fig. 1. Age distribution of subjects examined

had a volume of 1696 cubic millimeters (cylinder of 60 × 6 mm of internal diameter) and contained 3 g of glass beads, 0.5 mm in diameter. The blood was injected at a constant speed of 1 ml every 35 s, so that the contact time was 24 s. When necessary, the amount of citrate was corrected according to the hematocrit value (according to Hellem, 1960).

2. Platelet Aggregation by ADP and Collagen. Platelet aggregation has been studied by the method of Born on citrated platelet-rich plasma (PRP) arranging the number of the platelets to 300,000 c.mm by adding autologous platelet-poor plasma (PPP). Aggregation by ADP was determined employing ADP at very low concentration (final concentration 0.15 µM) in order to make evident a possible increased platelet reactivity to ADP. In evaluating platelet reactivity we considered both the entity of aggregation (expressed as percentage of optimal density, O.D.) and the maximal speed of aggregation per time unit (slope).

The collagen employed to induce platelet aggregation was a commercial purified collagen from bovine Achille's tendon at such concentration to give on a pool of control plasmas a lag time of 80-85 s before aggregation and a slope of 28-32 mm/min. The solution was then frozen in small amounts and its activity was checked weekly. We considered the lag time (in seconds) and the entity of the aggregation.

3. Aggregation at the 5th Minute of Plasma Mixtures. We investigated the disaggregation at the 5th minute (or residual aggregation at 5 min) in the presence of ADP 0.5 µM. The residual aggregation has been investigated also on plasma mixtures, namely on control PRP after addition of PPP of cerebrovascular patients. We designed as control a mixture of control PRP and control eterologous PPP.

4. *Nucleotides and ^{14}C-Serotonin Release*. The measurement of nucleotides and ^{14}C-serotonin release has been carried out on gel-filtered platelets on Sepharose 2B by a modification of the method of Tangen et al. (1971) in the presence of 8 μM ADP. The reagent concentration was higher than in the aggregation study to compensate the slightly lower activity of gel-filtered platelets. The release of total nucleotides has been measured by a modification of the method of Wolfe and Shulman (1970). The test is based on Mürer's (1968) observation that light absorbance at 260 μm is produced mainly by adenosine diphosphate and adenosine triphosphate. The concentration of adenine nucleotides was calculated by comparison with the absorbance of ATP in 0.5 M $HClO_4$. Results have been expressed as moles per 10^{11} platelets. The release of ^{14}C-serotonin has been measured according to Jerushalmy and Zucker (1971). The results have been expressed as percentage release of ^{14}C-serotonin.

5. *Platelet Survival*. Platelet survival time has been performed by both the ^{51}Chromium method of Aster and Jandl (1975) and by the ^{75}Se-seleniomethionine method of Najean and Ardaillou (1969).

Autologous platelets from 400-500 ml of blood were labeled with 100-150 μCi of ^{51}Cr. Blood samples for measurement of radioactivity in platelet concentrate were obtained beginning 3 h following reinfusion of labeled platelets and thereafter every 12 h during the first 3 days and daily during the following 7 days.

Platelet labeling with ^{75}Se-seleniomethionine was carried out by direct i.v. injection of 200 μCi of ^{75}Se-seleniomethione. Blood samples for determination of radioactivity were collected daily for 15 days. In order to calculate the utilization by platelets of ^{75}Se-seleniomethionine deriving from the catabolism of plasma proteins, we also collected a sample 20 days after the injection. The platelet radioactivity of this sample was subtracted from the radioactivity of each sample of the 15-day observation. We considered, as expression of platelet survival time, the time interval corresponding to the part of the curve of daily radioactivity plotted on arithmetic paper comprised between the 50% of the ascending branch and the 50% of the descending branch.

6. *Fibrinogen Turnover and Fibrinolytic Activity*. We studied fibrinogen turnover in 9 controls and in 13 CVP subjects contemporary with the evolution of platelet survival by ^{75}Se-seleniomethionine.

In the patients injected with ^{75}Se-seleniomethionine, the blood samples for determination of radioactivity was performed on the clot obtained from 3 ml of plasma clotted by thrombin and dissolved in 20% urea in NaOH 10%.

In order to investigate the *effects of heparin* on platelet survival time and on fibrinogen T/2 we used, in 6 controls and 13 CVP of ^{75}Se-seleniomethionine, ^{131}I-human fibrinogen. For this purpose we have injected human ^{131}I-fibrinogen. Blood samples were collected in siliconized tubes containing 3.8% natrium citrate (one part of citrate plus nine parts blood) and immediately centrifuged at 4° C. The radioactivity determination was per-

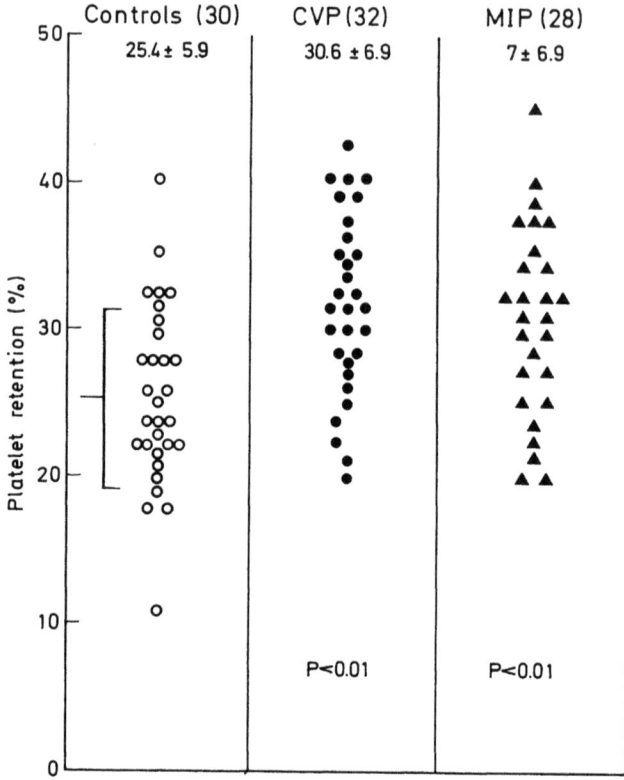

Fig. 2. Platelet adhesiveness

formed in the same way as for ^{75}Se-seleniomethionine. The radioactivity values (cpm) have been plotted on semilogarythmic paper. The *fibrinogen assay* has been performed according to Ratnoff and Menzie (1964). The *fibrinolytic activity* has been measured according to Milstone (1941) and expressed as fibrinolytic units according to Janusko and Dubinska (1965). The *FDP serum concentration* has been measured according to Merskey et al. (1966).

Results

1. Adhesiveness to Glass. Cerebrovascular patients showed a platelet adhesiveness to glass higher than that of controls (Fig. 2). The average adhesiveness was in controls 25.4 ± 6.9% ($p < 0.01$) in CVP. However, the increased retention of platelets by glass beads is not strictly specific, as also platelets from patients with myocardial infarction show a similar behavior (30.7 ± 6.9%). As can be seen from Figure 2, the values observed are widely diffuse and a wide overlapping between controls, CVP, and MI can be noted.

2. Platelet Aggregation by ADP and Collagen. Figure 3 shows that at a concentration of 0.15 μM ADP, an increased platelet aggregation in CVP can be observed. In these subjects the average change of O.D. was 20.8 ± 5.4%, against 17.6 ± 4.2% for controls

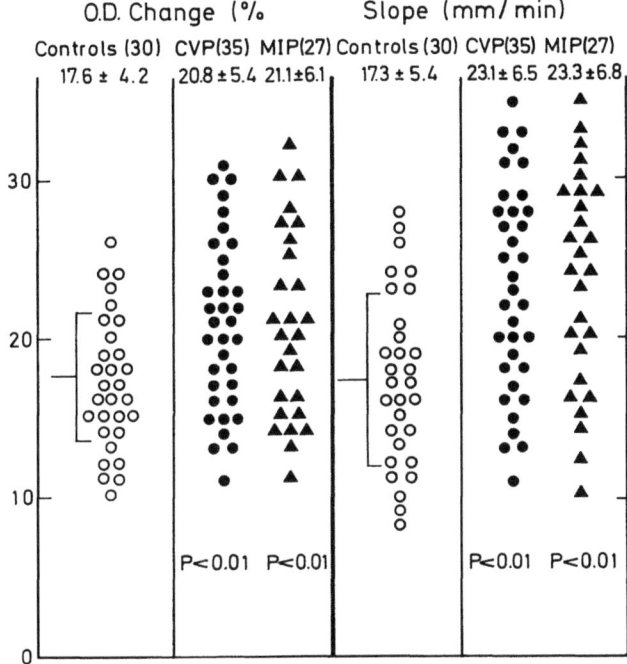

Fig. 3. Platelet aggregation by ADP 0.15 µM

(P < 0.01). The slope was also increased (23.1 ± 6.5 as compared to 17.3 ± 5.4; P < 0.01). Platelet aggregation was increased also in the MI patients. However, a wide overlapping of the values of platelet aggregation can be observed.

We did not find statistically significant differences between controls and CVP regarding the extent of aggregation in presence of collagen (Fig. 4). On the contrary, we were able to show in both CVP and in MIP a shortened lag-time. That is, the platelets of CVP aggregate earlier than those of controls, although the extent of aggregation is not significantly different in the two groups.

3. Aggregation at the 5th Minute on Plasma Mixtures. The study of the aggregation 5 min after the addition of ADP indicates that in CVP a decreased platelet disaggregation exists (Fig. 5). In fact, in the controls the aggregation by 0.5 µM ADP usually resulted in a reverse, coming back after 5 min to values of O.D. with a decrease from 4 to 13%. In only 2 out of 10 subjects did this remain up to 50%, indicating a persistent aggregation. On the contrary, in the group of the CVP, 11 showed an O.D. decrease up to 50% at the 5th min, and the other patients also showed at this time values higher than those of the controls (Fig. 5). The residual aggregation at the 5th min of control PRP increases after addition of PPP of CVP (Fig. 5)., i.e., the disaggregation decreases. On the contrary, addicted control eterologous PPP does not affect the disaggregation of the control PRP (Fig. 5, Column III).

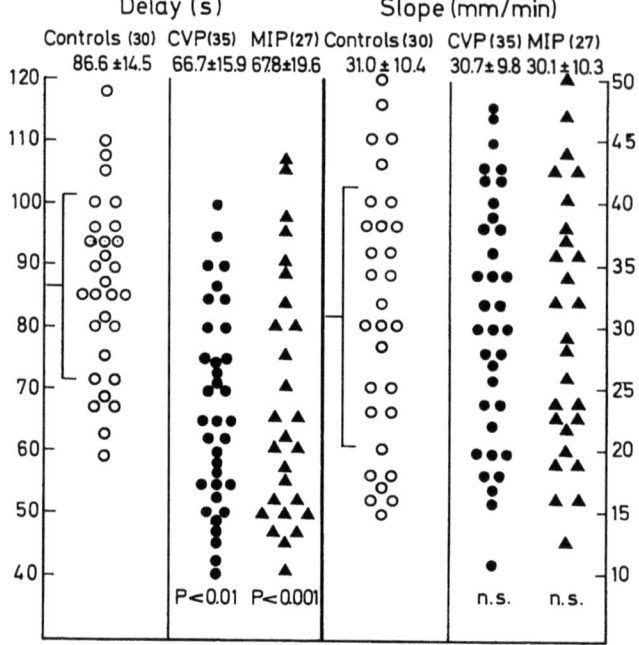

Fig. 4. Platelet aggregation by collagen

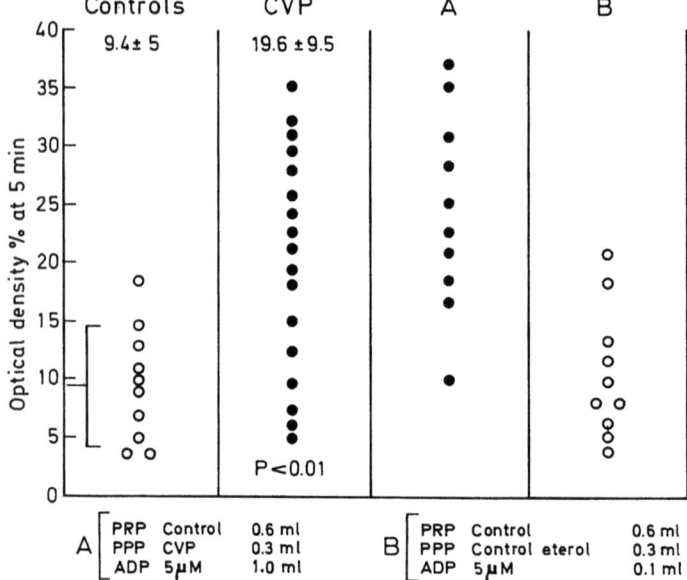

Fig. 5. Residual platelet aggregation 5 min after ADP 0.15 µM

4. Nucleotides and ^{14}C-Serotonin Release. As shown by Figures 6 and 7, respectively, the release of total nucleotides and of ^{14}C-serotonin did not result in a significant difference in controls and in CVP with 8 µM ADP. The mean nucleotide release for controls was 21.6 ± 16.3 µM/10^{11} platelets, whereas for CVP 25.8 ±20

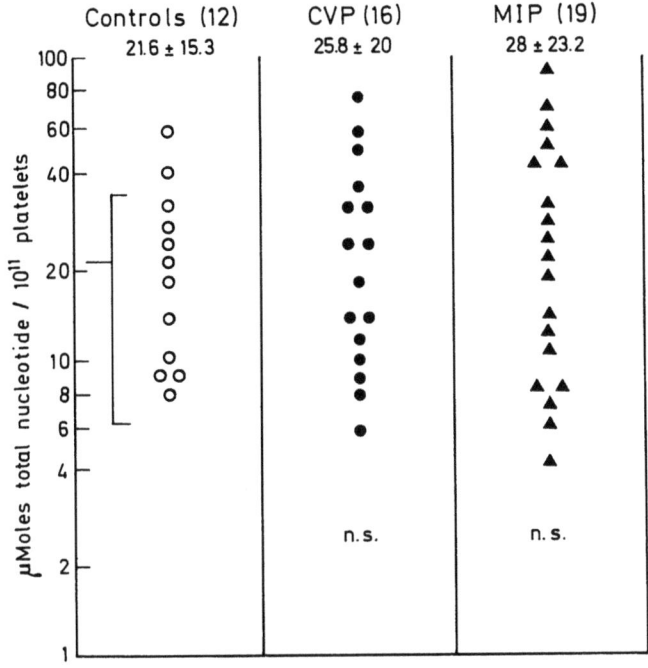

Fig. 6. Platelet nucleotide release by ADP 8 µM

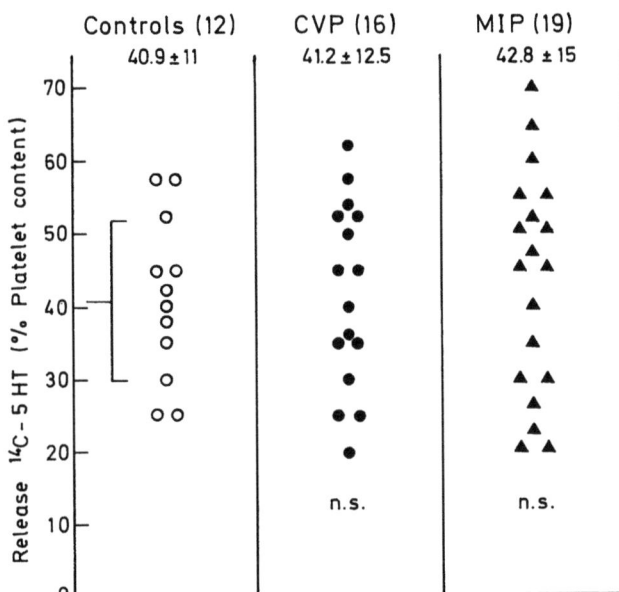

Fig. 7. Platelet release of ^{14}C-serotonin by ADP 8 µM

and for the MI patients 28 ± 23.2. Such values are not significantly different (P > 0.05). ^{14}C-serotonin release from control platelets was 40.9 ± 11%; in patients with CVP the release was 41.2 ± 12.5, and in patients with MI was 42.8 ± 15%. The differences do not statistically differ (P > 0.05).

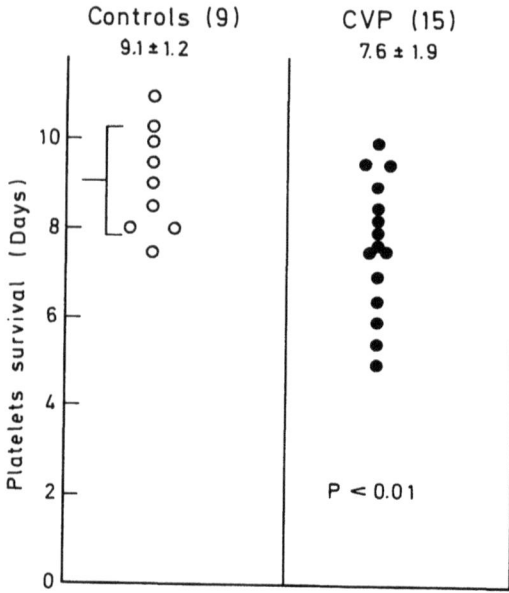

Fig. 8. Survival of ^{51}Cr-platelets

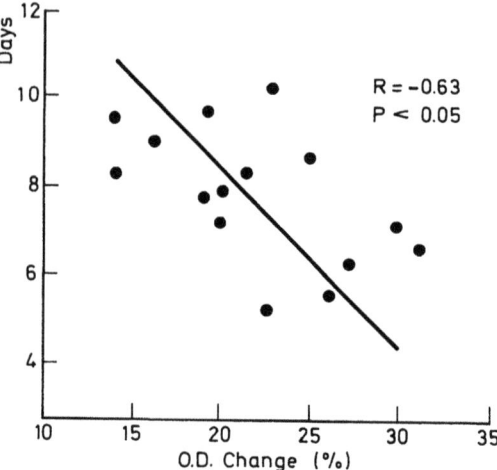

Fig. 9. Relationship between platelet aggregation by ADP 0.15 µM and ^{51}Cr-platelets survival in CVP

5. *Survival Time of Platelets*. In 9 controls and in 15 CVP of equivalent age we studied the survival of ^{51}Cr-platelets. Average platelet survival time for CVP was shortened (7.6 ± 1.9 days in respect to 9.1 ± 1.2 days; $P < 0.01$). However, several CVP showed survival times within the control range (Fig. 8). In the same subjects (Fig. 9) a negative correlation between platelet aggregation by 0.15 µM ADP and survival time can be observed ($r = 0.633$; $P < 0.05$).

The platelet survival time has resulted in a significant difference for CVP and for controls when also studied by ^{75}Se-seleniomethionine. The average platelet survival time was shortened in CVP (8.5 ± 2.1 days in respect to 10.2 ± 1.1 for controls; $P < 0.01$) (Fig. 10).

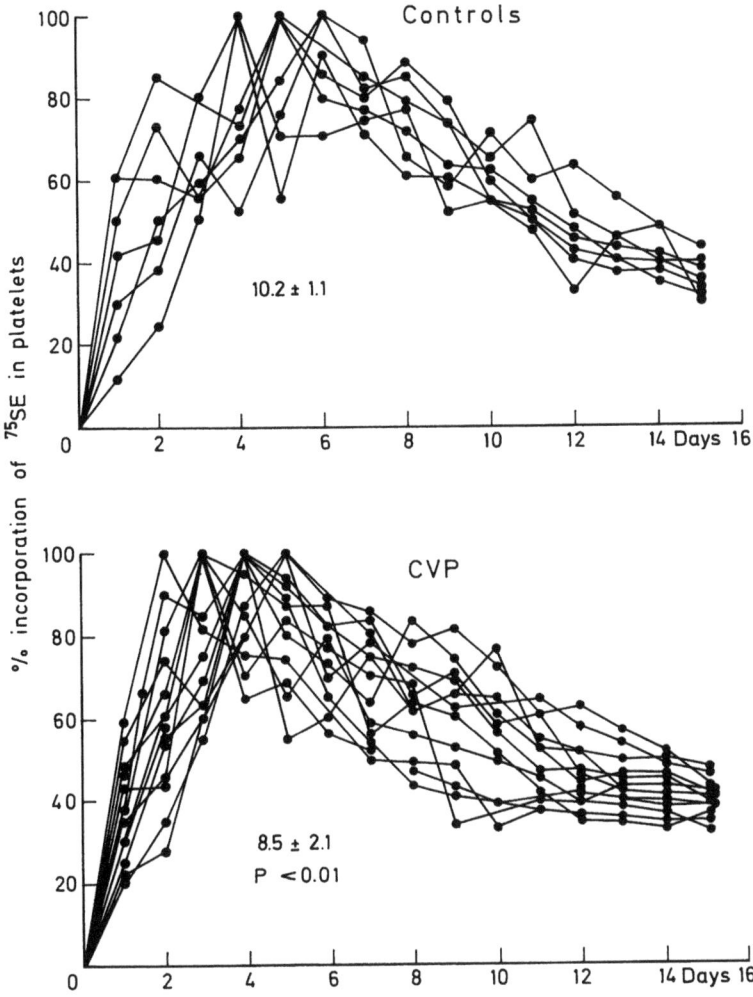

Fig. 10. ^{75}Se-seleniomethionine survival in controls and in CVP

6. *Fibrinogen Survival Time by* ^{75}Se-$Seleniomethionine$. In the subjects investigated for platelet survival time by ^{75}Se-seleniomethionine the determination of fibrinogen survival time has shown significant differences between CVP and controls (Fig. 11). Fibrinogen survival time resulted in 7.4 ± 0.9 days for CVP instead of 9.3 ± 1.8 for controls (P < 0.001). The concentration of plasma fibrinogen is not significantly different for two groups (Table 1). On the contrary, the average FDP serum concentration was increased in absence of any increase of plasmatic fibrinolysis activity (Fig. 12). In the CVP a good correlation can be observed between platelet and fibrinogen survival time (r = 0.783, P < 0.01), indicating a strict relationship between fibrinogen survival time and platelet consumption (Fig. 13).

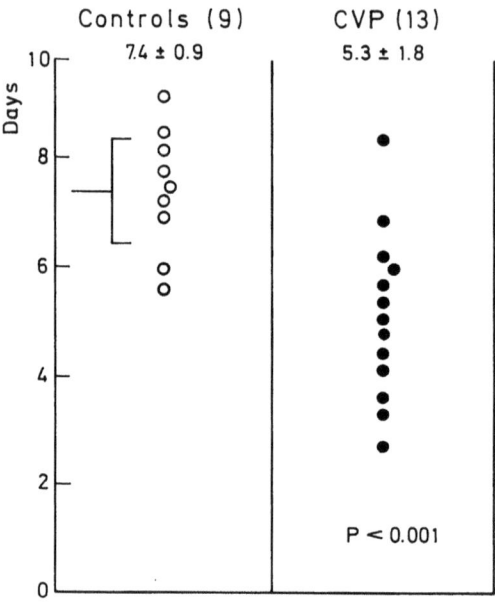

Fig. 11. ^{75}Se-seleniomethionine fibrinogen survival (days)

Table 1. Plasma fibrinogen concentration in patients studied by ^{75}Se-seleniomethionine (mg %)

Controls (9)	278 ± 48	
C.V.P. (13)	301 ± 61	N.S. (P > 0,05)

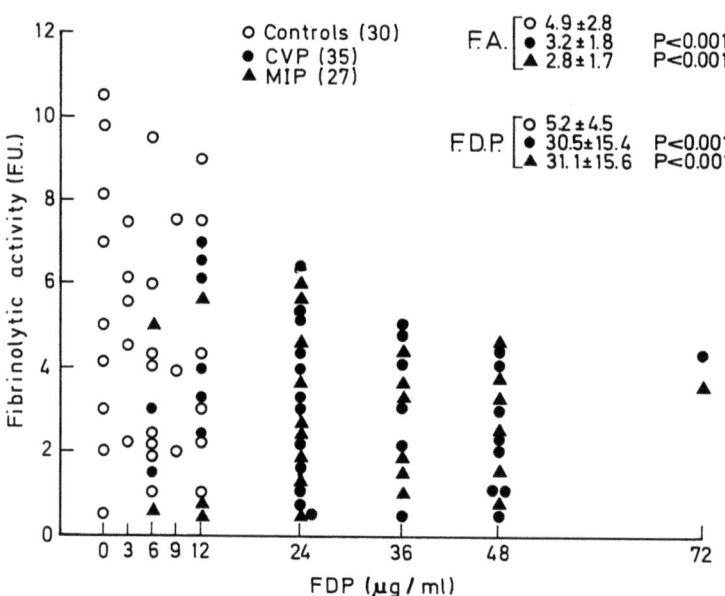

Fig. 12. Behavior of FDP concentration and fibrinolytic activity in controls, CVP and MIP

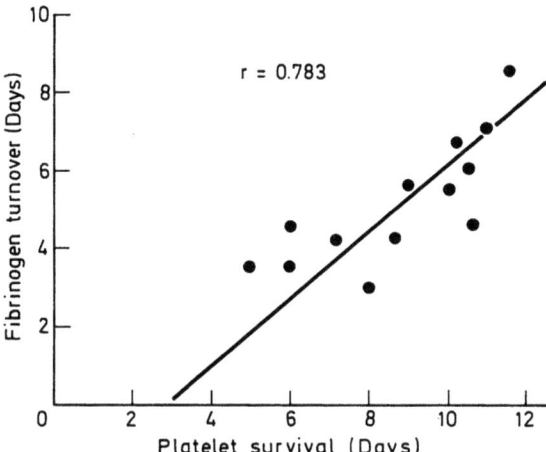

Fig. 13. Relationship between fibrinogen turnover and platelet survival

7. Effects of Heparin Administration on Platelet and Fibrinogen Turnover. In six controls and in nine CVP we studied ^{51}Cr-platelet survival time, ^{131}I-fibrinogen turnover, and serum FDP concentration before and during heparin administration (50 mg every 6 h, eV, for at least 4 days). Heparin administration has provoked a lengthening of ^{131}I-fibrinogen half-time in CVP and a decrease in FDP serum concentration in all the patients (Fig. 14). No detectable effect of heparin administration was observed in the controls. The ^{51}Cr-platelet decay curve was also flattened during heparin administration and the platelet survival time was lengthened (Fig. 15).

Discussion

The choice of an adequate control group is one of the most difficult problems in the study of platelet aggregation, as well as of blood clotting, in subjects with atherosclerosis. In fact, even in "normal" subjects, platelet aggregation shows wide fluctuations, and glass adhesiveness increases with age (Bucher et al., 1973). Moreover, the selection of the controls, although accurate, cannot rule out the possibility of selecting for the control group subjects with unrecognized atherosclerotic lesions. This fact can be an explanation of the frequent overlapping of the data for atherosclerotic patients and for control patients. In order to reduce, as possible, these inconveniences, we selected the control subjects for age. Our findings suggest that there is an increased platelet reactivity in subjects with chronic cerebrovascular disorders, expressed by an increased adhesiveness to glass and an increased aggregation in the presence of low concentrations of ADP, as well as by a lower lag time in the aggregation by collagen. An increased platelet adhesiveness-aggregation in these patients was previously reported by other authors (Neri Serneri et al., 1968; Gilroy et al., 1969; Geraud et al., 1972; Acheson et al., 1969; Danta, 1973). Danta (1973) reported that the second phase of aggregation is caused in these

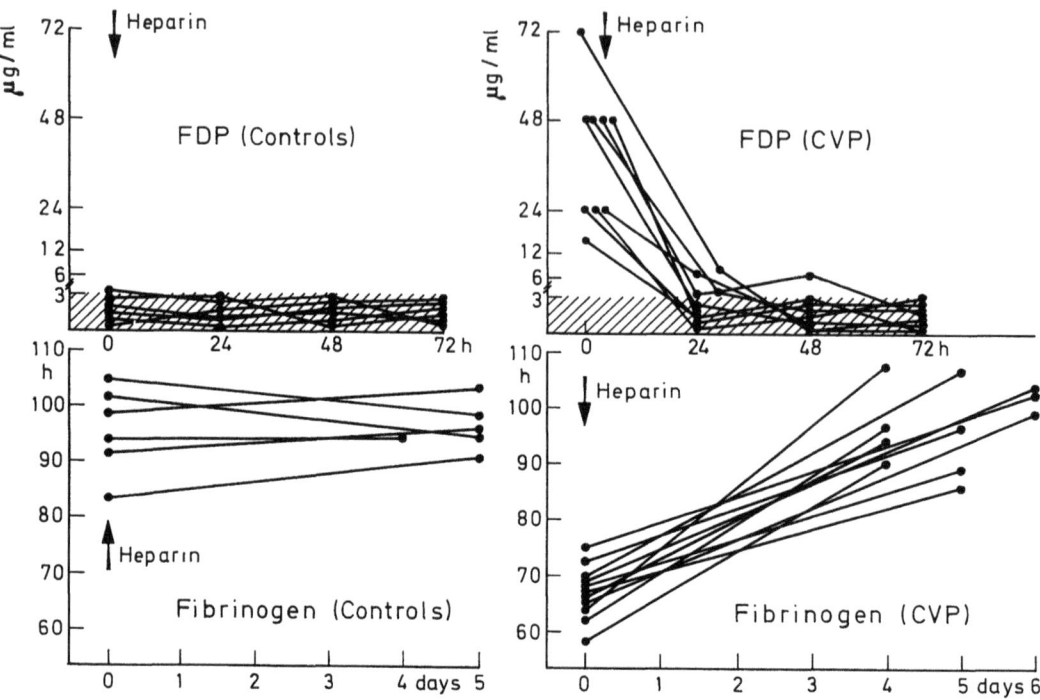

Fig. 14. Effect of heparin administration on ^{131}I-fibrinogen half-life and on FDP concentration

patients by lower ADP concentration than in controls, suggesting the occurrence of platelet hyperactivity in cerebrovascular patients.

However, the increased platelet reactivity is not specific for CVP, because MIP have also shown a very similar behavior. The average platelet survival time measured with both ^{51}Cr and ^{75}Se-seleniomethionine is shortened, although frequently CVP show a platelet survival time within the normal range. As pointed out by Murphy and Mustard (1962) and by Chakrabarti et al. (1968), and previously by us, it is difficult to designate as "normal" our controls without angiographic investigation. Therefore, our data, in this respect, are exposed to some criticism. Nevertheless, a shortened platelet survival time in subjects with coronary or peripheral atherosclerosis has been found also by other authors (Murphy and Mustard, 1962; Abrahamsen, 1968; Harker and Slichter, 1970; Steele et al., 1973; Salky and Dugdale, 1973). Interestingly, we found a positive correlation ($P < 0,0,5$) between shortened platelet survival time and increased aggregation by ADP.

Our findings do not allow us to state the factor or the factors responsible for increased platelet aggregation observed in CVP. The normal release of total nucleotides and ^{14}C-serotonine seems to exclude that platelet hyperreactivity is primitively related

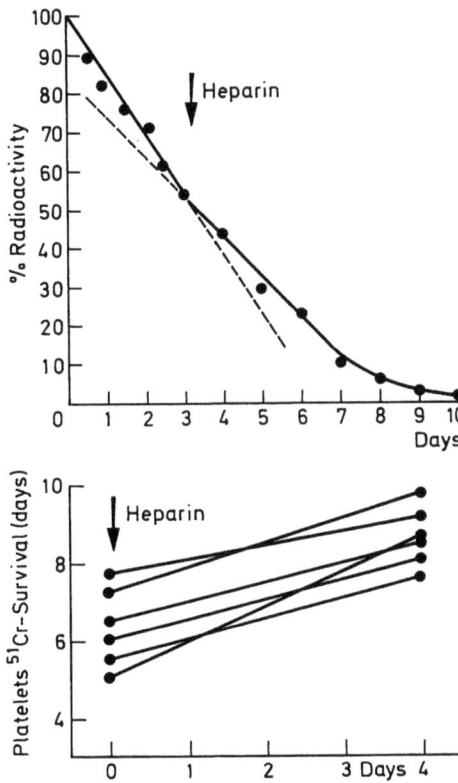

Fig. 15. Effect of heparin on ^{51}Cr platelet survival

to platelets themselves. On the other hand, the decreased disaggregation at the 5th min of control platelets after addiction of PPP and CVP suggests that platelet hyperreactivity is at least in part related to plasmatic factors. Such opinion is further supported by the shortened fibrinogen survival time in CVP and the increased serum FDP concentration, in absence of hyperfibrinolyses. These findings suggest a slow defibrination, which could take place by the thrombin-plasmin system as well as by the action of the other enzymes or factors able to modify or precipitate the fibrinogen molecule; between them, platelet factor 4, vaskulokinase, and, particularly, transglutaminase, present in atherosclerotic plaques (Laki, 1973) could be of importance. However, the normalization of the half-time of ^{131}-I-fibrinogen and of ^{51}Cr platelet survival and the decrease of serum FDP during heparin administration indicate, although they do not demonstrate it, an important role of the thrombinic system in the slow fibrin-coagulopathy of CVP. Probably, the shortening of survival time and the increase of FDP concentration are an expression of increased fibrin production. This hypothesis is further supported by the frequent association of these findings with thromboplastin and thrombin activation in atherosclerosis (Neri Serneri et al., 1970).

In conclusion, some of cerebrovascular patients show indirect evidences of chronic intravascular coagulation. Therefore, platelet hyperactivity is probably dependent on clotting activation, although other factors which we did not investigate cannot be excluded. Between them, an endogenous catecholamine increase, which could sensitize platelets to other aggregating agents (Haft et al., 1972) or the absence of an alfa globulin which perhaps inhibits platelet aggregation induced by collagen (Nossel et al., 1971).

References

Abrahamsen, A.F.: Platelet survival studies in man. Scand. J. Haemat. 5 (Suppl. 3), 1 (1968)

Acheson, J., Danta, G., Hutchinson, E.C.: Controlled trial of dipyridamole in cerebral vascular disease. Brit. med. J. 1, 614 (1969)

Aster, R.H., Jandl, H.: Platelet sequestration in man. I: Methods. J. clin. Invest. 43, 843 (1964)

Bucher, U., Robert, Y., Riedwyl, H.: Retention of platelets by glass beads. Variations with the age of the individuals. Thromb. Diath. haemorrh. (Stuttg.) 29, 671 (1973)

Chakrabarti, R., Hocking, E.D., Fearnley, G.R., Mann, R.D., Attwell, T.N., Jackson, R.: Fibrinolytic activity and coronary artery disease. Lancet 1, 987 (1968)

Danta, G.: Platelet aggregation in patients with cerebral vascular disease and in control subjects. Thromb. Diath. haemorrh. (Stuttg.) 29, 730 (1973)

Fisher, C.M.: Observations of the fundus oculis in transient monocular blindness. Neurology 9, 333 (1959)

Geraud, J., Rascol, A., Bierme, R., Boneu, B., Guiraud, B., Fernet, P., Geraud, G., David, J.: Troubles de la fonction plaquettaire et accidents ischémiques cérébraux. Perspectives thérapeutiques. Rev. neurol. 126, 31 (1972)

Gilroy, J., Barnhart? M.I.P., Meyer, J.S.: Treatment of acute stroke with Dextran 40. J. amer. Med. Ass. 210, 293 (1969)

Goldenfarb, P.B., Cathey, M.H., Zucker, S., Wilbur, P., Corrigan, J.J.: Changes in the hemostatic mechanism after myocardial infarction. Circulation 43, 538 (1971)

Haft, J.I., Gershengorn, K., Krantz, P.D., Oestreider, R.: Protection against epinephrine-induced myocardial necrosis by drugs that inhibit platelet aggregation. Amer. J. Cardiol. 30, 838 (1972)

Harker, C.A., Slichter, S.J.: Studies of platelet and fibrinogen kinetics in patients with prosthetic heart valves. New Engl. J. Med. 283, 1302 (1970)

Hellem, A.J.: The adhesiveness of human blood platelets in vitro. Scand. J. clin. Lab. Invest. 12, 51 (1960)

Januszko, T., Dubinska, L.: Estimation of activator of fibrinolysis by means of the euglobulin test. Acta med. pol. 6, 269 (1965)

Jerushalmy, Z., Zucker, M.B.: Some effects of fibrinogen degradation products (FDP) on blood platelets. Thromb. Diath. haemorrh. (Stuttg.) 25, 41 (1971)

Laki, K.: Langoustization of arteries: Circulation 48, Suppl. 4, 77 (1973)

McDonald, L.: Studies on Blood Coagulation and Thrombosis and on the Action of Heparin in Ischaemic Heart Disease. Cambridge, England: Cambridge University Press 1964

McDonald, L., Edgill, M.: Changes in coagulability of the blood during various phases of ischaemic heart disease. Lancet 1, 1115 (1959)

Merskey, C., Kleiner, G.J., Johnson, A.J.: Quantitative estimation of split products of fibrinogen in human serum, relation to diagnosis and treatment. Blood 28, 1 (1966)

Milstone, H.: A factor in normal human blood which participates in streptococcal fibrinolysis. J. Immunol. 42, 109 (1941)

Mürer, E.H.: Release reaction and energy metabolism in blood platelets with special reference to the burst in oxygen uptake. Biochem. biophys. Acta 162, 320 (1968)

Murphy, E.A., Mustard, J.F.: Coagulation tests and platelet economy in atherosclerotic and control subjects. Circulation 25, 114 (1962)

Naimi, S., Goldstein, R., Proger, S.: Studies on coagulation and fibrinolysis of the arterial and venous blood in normal subjects and patients with atherosclerosis. Circulation 27, 904 (1963)

Najean, Y., Ardaillou, N.: The use of 75 Se-Methionine for the in vivo study of platelet kinetics. Scand. J. Haemat. 6, 395 (1969)

Neri Serneri, G.G., Silvestrini, E., Paoletti, P., Masotti, G.: Studi sulle sindromi trombofiliche. V: L'adesività e l'aggregazione delle piastrine nella malattia aterosclerotica. Riv. clin. Med. 68, 835 (1968)

Neri Serneri, G.G., Paoletti, P., Silvestrini, E., Gensini, G.F., Masotti, G., Abbate, R.: La sindrome trombofilica della malattia ateromasica: aspetti patogenetici e semeiologici. G. Aterosc1. 8, 409 (1970)

Nestel, P.S.: A note on platelet adhesiveness in ischemic heart disease. J. clin. Path. 14, 150 (1961)

Nossel, H.L., Wilner, G.D., Drillings, M.: Inhibition of collagen-induced platelet aggregation by normal plasma. J. clin. Invest. 50, 2168 (1971)

Pfleiderer, T., Rucker, G.: Über Kaliumgehalt und Adhäsibität der Thrombocyten von gesunden Menschen und Patienten mit obliterierender Gefäßsklerose. Klin. Wschr. 42, 82 (1964)

Ratnoff, O.D., Menzie, C.: Estimation of fibrinogen in small samples of plasma. In: Blood Coagulation, Hemorrhage and Thrombosis. Tocantins, M.L., Kazal, L.A. (eds.). New York: Grune and Stratton 1964, p. 121

Ross Russel, R.W.: Observations on the retinal blood vessels in monocular blindness. Lancet 2, 1422 (1961)

Salky, N., Dugdale, M.: Platelet abnormalities in ischaemic heart disease. Amer. J. Cardiol. 32, 612 (1973)

Steele, P.P., Weily, H.S., Davies, H., Centon, E.: Platelet function in coronary artery disease. Circulation 48, 1194 (1973)

Stormorken, H.: Platelet adhesiveness in coronary heart disease. Evaluation of the platelet rich plasma-ADP method. Acta med. scand. 188, 339 (1970)

Tangen, O., Berman, H.J., Marfey, P.: Gel filtration: a new technique for separation of blood platelets from plasma. Thromb. Diath. haemorrh. (Stuttg.) 25, 268 (1971)

Wolfe, S.M., Shulman, N.R.: Inhibition of platelet energy production and release reaction by PGE_1, theophylline and cAMP. Biochem. biophys. Res. Commun. 41, 128 (1970)

Arterial Hypertension and Platelet Aggregation in the Pathogenesis of Cerebrovascular Diseases

S. Lentini, E. Bologna, and C. Pirro

The most important risk factor clearly identifiable in subjects predisposed to cerebrovascular accident (CVA), of both sexes and all ages, is arterial hypertension even if it be only mild. In fact, CVAs are five times more frequent in hypertensive subjects than in normotensives and are the second most frequent cause of death after coronary diseases (Froment and Froment, 1969; Kannel, 1971). The results of the Framingham study confirm the greater risk of atherosclerotic lesions in hypertensive subjects and demonstrate that the incidence of the arterial complication, thrombotic cerebral infarction, is that most closely related to the increase in arterial pressure (Gordon and Kannel, 1972).

Examining the incidence of atherosclerotic lesions in 500 consecutive autopsies, Wilkins et al. (1959) discovered that the main differences between normotensive and hypertensive subjects were to be found in the cerebral arteries, especially in the middle cerebral artery, the posterior cerebral artery, and the basilar cerebral artery (Wilkins et al., 1959). In Figure 1, taken from the above-mentioned data, it can be seen how the incidence of atherosclerotic lesions, in all arterial regions, is greater in hypertensive patients than in normotensives. It was also demonstrated that this difference is particularly enhanced in the coronary region and, to an even greater extent, in the cerebral region - in the middle cerebral arteries, the posterior cerebral arteries, and the basilar cerebral arteries.

The total incidence of CVAs, like that of other complications in arterial hypertension (AH), is lowered by antihypertensive treatment although cerebral hemorrhage is most favorably affected and has now become an exception in adequately treated patients. Thrombosis, on the other hand, is influenced to a lesser degree and is, at present, the most frequent cause of CVA (Aurell and Hood, 1964; Kannel, 1971; Leishamen and Oxon, 1963; Pickering, 1972; Smirk and Hodge, 1963; VA Cooperative Study Group on Hypertensive Agents, 1970). In fact, comparing autopsy results for hypertensive subjects whose death was due to cerebral ictus it was found that, in the thirties, hemorrhagic ictus was three times more frequent than thrombotic ictus while, recently, the latter has become more frequent (Yates, 1964).

These epidemiologic differences, probably due to antihypertensive therapy, may be explained by numerous factors, the first being the pathologic conditions at the root of CVA. In nonelderly pa-

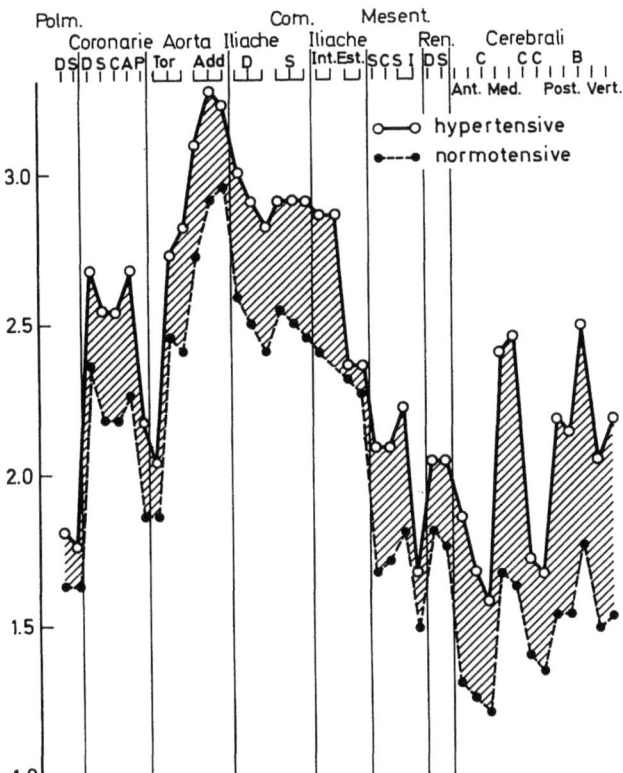

Fig. 1. Diffusion and severity of atherosclerotic lesions in arterial districts of normotensive (- - -) and hypertensive (———) subjects (from data of Wilkins et al., 1959)

tients, in particular, cerebral hemorrhage is due to the rupture of congenital or acquired microaneurysms, located in the small-caliber cerebral arteries especially in the perforating arteriolae of the basal ganglia and the subcortical region which had a diameter of less than 1 mm (Cole and Yates, 1967; Russel, 1963). Thrombosis, on the other hand, mainly affects the major arteries where the lumen is narrowed by atherosclerotic plaques, thus conferring a tendency to thrombotic occlusion.

Cerebral hemorrhage is directly dependent on the increase in pressure levels while the relationships between hypertension and thrombosis, although undeniable, are indirect (Pickering, 1972), since caused by atherosclerotic vascular lesions.

The numerous etiopathogenic factors of atherosclerosis account for the tendency of these alterations to progress with age even when AH is adequately treated. Diversely, where the effect of antihypertensive therapy on cerebral circulation is concerned, a distinction must be made between hypertensive subjects whose cerebral arteries are still undamaged and hypertensive subjects whose arteries have already been narrowed by atherosclerotic alterations. In the first group, antihypertensive therapy can prevent atherosclerosis but, in the second, the reduction in pressure levels, even though it may prevent further aggravation, can certainly cause ischemia in a cerebral region and encourage thrombosis in an already damaged artery.

It has also been demonstrated that the lower pressure limit necessary for an efficient self-regulation of the cerebral circulation (i.e., for the maintenance of adequate cerebral blood flow) is much higher in hypertensive subjects than in normotensives (Strandgaard et al., 1973). This explains the lower compensatory capacity of hypertensive subjects when sudden falls in arterial pressure occur.

The most well-known mechanism by which AH induces or favors atherosclerosis and cerebral thrombosis is, without doubt, related to an increase in the pulsation, distention, and shearing stress of the arterial walls.

The increase in arterial wall pulsation in hypertensive subjects has already been adequately demonstrated and may easily be documented by the simple examination of the classic patterns of maximum and minimum arterial pressure measured in normotensive subjects and in patients affected by essential benign hypertension and essential malignant hypertension, taken from Pickering (Fig. 2). It is obvious that there was considerable increase in arterial pulsation in essential hypertension and particularly in malignant hypertension when compared with normotensives and that this increase is constant at all times, persisting even during sleep.

This real mechanical trauma can cause alterations in the trophism of the arterial tunic by acting both directly on cellular elements and by hampering microcirculation and vasa vasorum circulation and eventually causing functional and structural alterations of the arterial intima producing dystrophic lesions, lipidic infiltrations, and fibrosis (Fry, 1973). In their initial phase, these alterations correspond to the cerebral arterial lesion which may be considered typical of hypertensive subjects, i.e., the "lipohyalinosis." This lesion weakens the arterial wall and is the main cause of its rupture and, therefore, of the formation of focal intracerebral hemorrhages. In the more advanced stages segmental occlusion of the arterial lumen may occur (Sandok and Whisnant, 1974).

To this mechanism must be added the variations in lateral pressure related to local modifications in the blood flow rate. In fact, where the blood flow rate is greater owing to artery conformation (external side of the angulations, medial walls of the bifurcations), where congenital stenosis is present or, more frequently, due to preexisting atherosclerotic and thrombotic lesions, a "suction" in the direction of the intima is created and consequent parietal dystrophy occurs (Texon, 1971).

Others, as well as ourselves, have observed that arterial hypertension is generally accompanied by an increase in platelet adhesiveness and aggregation (Carleo and Cioffi, 1970; Coccheri

---

Fig. 2. Blood pressure values registered at 5 min intervals for 24 h in normotensive (a), essential hypertensive (b), and malignant hypertensive (c) patients (from Pickering: "High Blood Pressure"," London: Churchill 1974)

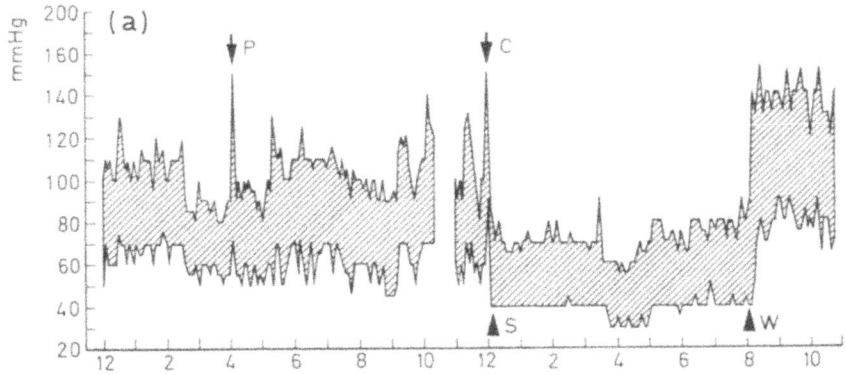
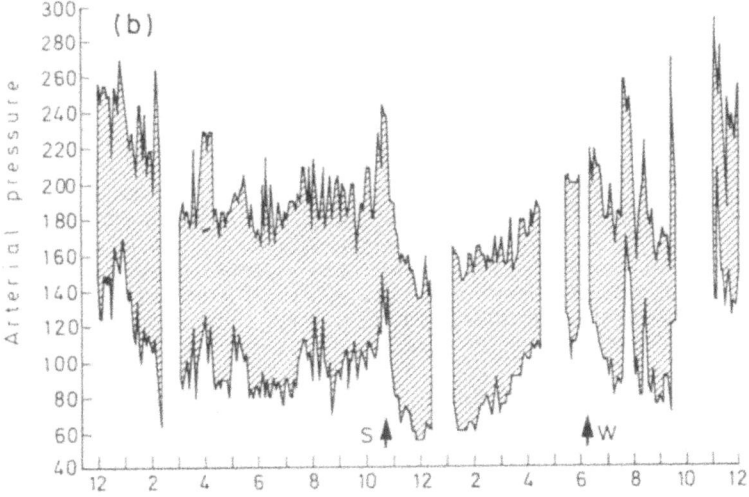
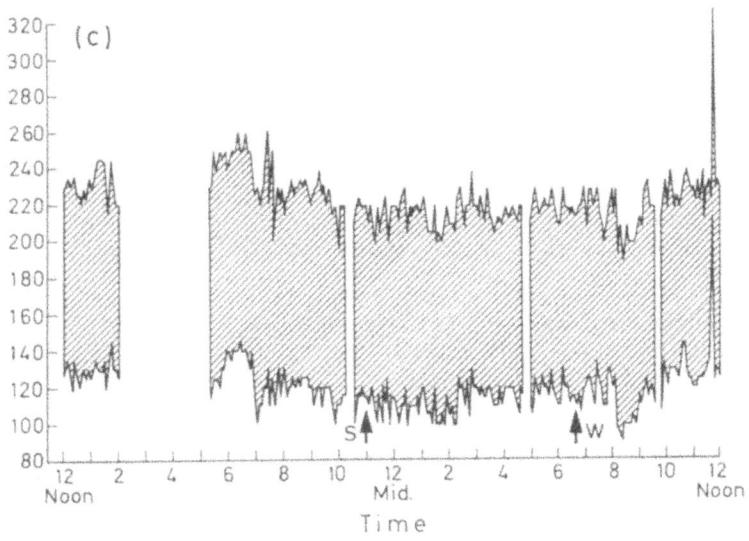

Platelet aggregation (degree)	Normal subjects	Hypertensives Essential	Hypertensives Secondary
V	0	44	30 %
IV	2	32	41 %
III	16	14	23 %
II	32	7	6 %
I	50	3	0 %

Fig. 3. Platelet aggregation (Breddin method II) n normotensive and in hypertensive subjects

and Fiorentini, 1971; Poplawski et al., 1968). The notion of the fundamental role played by the platelets in the thrombotic process and the interdependent relationship between this and atherogenesis lead us to believe that this phenomenon may also be of major importance in the genesis of the thrombotic alterations occurring in the arterial system and particularly in the cerebral arteries of hypertensive subjects.

With extensive case material we were able to demonstrate that the increase in platelet aggregation is present even in the early phases of arterial hypertension and also in young hypertensive subjects with no clinical signs of target organ involvement particularly in the cerebral artery circulation. Moreover, this phenomenon is independent of the etiopathogenic nature of hypertension since it is manifest both in essential and secondary hypertension (Lentini, 1974; Lentini and Bologna, 1974).

The results of our investigations may be seen in Figure 3. The aggregation levels are distributed in accordance with the five grades of the Breddin test, the fourth and fifth being pathologic. It may be observed that there is an evident difference in behavior between normotensives and hypertensives. In fact, the latter, independently of the essential or secondary nature of hypertension, in the majority of cases manifest pathologic aggregation levels.

Arterial hypertension may cause platelet aggregation to increase through various mechanisms: (1) hemodynamic: creating or accentuating the turbulence of the flow and formation of vortexes, (2) bioelectric: modifying the potential of the arterial walls in a positive sense through an increase in cations (Na^+) thus favoring adhesion since the platelets are electronegative, (3) endocrine: through the aggregation effect of certain endogenous substances such as catecholamines and angiotensin which increase

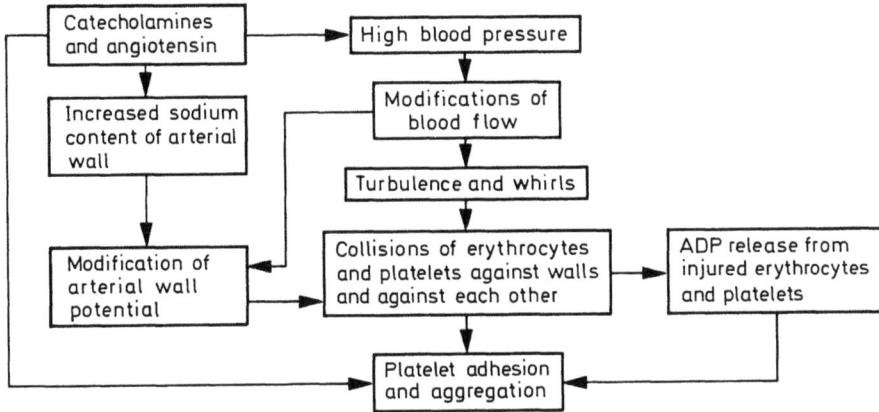

Fig. 4. Schematic drawing relating arterial hypertension to the increase in platelet aggregation

in some types of arterial hypertension. In Figure 4 we schematically show the ascertained and presumed mechanism relating arterial hypertension to the increase in platelet aggregation.

Immediately following severe obstructive arterial lesions, a considerable drop in blood flow rate may occur and this might favor the aggregation of platelets circulating on atherosclerotic plaques (Richardson, 1973).

The hemodynamic mechanism is particularly important in the genesis of cerebrovascular diseases since it predominantly affects the arteries which are least sustained by the surrounding structures, such as the cerebral arteries, especially regarding dilatations, bifurcations, angulations, and ramifications. In fact, atherosclerosis and thrombosis occur more frequently and with decreasing incidence in the carotid sinus, the paraclinoid portion of the internal carotid artery where the internal carotid divides into the anterior and middle cerebral arteries, in the union of the vertebral arteries to form the basilar trunk, and in the bifurcations of the basilar artery (Fisher, 1953).

Given the importance of platelet intervention in the genesis of atherosclerosis and arterial thrombosis, the increased tendency toward aggregation in hypertensive patients raises two problems of a practical nature. Will the normalization of arterial pressure, as such, also normalize platelet aggregation? Among the antihypertensive drugs are there any which possess antiaggregating activity independently of their action on arterial hypertension? Our present investigations are oriented toward a solution to these two problems.

In the first series of investigations conducted on 75 patients with essential arterial hypertension, we observed that the pharmacologic normalization of arterial pressure is accompanied by the normalization of platelet aggregation only where target organs, especially the cerebral vascular system, are either not

Table 1

Class of arterial hypertension	n	Blood pressure normalization (150/90 or less)	PAT normalization
I	27	27	26
II	23	18	7
III	14	7	none
IV	11	2	none

involved at all or are involved only to a slight extent. This can be seen in Table 1 where our hypertensive subjects have been divided into four groups in conformity with a well-known classification based on the presence, and on the degree, of involvement of the target organs. Only in the first group, where vascular damage had not occurred, was there a parallel normalization in arterial pressure and platelet aggregation. This correlation was poor in group 2 and completely lacking in groups 3 and 4.

Clinically these results are one more argument in favor of treating hypertension in its early phases, i.e., when the increase in platelet aggregation is not yet conditioned by arterial wall lesions but depends only on functional characteristics related to hypertension.

Information is very scarce regarding possible direct action on platelet aggregation by hypertensive drugs. It appears that reserpine is the only antihypertensive drug studied from this point of view.

We, therefore, began a systematic study of this problem examining the in vitro and in vivo effect of some antihypertensive drugs.

The in vivo studies were conducted both by acute intravenous administration and in long-term oral therapy. In the in vitro tests the substances being examined were added to platelet-rich plasma (PRP) in such quantities as to determine concentrations comparable with those reached during in vivo treatment at average doses.

The Born and Breddin methods were used to study platelet aggregation both in the long-term in vivo studies and in the in vitro tests while, in the acute in vivo studies, only the Born method was used. With this latter method we used ADP, epinephrine, and collagen as aggregating agents (final concentrations in the PRP were µM 2, µM 1, and µg 5, respectively).

Medical literature has shown that, in the animal, *reserpine* is capable of inhibiting ADP aggregation and adhesion to glass and it also obstructs the growth of experimental thrombi (Zweifler, 1967) while, in man, we ourselves had previously demonstrated

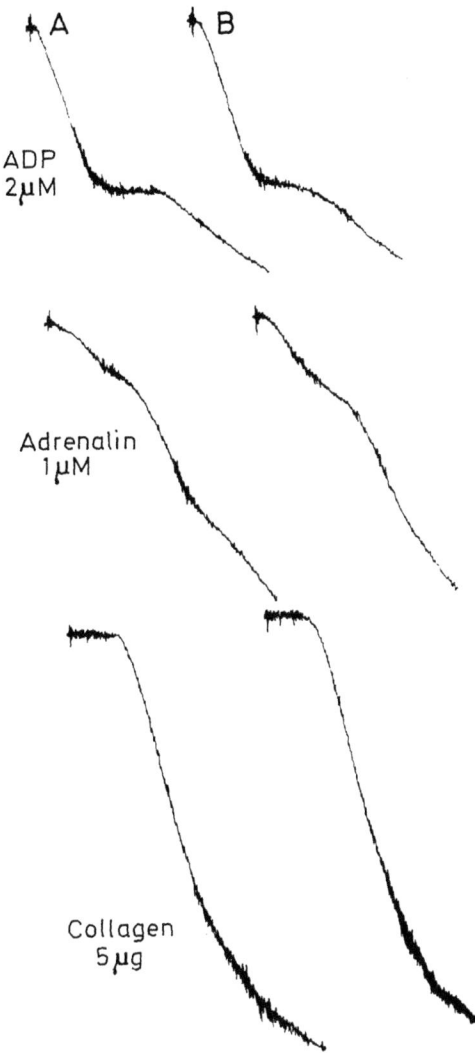

Fig. 5. Platelet aggregation induced by ADP, adrenaline, and collagen in native PRP (A) and in PRP incubated with reserpine, 2.10^{-9} (B)

that the administration of 0.30 mg daily of this drug does not modify platelet adhesiveness to glass (Bologna, 1969b).

The results of our present investigations (Figs. 5, 6, and Table 2) are not univocal. It can be seen that the addition of reserpine in vitro does not cause alterations worthy of note (Fig. 5) while with intravenous administration there is a net reduction in response, especially evident in the ADP test and less so in the collagen test (Fig. 6). Long-term oral administration (0.30 mg daily) partially inhibits aggregation, according to Breddin, and in ADP and epinephrine aggregation, according to Born (Table 2).

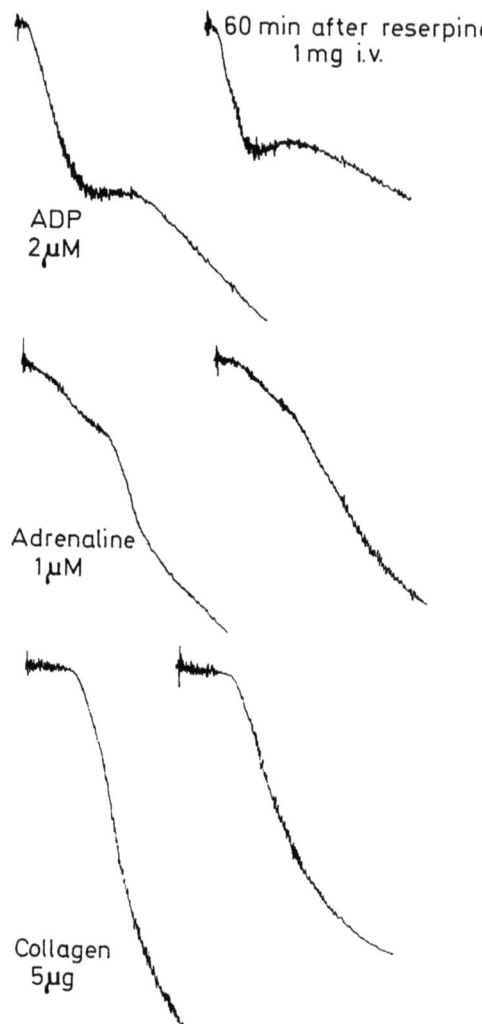

Fig. 6. Platelet aggregation induced by ADP, adrenaline, and collagen in PRP from a subject before and 60 min after i.v. administration of reserpine, 1 mg

Table 2. Inhibition of platelet aggregation during long term treatment with antihypertensive drugs

Drug	PAT	ADP	Adrenaline	Collagen
Clonidine mcg 450/die	+	−	−	−
Reserpine mg 0,30/die	±	+	+	−
Methyldopa mg 750/die	−	−	−	−
Guanethidine mg 50/die	+	+	±	−

+ = strong inhibition; ± = moderate inhibition; − = no inhibition

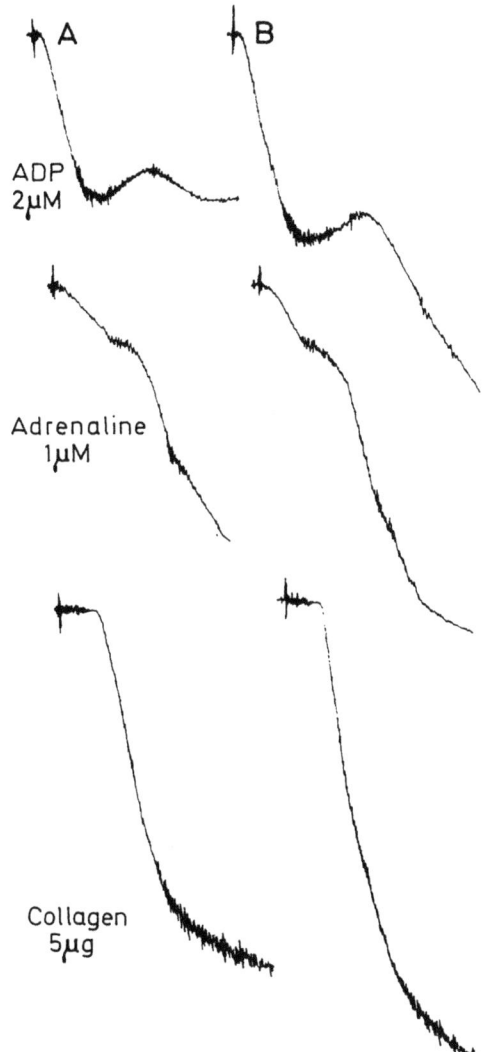

Fig. 7. Platelet aggregation induced by ADP, adrenaline, and collagen in native PRP (A) and in PRP incubated with clonidine, 2.10^{-9} (B)

The reaction on platelet aggregation of *clonidine*, which is one of the most recent and most active antihypertensive drugs, differs according to which test is being done. When the drug is added to PRP in vitro there is a constant increase in aggregation with all three aggregating agents - ADP, epinephrine, and collagen (Fig. 7). Diversely, in the acute in vivo studies, the effect observed is exactly the opposite since there is a diminished response to the aggregating agents. The reduction is most evident when ADP is used (Fig. 8). We feel it opportune to emphasize that the phenomenon observed in vivo is completely independent of variations in arterial pressure. For example, in the case seen in this figure, the patient is a young normotensive whose pressure did not alter during the course of the study.

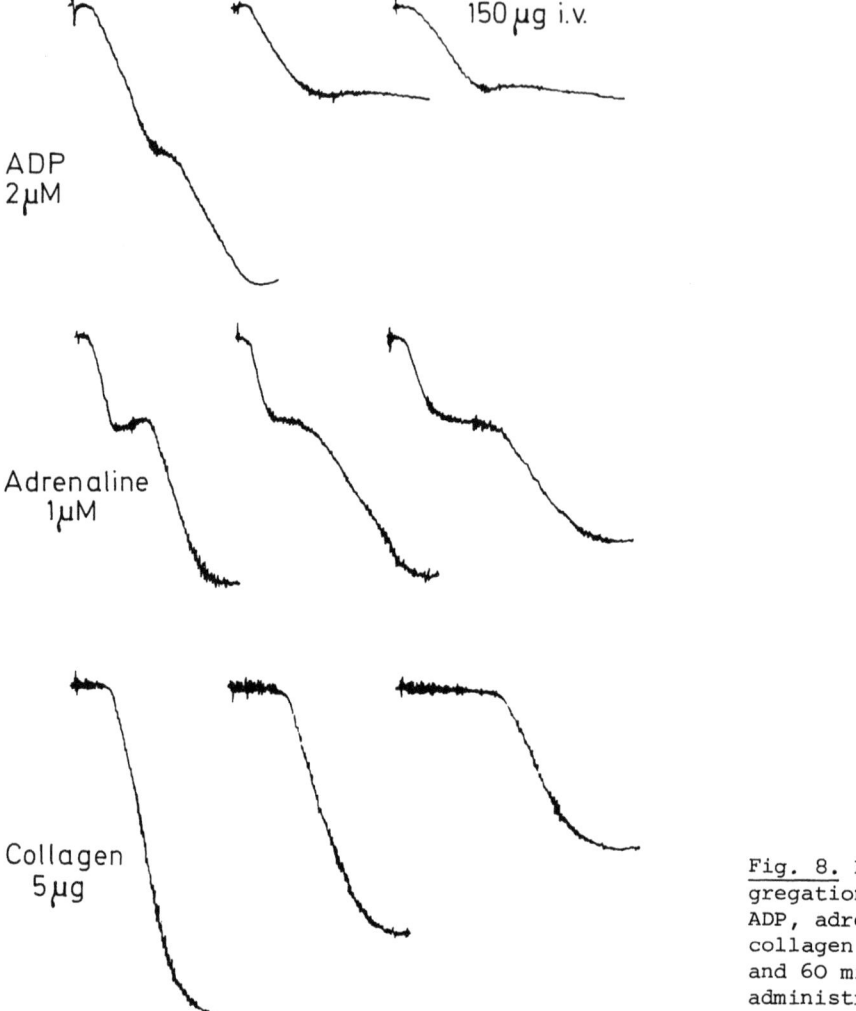

Fig. 8. Platelet aggregation induced by ADP, adrenaline, and collagen before, 30, and 60 min after i.v. administration of clonidine, 150 mg

Finally, long-term oral administration at usual average doses (450 mg daily) almost always resulted in normalization using the Breddin test while, with the Born method, it did not notably influence aggregation (Table 2). As we know, clonidine is an imidazolinic derivative and, besides its antihypertensive effect on the central nervous system, it also produces an adrenergic-stimulating action on peripheral organs with sympathetic innervation. This action, which is independent of endogenous norepinephrine, is exclusively exerted on the alpha-adrenoreceptors (Hoefke and Kobinger, 1966; Kobinger, 1973). It explains the transient increase in arterial pressure which is often to be seen after intravenous administration of the drug and on interruption of long-term oral treatment.

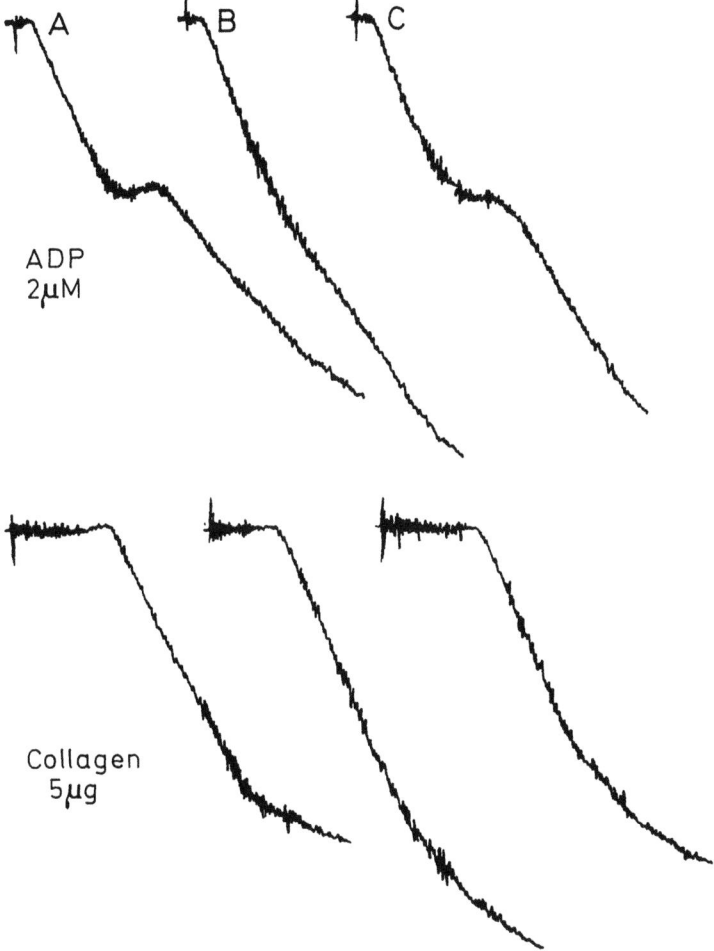

Fig. 9. Platelet aggregation induced by ADP and collagen in native PRP (A), in PRP incubated with clonidine, 2.10^{-9} (B), and in PRP incubated with clonidine, 2.10^{-9} plus phentolamine, 7.10^{-5} (C)

The increase in aggregation which clonidine induces in vitro is probably related to the stimulation which this substance exerts on the alpha-receptors in the platelet membrane (Mills and Roberts, 1967). The addition of phentolamine does, in fact, prevent the increase in ADP and collagen aggregation induced by clonidine (Fig. 9). Less easily explainable is the antiaggregation effect observed in the acute in vivo studies. Hypothetically speaking, it could be related to the inhibiting effect of clonidine on the generation of angiotensin II (Frohlich, 1974), which is known to possess proaggregant activity (Bologna, 1969a).

Two other antihypertensive drugs examined for their platelet aggregation effect were *methyldopa* and *guanethidine*. Our initial results indicate that methyldopa does not affect aggregation in long-term oral administration (750 mg daily) while, in vitro, it partially inhibits epinephrine and collagen aggregation. Diversely, guanethidine, which has no notable effect in vitro, considerably inhibits aggregation after administration in vivo with the Breddin method (50 mg daily). With the Born method, it inhibits ADP aggregation and, to a lesser extent, also epinephrine (Table 2).

Besides studying true antihypertensive drugs, it might be useful to examine the possible effect on aggregation of *saluretics* since, as a rule, they are also employed in the treatment of arterial hypertension.

It has already been observed that furosemide in vitro inhibits collagen aggregation, the first phase of ADP aggregation and the second wave of epinephrine aggregation. These effects, however, occur using only very high concentrations, much higher than those which, in fact, can be employed in therapy (Richardson, 1973).

However, even if other saluretics should show antiaggregating activity in vitro at concentrations similar to those which may be employed in therapy, it must be remembered, when considering possible clinical application, that diuretic treatment is accompanied by a higher incidence of thromboembolic complications. These are probably related to the hemoconcentration caused by hydrosaline depletion and the consequent plasmatic concentration increase of coagulation factors (Egeberg, 1963; Gormsen, 1954), also shown by the diuretic-induced alterations on the thromboelastogram and on Quick time (Deutsch and Kock, 1967).

A comprehensive judgment on the opportunity of selecting the antihypertensive drugs also on the basis of their antiaggregating platelet activity is not yet possible. This derives from the fact that the methods for studying platelet aggregation differ considerably from one another and have no definite clinical correspondence.

References

Aurell, M., Hood, B.: Cerebral hemorrhage in a population after a decade of active antihypertensive treatment. Acta med. scand. 176, 377 (1964)
Bologna, E.: Effetti dell'angiotensina sull'aggregazione piastrinica. Boll. Soc. ital. Biol. sper. 45, 1030 (1969a)
Bologna, E.: Effetti della reserpina sul numero della piastrine circolanti e sulla adesività piastrinica in vitro. Boll. Soc. ital. Biol. sper. 45, 1035 (1969b)
Carleo, R., Cioffi, L.A.: Adesività piastrinica nell'ipertensione. Rass. Med. sper. 17, 89 (1970)
Coccheri, S., Fiorentini, P.: Platelet adhesiveness and aggregation in hypertensive patients. Acta med. scand., Suppl. 525, 273 (1971)
Cole, F.M., Yates, P.O.: Comparative incidence of cerebrovascular lesions in normotensive and hypertensive patients. Neurology 18, 255 (1968)
Deutsch, E., Kock, M.: Die Wirkung von Furosemid auf die Blutgerinnung. Wien. med. Wschr. 117, 498 (1967)

Egeberg, O.: The effect of oedema drainage on the blood clotting system. Scand. J. clin. Lab. Invest. 15, 14 (1963)

Fisher, C.M.: The arterial lesions underlying lacunes. Acta neuropath. (Berlin) 12, 1 (1969)

Fry, D.L.: Response of the arterial wall to certain physical factors. In: Atherogenesis: Initiating Factors. Amsterdam: Elsevier 1973

Frohlich, E.D.: Inhibition of adrenergic function in the treatment of hypertension. Arch. intern. Med. 133, 1033 (1974)

Froment, A., Froment, R.: L'hypertension arterielle permanente. Données statistiques, évolution, complications et pronostic. Rev. Prat. 19, 207 (1969)

Gordon, T., Kannel, W.B.: Predisposition to atherosclerosis in the head, heart, and legs. The Franingham Study. J.A.M.A. 221, 661 (1972)

Gormsen, J.: Does rapid dehydration in cardiac decompensation produce thromboembolic complications? Acta med. scand. 148, 61 (1954)

Hoefke, W., Kobinger, W.: Hypertension: mechanisms and management. Arzneimittel Forsch. 16, 1038 (1966)

Kannel, W.B.: Current status of the epidemiology of brain infarction associated with occlusive arterial disease. Stroke 2, 295 (1971)

Kobinger, W.: In: Hypertension: Mechanisms and Management. Onesti, G. et al. (eds.). New York: Grune and Stratton 1973

Leishaman, A.W.D., Oxon, D.M.: Merits of reducing high blood pressure. Lancet I, 1284 (1963)

Lentini, S.: Nuove prospettive sulla patogenesi del danno vascolare da ipertensione arteriosa: il rudo dell'aggregazione piastrinica. Atti Accad. Lancisiana, XVIII/1, 37-62 (1974)

Lentini, S., Bologna, E.: Platelet aggregation and drugs. Symposium on platelet aggregation and drugs. Roma, March 7-8. London: Academic Press 1974, p. 63

Mills, D.C.B., Roberts, G.C.K.: Effects of adrenaline on human blood platelets. J. Physiol. (Lond.) 193, 443 (1967)

Pickering, G.: Hypertension. Definitions, natural histories and consequences. Amer. J. Med. 52, 570 (1972)

Poplawski, A., Skorulska, M., Niewiarowski, S.: Increased platelet adhesiveness in hypertensive cardiovascular disease. J. Atheroscl. Res. 8, 721 (1968)

Richardson, P.D.: Effect of blood flow velocity on growth rate of platelet thrombi. Nature (Lond.) 245, 103 (1973)

Rossi, E.C., Levin, N.W.: Inhibition of ADP-induced platelet aggregation by furosemide. J. Lab. clin. Med. 81, 140 (1973)

Russel, R.W.R.: Observations on intracerebral aneurysms. Brain 86, 425 (1963)

Sandok, B.A., Whisnant, J.P.: Hypertension and the brain. Arch. intern. Med. 133, 947 (1974)

Smirk, F.H., Hodge, J.V.: Causes of death in treated hypertensive patients: based on 82 deaths during 1959-1961 among an average hypertensive population at risk of 518 persons. Brit. med. J. II: 1221 (1963)

Strandgaard, S., Olesen, J., Skinjøj, E., Lassen, N.A.: Autoregulation of brain circulation in severe arterial hypertension. Brit. med. J. I, 507 (1973)

Texon, M.: Atherosclerosis: its hemodynamic basis and implications. Med. Clin. N. Amer. 58, 257 (1971)

Veterans Administration Cooperative Study Group on Antihypertensive Agents. J.A.M.A. 202, 1028 (1967)

Veterans Administration Cooperative Study Group on Antihypertensive Agents. J.A.M.A. 213, 1143 (1970)

Wilkins, R.H., Roberts, J.C., Moses, C.: Autopsy studies in atherosclerosis, III. Circulation 20, 527 (1959)
Yates, P.O.: A change in the pattern of cerebrovascular disease. Lancet I, 65 (1964)
Zweifler, A.J.: Impairment of platelet function and thrombus growth in reserpine-treated rabbits. J. Lab. clin. Med. 70, 16 (1967)

Smoking, Cerebrovascular Diseases and Platelet Functions*

G. Kauchtschischvili, G. Grignani, P. Bo, G. Gamba, and G. Nappi

A relationship between predisposition to thromboembolia (particularly in cerebrovascular diseases) and platelet functions is not yet well established, neither is the effect of cigarette smoking on platelets (Hawkins, 1972; McDonald et al., 1972; Deutschinoff, 1971; Koidakis, 1971).

We have studied 118 patients, males and females, between ages of 14 and 90 years with high thromboembolic risk (more than 12 degrees, Table 1). The ADP-induced platelet aggregation was studied in 113 cases, the adrenalin-induced platelet-aggregation in 30, and the platelet-adhesiveness in vitro in 48. We found an increased response to ADP in 26%, to adrenalin in 37%, and increased adhesiveness in 25% of cases.

Age does not seem to particularly affect platelet functions (Table 2). Indeed, we found an increased platelet activity only in 25% of these patients.

Smoking influence was also studied. Subjects were divided into three groups: first, nonsmoking patients with cerebrovascular diseases (22 cases); second, heavy smoking patients with cerebrovascular diseases (17 cases); third, young heavy smokers (11 cases). (See results in Table 3.)

In order to examine the immediate influence of smoking on platelets we used the two Kauffman cigarette tests (1960) in 16 of our subjects. They were divided in two further groups: young smokers (aged 20-40 years) and old cerebrovasculopathic smokers (55-70 years).

Results: in young subjects there was a significant increase of platelet aggregation and adhesiveness (78%), whereas in old cerebrovasculopathic patients platelet response to the smoking test was less evident.

According to our former studies about smoking and cerebral circulation (Currò Dossi et al., 1973; Kauchtschischvili et al., 1972, 1973), we can suggest that young subjects are more sensitive to cigarette smoking than older subjects.

---

*This article has been published in "L'aggregazione piastrinica nella patogenesis della vasculopatie cerebrali" abstract of the Round Table of Rome 30/31.10.1974. Edition: Boehringer Ingelheim, Firenze

Table 1

Risk-factors	Risk-degrees
1. Heredity	
Myocardial infarction	1
Angina pectoris	1
Hypertension	1
Diabetes	1
2. Life habits	
Smoking (more than 20 cigarettes in a day)	2
Hypernutrition	2
Sedentary life	1
Heavy responsibility profession	1
3. Arterial diastolic pressure	
95-105 mm Hg	2
105-115 mm Hg	3
more than 115	4
4. Glycosuria	2
5. Uricemia: more than 7 mg%	2
6. Eyes: corneal arc	3
7. Skin: xantomas	4
8. Trigliceridemia: more than 150 mg%	3
9. Cholesterolemia: more than 250 mg%	4

Total risk degrees	Risk
More than 12	Very high
From 9 to 12	Middle
From 5 to 8	Light
From 0 to 4	Practically null

Table 2. Platelet functions in 118 patients with high thromboembolic risk

	ADP-induced platelet-aggregation			Total	Adrenalin-induced platelet-aggregation			Total	Platelet adhesiveness			Total
	Incr.	Normal	Decr.		Incr.	Normal	Decr.		Incr.	Normal	Decr.	
14-30 years[a]	2	–	–	2	1	1	–	2	1	1	–	2
31-50 years	1	1	–	2	2	–	–	2	–	2	–	2
51-70 years	16	44	5	65	2	10	–	12	5	18	1	24
%	24.6	67.6	7.8	100					21	75	4	100
71-90 years	10	31	3	44	6	8	–	14	6	13	1	20
%	22.7	70.4	6.9	100					30	65	5	100
Total	29	76	8	113	11	19	–	30	12	34	2	48
%	26	67	7	100	37	63	–	100	25	71	4	100

[a] Two young diabetic patients

Table 3. Effect of smoking on platelet functions

	ADP-induced platelet-aggregation			Total	Adrenalin-induced platelet aggregation			Total	Platelet Adhesiveness			Total
	Incr.	Normal	Decr.		Incr.	Normal	Decr.		Incr.	Normal	Decr.	
Non-smoking cerebro-vasc. patients	6	13	3	22	2	3	–	5	4	17	2	23
%	27	59	14	100	40	60	–	100	18	74	8	100
Heavy smoking cerebrovasc. patients	3	14	–	17	7	13	–	20	7	13	–	20
%	22	78	–	100	35	65	–	100	35	65	–	100
Young heavy smokers	3	6	2	11	–	5	–	5	1	5	5	11
%	27	55	18	100								

References

Currò Dossi, B., Nappi, G., Rognone, F., Kauchtschischvili, G.: Acute effects of smoking on cerebral circulation in smokers and non smokers. Acta geront. (Milano) 23, 75 (1973)

Deutschinoff, A.: Plaquettes et fonctions plaquettaires au cours de consommation des cigarettes, du vin et des d'oignons. The second Mediterranean Congress on Thromboembolism, Istanbul, Oct. 1971

Hawkins, R.I.: Smoking, platelets and thrombosis. Nature (Lond.) 236, 450 (1972)

Kauchtschischvili, G., Nappi, G., Fici, F.: Aspetti funzionali del circolo cerebrale nel fumatore, nell'etilista cronico, nell'anziano sotto l'influenza di stimoli vasoattivi (fumo, isossisuprina). XIX Congr. Nazionale Soc. It. Geront. Parma, Oct. 1972

Kauchtschischvili, G., Nappi, G., Bo, P., Bono, G.: Fumo e circolo cerebrale. Acta geront. (Milano) 23, 282 (1973)

Kauffman, H.: Bull. Acad. Nat. Med. 123, 31 (1960)

Koidakis, A., Maudalaki, T., Georgiou, B., Antonakis, E.: L'influence de fumer sur l'agrégation des plaquettes. The second Mediterranean Congress on Thromboembolism, Instanbul, Oct. 1971

McDonald, T.P., Woodard, P., Cotrell, M.: Effect of nicotine on clot retraction of rat blood platelets. Blood 40, 595 (1972)

Contribution of Platelet Aggregation and Serotonin Release to Progressive Cerebral Infarction[1]

J. S. Meyer, K. M. A. Welch, and J. Buckingham[2]

Prevention and amelioration of acute cerebral infarction resulting from occlusion of the aortocerebral arterial circulation is best effected by maintenance of tissue metabolic viability by enhancing the cerebral collateral circulation. This may be achieved by several therapeutic approaches but let us consider here the role of components of the blood in the ischemic zone, particularly the platelet. Certain conditions contribute to impairment of the cerebral collateral circulation and hence are responsible for the diminished perfusion so commonly found after ischemia when occlusion of major cerebral vessels is relieved, which has sometimes been termed the "no reflow" phenomenon (Ames et al., 1968). Conditions contributing to "no reflow" include intraluminal aggregation of blood elements, swelling of perivascular astrocytes due to edema formation with vascular compression and vasospasm (Meyer, 1961) (Fig. 1). This paper examines in detail the possible participation of the platelet in such adverse pathophysiologic conditions that may contribute to progression of cerebral infarction.

Using the Forbes window technique, observations have been made in primates on changes in pial circulation overlying areas of cortical ischemia produced by occlusion of the middle cerebral artery (MCA) (Meyer, 1958; Waltz and Sund, 1967) (Fig. 2). Platelet aggregation was frequently observed, most commonly in terminal venules (20-200 µ/decimeter) preceded by hemoconcentration, stasis, and red cell clumping. Permanent platelet thrombi and emboli were recognized and recovery of collateral flow was often accompanied by fragmentation or disintegration of platelet aggregates in areas of spreading cortical pallor and in some instances this was seen at sites distal to platelet thrombi and emboli lodged in small terminal vessels. Spasm of leptomeningeal arterioles with cortical pallor was observed and is best accounted for by release of a vasoconstrictor agent such as serotonin from the aggregated platelets and/or from ischemic brain tissue

---

[1] This work was supported by Grant NS 09287 from the National Institute of Neurological Diseases and Stroke, and in part by Grant RR 00350 from the General Clinical Research Centers Branch, Division of Research Resources, National Institutes of Health, Bethesda, Maryland 20014

[2] James Buckingham was awarded a Student Clerkship from the American Heart Association, Stroke Council

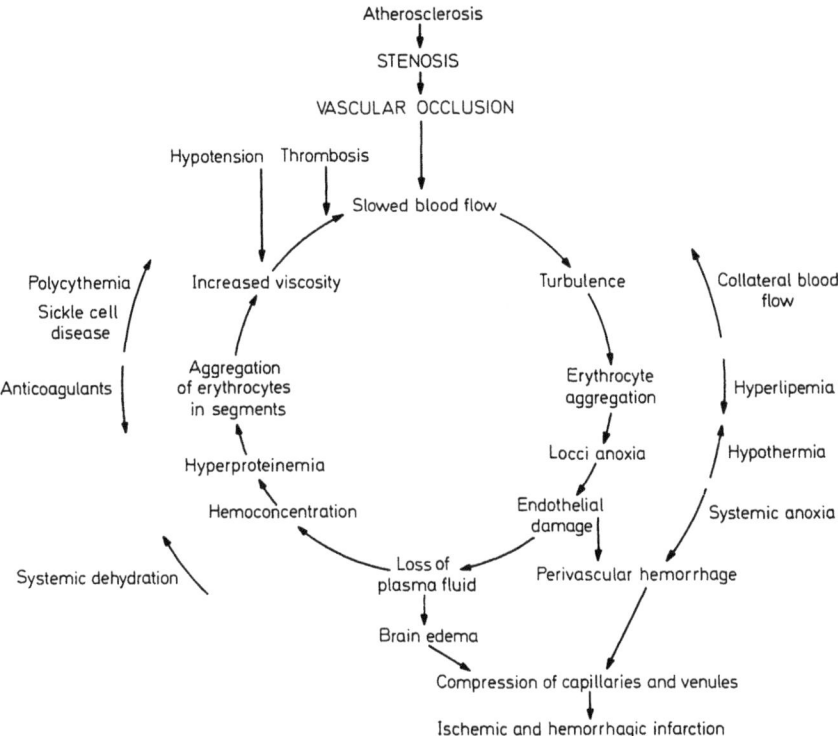

Fig. 1. The train of events that may result from atherosclerosis, cerebrovascular stenosis, or occlusion. The cycle shows various physiologic and pathologic factors that determine the extent of cerebral ischemia or infarction. Once blood flow is significantly reduced and ischemic anoxia is severe enough to cause endothelial damage, a potential vicious cycle is initiated. (After Meyer, J.S.. In: Pathogenesis and Treatment of Cerebrovascular Disease, Springfield, Ill.: Thomas 1961, p. 87

itself since many neurons are rich in serotonin as an important neurotransmitter.

Estimations of regional cerebral blood flow (rCBF) in primates by means of thermistor electrodes plus oxygen and hydrogen cathodes directly implanted in cortical tissue as well as by the antipyrine method (Yamaguchi et al., 1971) have permitted more exact observation of no reflow, "uneven perfusion", or "poor reflow" after removal of temporary MCA occlusion. Figure 3 shows that areas of reflow and no reflow may result after temporary MCA occlusion in the territory of its supply. The rapid, spontaneous, and complete reversal of the region of no reflow in the parietal ischemic zone (electrode 2) seen in this illustration can best be explained by sudden disintegration of intraluminal platelet thrombi or release of spasm in a major branch of the MCA. Both these events have been seen to occur in the leptomeningeal vessels after MCA occlusion and release. Note the subsequent reactive hyperemia and abolition of regional autoregulation in the previously ischemic area.

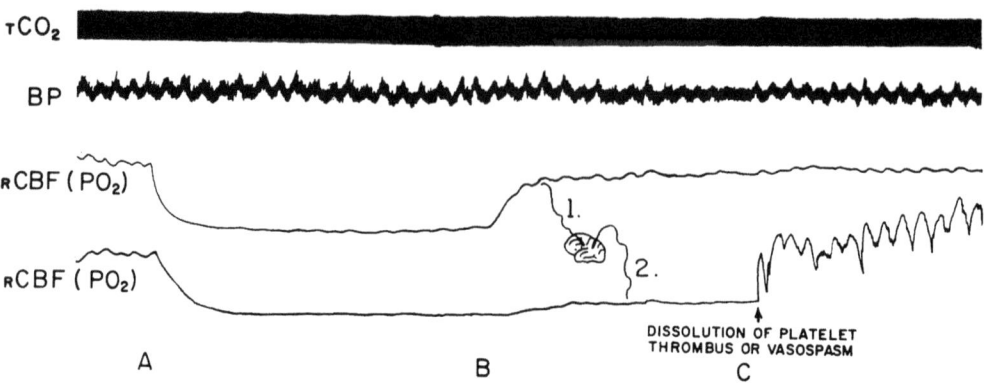

Fig. 2. Shows transient clumping of platelets in small pial vessels following vascular occlusion in the cat. (A) Macrophotograph (400 ×) 6 h after occlusion of middle cerebral artery and its principle collateral vessels. Upper arrow points to clump of platelets, lower arrow points to white cells sticking to walls. (B) Same field 3 min later. Arrow points to same clump of platelets that has moved along the vessel and increased in size. (After Meyer, J.S.: Localized changes in properties of the blood and effects of anticoagulant drugs in experimental cerebral infarction. New Engl. J. Med. 278, 151-159 (1968))

Measurements of cerebral arteriovenous (A-V) differences for whole blood serotonin in baboons revealed an initial release of serotonin from ischemic brain into cerebral venous blood following bilateral vertebral artery occlusion (Meyer et al., 1974). However, this was brief and was followed 30-60 min after release of temporary total cerebral ischemia by a markedly increased positive cerebral A-V difference (i.e., brain uptake) for serotonin as illustrated in Figure 4A. The latter results suggest that after near total arrest of the cerebral circulation, platelet aggregation occurs in the microcirculation of ischemic brain with release of serotonin which then moves into the brain across a damaged blood brain barrier.

Clinical studies have shown increased platelet aggregates in the circulating blood of patients with acute cerebral infarction (Barnhart et al., 1970). The presence of such aggregates bore qualitative relation to recovery or progression of cerebral infarction. Measurements of serotonin in cerebrospinal fluid (CSF) of patients with recent cerebral hemispheric infarction revealed elevated levels in the early stages after infarction, these abnormal levels subsiding to almost zero 3 weeks after the ictus also coinciding with improvement in the patients' neurologic status (Meyer et al., 1974) (Fig. 5). The elevated serotonin levels were found to correlate inversely with concurrent measurements of hemispheric blood flow (HBF) measured in both cerebral hemispheres, HBF improving as CSF serotonin levels fell 2-3 weeks after the infarction (Fig. 4B). The source for increased serotonin levels in CSF may be release or displacement of the neurotransmitter from ischemic brain tissue and/or release from intravascular aggregates of blood platelets in the

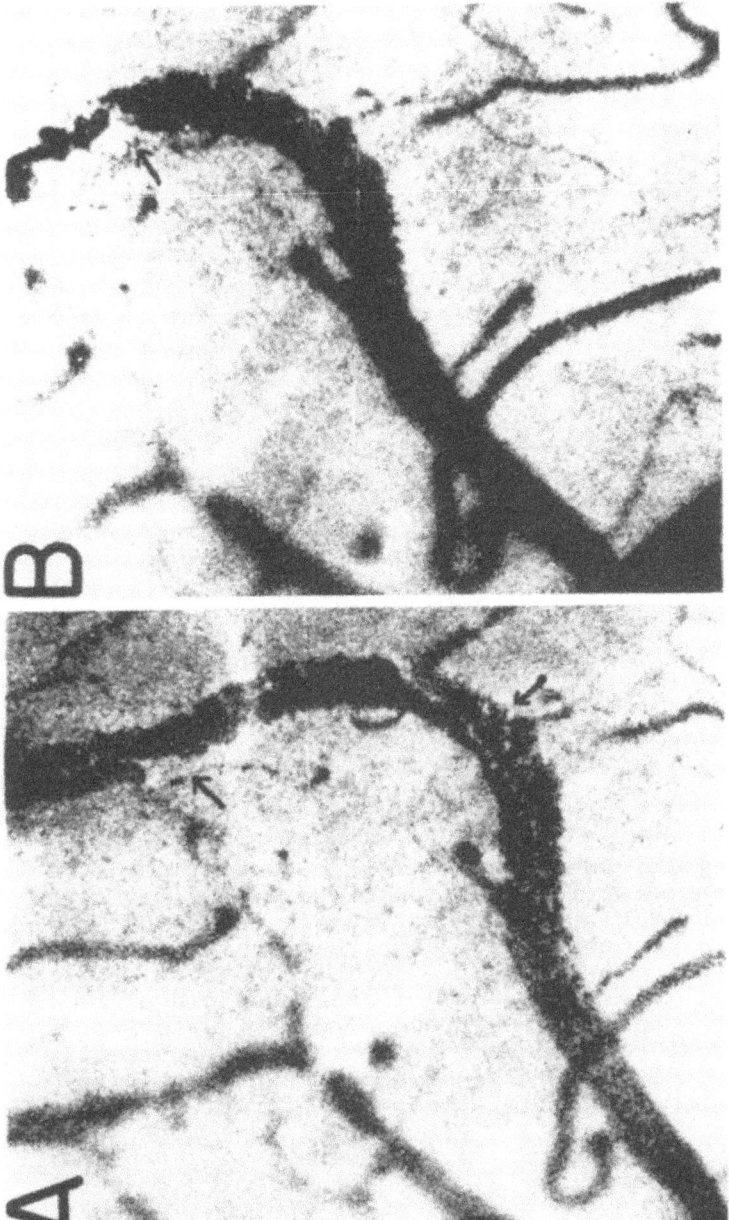

Fig. 3. Shows areas of reflow and no reflow in brain tissue after occlusion, marked A and release, marked B of the middle cerebral artery in the baboon. Electrode 1 shows prompt reflow but electrode 2 shows nonreflow. Note that at the point marked C a sudden spontaneous recovery of flow (due to dissolution of a platelet thrombus) occurs followed by hyperemia or spasm and loss of regional autoregulation so that regional flow fluctuates directly with changes in blood pressure. (rCBF = regional cerebral blood flow as measured by directly implanted oxygen electrodes. TCO_2 = end-tidal CO_2 and BP = blood pressure)

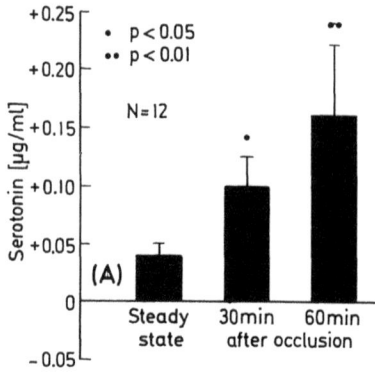 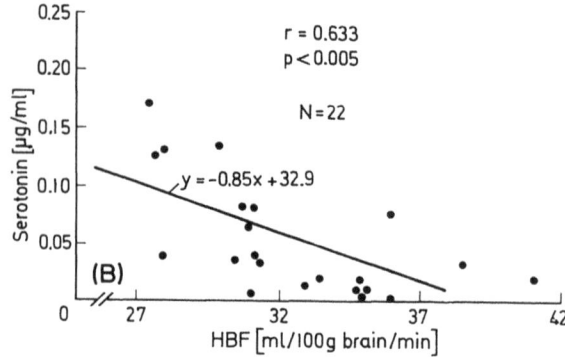

Fig. 4. (A) Shows widening of positive cerebral A-V differences for whole blood serotonin at 30 and 60 min following temporary occlusion and release of all four major extracranial arteries. Asterisks show significance of change from steady state values using the paired t-test. (B) Shows inverse correlation of levels of serotonin in CSF with hemispheric blood flow of infarcted hemisphere in patients with cerebral infarction. In normal individuals CSF serotonin values are zero or near zero

ischemic cerebral microcirculation with subsequent movement across the damaged blood brain barrier into brain tissue and CSF, such as has been shown to occur with cerebral catecholamines (Fig. 5). After displacement from its normal sites, serotonin may cause cerebral vasoconstriction and adversely affect brain perfusion. Studies in human and animal models indicate that both mechanisms (i.e., platelet release and release from brain parenchyma) play a part in the progression of cerebral infarction. Furthermore, the results of these investigations suggest that the vasoconstrictor action of serotonin may contribute to reduction of blood flow not only in the infarcted zone but also in the nonischemic cerebral hemisphere, i.e., diaschisis, thus causing impairment of collateral flow and progression of cerebral infarction. Use of antiserotonin or brain serotonin depleting agents in the prevention of the adverse pathophysiologic sequelea of cerebral infarction is currently under investigation and appears to be a promising new therapeutic approach to the problem of cerebral infarction.

In order to clarify the possible role of serotonin in the pathogenesis of cerebral infarction, a series of 233 gerbils were subjected to occlusion of one carotid artery which regularly produces hemiplegia in this animal under hypercapnic hypothermic anesthesia. One group received P-Chlorophenylalanine (PCPA) and one did not. In the PCPA-treated group, stroke was reduced from 42 to 23% (Table 1). Since PCPA principally blocks serotonin synthesis it is concluded that depletion of serotonin and its release into infarcted brain benefits the course of cerebral infarction in some way.

Measurement of brain swelling between the two groups was not significantly different, so that the benefit is not apparently related to reduction of cerebral edema (Fig. 6).

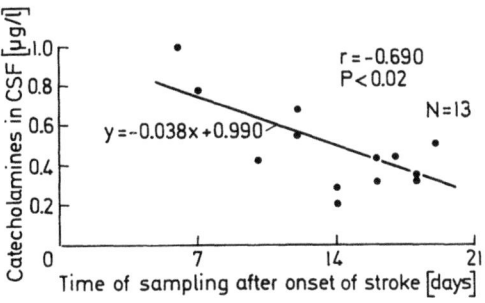
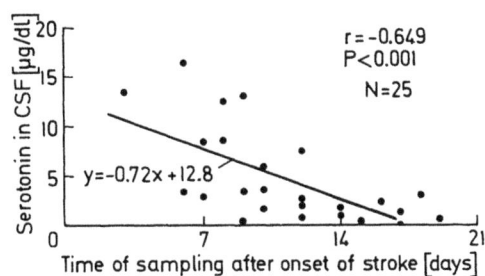

Fig. 5. Shows that serotonin and catecholamine levels in the CSF are increased in acute cerebral infarction but decrease with the passage of time, serotonin approaching zero levels 3 weeks after infarction. (Meyer et al.: Disordered neurotransmitter function demonstrated by measurement of norepinephrine and 5-hydroxytryptamine in CSF of patients with recent cerebral infarction. Brain 1974, in press)

Table 1. Comparison of effect of three separate anesthetic procedures and P-chlorophenylalanine on stroke incidence rate in gerbil

Procedure	Ether	Pentobarbital	Hypercapnic hypothermia (HH)	HH + PCPA
Stroke incidence	44%	42%	42%	23%
Number	45	29	130	103

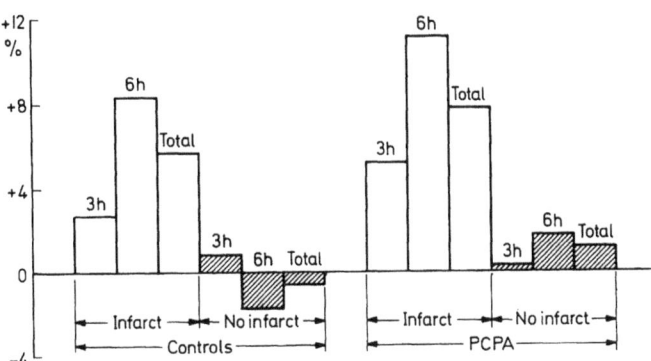

Fig. 6

References

Ames, A., III, Wright, R.L., Kowada, M., Thurston, J.M., Magno, G.: Cerebral ischemia. II. The "no reflow" phenomenon. Amer. J. Path. $\underline{52}$, 437-453 (1968)

Barnhart, M.I., Gilroy, J., Meyer, J.S.: Dextran 40 in cerebrovascular thrombosis. In: Platelet Adhesion and Aggregation in Thrombosis: Countermeasures. Mammen, E.F., Anderson, G.F., Barnhart, M.I. (eds.). Transactions of the 18th Annual symposium on Blood, Wayne State University School of Medicine, Detroit, Michigan. Stuttgart and New York: F.K. Schattauer Verlag 1970, pp. 321-342

Meyer, J.S.: Localized changes in properties of the blood and effects of anticoagulant drugs in experimental cerebral infarction. New Engl. J. Med. $\underline{278}$, 151-159 (1958)

Meyer, J.S.: Changes in cerebral blood flow resulting from vascular occlusion. In: Pathogenesis and Treatment of Cerebrovascular Disease. Fields, W.S. (ed.). Springfield, Ill.: Charles C. Thomas 1961, pp. 1-25

Meyer, J.S., Welch, K.M.A., Okamoto, S., Shimazu, K.: Disordered neurotransmitter function demonstrated by measurement of norepinephrine and 5-hydroxytryptamine in CSF of patients with recent cerebral infarction. Brain, 1974, in press

Waltz, A.G., Sundt, T.M., Jr.: The microvasculature and microcirculation of the cerebral cortex after arterial occlusion. Brain $\underline{90}$, 681-696 (1967)

Yamaguchi, T., Waltz, A.G., Okazaki, H.: Hyperemia and ischemia in experimental cerebral infarction: Correlation of histopathology and regional blood flow. Neurology $\underline{21}$, 565-578 (1971)

Effect of Agents which Modify Platelet Aggregation and/or Coagulation on Experimental Platelet Embolism and Intravascular Coagulation

M. G. Bousser, L. Bara, R. J. Prost, and M. Samama

Introduction

There is growing evidence that platelet microemboli play a role in the pathogenesis of cerebral transient ischemic attacks in their occlusive variety. However, most atherosclerotic completed strokes, and strokes in evolution, are related to the formation of a mixed thrombus on atheromatous lesions situated in the internal carotid, vertebral, or basilar arteries. This primary thrombus, which contains fibrin, can either give rise to platelet fibrin thrombi or become occluded and then provoke the formation of a propagated red thrombus which can itself obturate important collateral channels or embolize. It is therefore obvious, and already well known, that platelet aggregation and coagulation are concurrent and both play a vital role in the pathogenesis of most ischemic strokes in atheromatous patients.

The aim of this study was to compare the effects on thrombogenesis and on platelet aggregation of agents which can decrease or increase platelet aggregation and/or coagulation: two classical methods for inducing experimental thrombosis in rabbits were used:

1. The platelet thrombus formation and embolization in cortical arteries (Honour and Ross Russel, 1962).

2. The intravascular coagulation in jugular veins (Wessler, 1952).

Six agents were used: three which might potentially decrease thrombogenesis: aspirin, prostaglandin E1, heparin, and three potential thrombosis-promoting substances: adrenaline, ADP, and ellagic acid.

The results obtained are presented here together with their potential clinical implications.

Materials and Methods

Experimental Platelet Thrombosis was studied in the rabbit cerebral cortical arteries exposed and observed microscopically, as described by Honour and Ross Russel (1962). Only major injuries were inflicted: an artery, 100-200 μm in diameter, is pinched firmly with a needle pointed ophthalmic forceps, producing transient external bleeding. When the bleeding has stopped, a mural thrombus forms at the site of the injury, grows, becomes occlusive, and is then rapidly detached. This thrombus formation and embolization goes on in a steady succession for a variable length of time, sometimes several hours. This steady succession can be impaired by prior administration of a potentially antithrombotic agent.

After numerous experiments with this method, it was decided:

1. To study only injuries inflicted after or during the administration of the tested substance.

2. To study, for each substance, a number of rabbits and a number of injuries (two to four) sufficient to give statistical validity (unpaired student "t" test).

3. To use for each group of rabbits the following criteria:

    a) The number of injuries giving respectively zero (inactive injuries), one to five, and more than five emboli.

    b) The mean initial bleeding time.

    c) The mean initial "embolization time" calculated as the interval between the end of the initial bleeding and the embolization of the first thrombus.

    d) The frequency of rebleedings.

Two groups of experiments were performed in two different laboratories by two different persons: in the first, aspirin and PGE_1 were tested on at least 10 rabbits each and compared to a control group of 13 rabbits; the anesthetic used was urethane, in the second, all other agents were tested on at least 5 rabbits each and compared to a second control group of 10 rabbits; the anesthetic used was intravenous nembutal (25 mg/kg).

Platelet Aggregation was usually tested in those rabbits used for experimental platelet thrombosis. It was studied by modification of the Born (1962) and O'Brien (1962) optical density techniques on citrated PRP prepared by a slow centrifugation of blood samples, taken immediately before, 15, and 30 min after the injection, or after the beginning of the infusion of the tested substance.

1. Three clumping agents were used: (1) ADP (0.16 µg/ml), (2) a mixture of ADP (0.16 µg/ml) and adrenaline (4.1 µg/ml); (3) collagen (66 µg/ml.

2. The parameters used were: the percentage maximal fall in OD, the velocity, expressed as the fall of OD in 1 min and, in collagen-induced aggregation, the latent period expressed in seconds.

3. Each rabbit in this study was its own control. Statistical evaluation was done by the paired student "t" test.

Intravascular Coagulation was studied by the method of Wessler (1952) modified as to successively isolate four different venous segments: each external jugular vein and its external bifurcation branch. To test a potential thrombosis-promoting substance, ligations were respectively done 15 s, 2 min, and 5 min after injection of this agent. In animals pretreated with a potential antithrombotic substance, thrombosis was induced by the injection of human serum (Wessler, 1953). Each isolated segment was opened after 10 min and examined for thrombosis.

Aspirin

Experimental Platelet Thrombosis (Table 1). Aspirin (50-200 mg/kg) was injected i.v. in 15 rabbits:

1. 47 injuries were inflicted: 15 of which were inactive;
2. Embolization time was significantly increased;
3. Bleeding time and frequency of rebleedings were normal.

Platelet Aggregation

1. ADP-induced aggregation was not changed.
2. ADP-adrenaline-induced aggregation was moderately decreased (32%).
3. Collagen-induced aggregation was markedly decreased (58% fall in OD) and slowed (38% fall in velocity).

Intravascular Coagulation was studied in six rabbits:

1. Aspirin (150 mg/kg) constantly failed to inhibit serum-induced thrombus formation.

Summary: Aspirin decreased platelet thrombus formation, decreased and slowed collagen, and to a lesser extent, mixture-induced aggregations failed to prevent serum-induced venous thrombosis.

Prostaglandin E1

Experimental Platelet Thrombosis (Table 2). PGE1 infused in 12 rabbits, at rates of 1-5 µg/kg/min, was found to be a very powerful inhibitor of thrombus formation:

1. All 15 injuries inflicted during the infusion were totally inactive for as long as PGE1 was being infused.
2. Bleeding time and frequency of rebleedings were unchanged.

Platelet Aggregation (Table 3) induced by all agents, was significantly decreased but not totally inhibited. Platelet count was not modified.

Intravascular Coagulation (Table 4) was induced in five rabbits 10 min after the onset of the infusion of PGE1 (5 µg/kg/min):

1. At 15 s and 1 min thrombosis was present in 4, absent in 1.
2. At 2 min thrombosis was present in 3, absent in 2.

This very weak inhibition is not significant because of the small number of animals studied.

Summary: PGE_1 totally inhibited platelet thrombus formation, partially inhibited platelet aggregation induced by all three agents and failed to prevent serum-induced venous thrombosis.

Heparin

Experimental Platelet Thrombosis (Table 5)

1. Heparin (1000 U/kg) was injected in four rabbits and did not change the pattern of behavior of injuries.
2. Heparin (4000 U/kg) was injected in only two rabbits: Thrombus formation, bleeding time, and embolization time were normal but rebleedings were frequent and occurred at the site of the injury nearly every time white bodies embolized. Moreover, white bodies looked small and instable with small fragments being detached, and they often embolized before becoming occlusive.

Platelet Aggregation. At both doses, the only change was a decrease and delay in collagen-induced aggregation, 15 and 30 min after injection.

1. Platelet count was slightly, but not significantly decreased 2 min after injection.

Intravascular Coagulation. Heparin at doses as low as 500 U/kg was found to constantly inhibit serum-induced intravascular coagulation.

Summary: Heparin at high doses (1) did not prevent platelet thrombus formation but decreased the stability of white bodies and hemostatic plugs, (2) decreased and delayed collagen-induced aggregation, and (3) constantly prevented serum-induced venous thrombosis.

Adrenaline (1 µg/kg)

1. Did not modify the pattern of behavior of injured cortical arteries.
2. Failed to alter mixture and collagen-induced aggregation but did enhance ADP-induced aggregation, 15 and 30 min after injection.
3. Did not induce thrombosis in venous segments.

ADP

Experimental Platelet Thrombosis

1. ADP (30 mg/kg) was injected intravenously for 3 min into three rabbits, nine injuries were inflicted less than 40 min after injection. The only change was decrease in thrombus formation with four inactive injuries and five producing one to five white bodies.

Platelet Aggregation (Table 6)

1. 5 to 10 min after injection, aggregation induced by all three agents was significantly decreased and the latency of collagen-induced aggregation increased.
2. 30 min after injection, the mixture-induced aggregation was still inhibited, whereas the others were back to normal.

Platelet Count was significantly decreased 30 s after injection.

Intravascular Coagulation. ADP (30 mg/kg) injected in six rabbits, and (12 mg/kg/h), infused in six rabbits, always failed to induce thrombosis.

Summary: ADP (1) decreased thrombus formation in cortical arteries, (2) 5-10 min after injection, decreased aggregation-induced by all three agents, (3) failed to induce thrombosis in isolated venous segments.

Ellagic Acid

Experimental Platelet Thrombosis (Table 7). Ellagic acid (10 mg/kg) was injected i.v. into 3 rabbits. There were 14 injuries inflicted, eight of them less than 20 min after the injection and others later. These two groups of injuries had very different patterns of behavior.

The earliest injuries produced very few emboli: one produced none, five produced one, and two produced two. Although the injuries stopped bleeding normally 41.6 (SD: 38.5, $P < 0.90$) with a mural thrombus appearing rapidly, the occlusive thrombi which were formed, instead of embolizing rapidly, used to stay in the arteries for several minutes (317.5) (SD: 208, $P < 0.001$) which is a very rare phenomenon in controls.

By contrast, injuries inflicted more than 20 min after injection had a normal behavior.

Ellagic acid (2 mg/kg/min) was infused into three rabbits. Again early and late injuries had different patterns of behavior, but this time early injuries behave normally, whereas late injuries produced these unusually stable white bodies, whose embolization time was very significantly increased 321.4 (SD: 202, $P < 0.01$).

There was one rebleeding in the first group of injuries and two in the second.

Platelet Aggregation and Platelet Count (Table 8)

1. Mixture and collagen-induced aggregations were very significantly decreased and slowed down 5 min after injection ($P < 0.001$).
2. ADP-induced aggregation was decreased but not significantly ($P < 0.10$). Platelet count was also decreased ($P < 0.02$).

Intravascular Coagulation

Ellagic acid (5 mg/kg) was injected i.v. into 15 rabbits. Thrombosis was present in 10 segments, 15 s after the injection, in 14 segments at 2 min after the injection, and in 8 segments at 5 min after the injection. Glass and silicone coagulation times were significantly decreased 2 and 5 min after the injection, but were back to normal 15 min after (Table 9).

Summary: Ellagic acid (1) decreased thrombus formation in arteries but increased thrombus stability, (2) decreased aggregation induced by all three agents, (3) shortened coagulation time and induced thrombosis in venous segments.

Discussion and Conclusion

The results of these experiments have led to a better understanding of the events occurring in these two models of experimental thrombosis.

Thrombosis in isolated venous segments was induced by activators of coagulation (ellagic acid, which is an activator of Hageman factor (Ratnoff and Crum, 1964), but not by aggregating agents (ADP-adrenaline). Similarly, serum-induced thrombosis was inhibited by heparin but not by inhibitors of platelet aggregation such as PGE_1 and aspirin. This confirms that thrombus formation is, in this method, very similar to intravascular clotting, and therefore constitutes a good model to study the effect of drugs on the prevention of venous thrombosis. In this respect, although it was always difficult to draw from animal studies, valid conclusions about human pathophysiology, these results are in agreement with those of the Medical Research Council (1972) which were unable to show any effect of aspirin in the prevention of human venous thrombosis. This, together with the proven efficacy of anticoagulants, suggests that antiaggregant drugs should not be used alone in the prevention of human venous thrombosis.

Thrombosis in injured cortical arteries seems to be a much more complex phenomenon since all agents tested had different effects on the various criteria used.

Initial bleeding time was not significantly affected by any of these agents. This is probably due to the fact that bleeding time varies widely from injury to injury because the stimulus itself - pinching an artery - is most variable. This leads to a very high standard deviation of the mean of bleeding times and explains the apparent lack of effect of agents like aspirin. However, it is of interest, that, despite a remarkable inhibitory effect on thrombus formation, PGE_1 was not found to increase the bleeding time as already mentioned by Emmons et al. (1967). This contrasts with the results obtained by Kinlough et al. (1970) in transsected mesenteric vessels of rabbits injured at 10 µg/kg/min which is over twice as high as the dose we used. Although the use of PGE_1 itself in man is not possible, it is important to show that a substance is able, at least in rabbits, to totally prevent platelet thrombus formation without significantly enhancing the primary bleeding time.

Initial visible thrombus formation - which is the reverse of the number of inactive injuries - was decreased by aspirin and ADP, totally inhibited by PGE_1, but unchanged by heparin, adrenaline, and ellagic acid.

These results are in agreement with literature as far as aspirin, PGE_1 and heparin are concerned and they correlate well with the histologic observation that white bodies are formed mainly of platelets (Honour and Ross Russell, 1962). This experimental situation seems to be very similar to the formation in retinal or in cortical arteries of platelet aggregates, sometimes responsible for monocular blindness or transient ischemic attacks. This seems to be confirmed by some reports showing the efficacy of aspirin in the prevention of these two conditions in man (Harrison et al., 1971; Mundahl et al., 1972; Dyken et al., 1973).

The absence of potentiation by adrenaline might be due either to the fact that aggregation - except ADP-induced aggregation - was not much increased as already shown by Sinakos and Caen (1967), or to the method itself in which a decrease in thrombus formation is easy to prove whereas an increase might not be demonstrable, at least with the parameters used.

The inhibitory effect of ADP on thrombus formation contrasts with the results of numerous experiments in which ADP has been found to induce the formation of platelet aggregates. But it has been shown that the formation and the resulting changes are maximal 30 s after injection (Jørgensen et al., 1970; Kobayashi and Didisheim, 1973). Later, as shown in our experiments, there is a profound inhibition of platelet aggregation, probably due to the formation of adenosine, an ADP degradation product which is known to inhibit platelet aggregation and to decrease the formation of platelet thrombi in injured rabbit cortical arteries (Born et al., 1964).

Thrombus formation, which therefore seems to be similar to platelet aggregation in vivo, was compared to platelet aggregation studied in vitro on blood samples collected at the same time. Though there was a good correlation between these two phenomenon, some discrepancies could be noted, particularly in PGE_1-treated animals: some animals had a complete inhibition of thrombus formation in spite of a normal aggregation in vitro. This emphasizes once more the need to study experimental thrombosis as well as in vitro tests of human platelet behavior in the assessment of potential antithrombotic agents.

Initial embolization time was not modified by heparin and ADP, but was increased by aspirin and ellagic acid. But whereas, in aspirin-treated animals white bodies once formed, broke off from the injured vessel wall, in ellagic acid-treated animals, they remained in arteries. The increase in embolization time has thus two different meanings: (1) in the aspirin group, it is the formation of thrombi which seemed to be delayed; (2) in the ellagic acid group, thrombus formation looked normal but there was a stabilization of white bodies. The initial embolization time should therefore be divided in two parameters: (1) thrombus formation time in which platelet aggregation plays the major role, (2) occlusion time in which coagulation is the main factor.

The frequency of rebleedings was not modified by aspirin and PGE_1 but was severely increased by heparin. This suggests that thrombus stability depends mainly on coagulation mechanism, and that fibrin must therefore be present very early in white bodies, although platelets without fibrin could be seen be electronmicroscopy by Honour and Ross Russel (1962). The fact that ellagic acid increases plug stability might lead to the development of an interesting experimental method with which other studies are under way and which involves the four main factors occurring in thrombosis: (1) blood flow; (2) endothelial injury; (3) platelet adhesion/aggregation; and (4) coagulation.

Table 1. Aspirin. Experimental platelet thrombosis

	Number of injuries				Initial embolisation time		
	Total	Inactive	Producing 1.5 W. Bodies	Producing > 5 W.B.	Mean	SD	p
Control group 13 rabbits	45	1	14	30	109.6	51.7	
Aspirin group 50-200 mg/kg 15 rabbits	47	15	22	10	198	137.7	< 0.001

Initial bleeding time and frequency of rebleedings: normal

Table 2. Prostaglandin E1. Experimental platelet thrombosis

	Number of injuries		Initial bleeding time			Rebleedings
	Total	Inactive	Mean	SD	p	
13 controls	45	1	18,5	11,4		5
12 PGE_1 1-5 µg/kg/min	15	15	24.2	23.1	< 0.40	0

Table 3. Prostaglandin E1 and platelet aggregation

Agent		Before infusion		During infusion		p
		Mean	SD	Mean	SD	
ADP 0.16 µg/ml	OD	31.11	5.33	19.22	4.82	< 0.01
ADP/adrenaline 0.16/4.1	OD	48.55	4.80	32.88	5.63	< 0.001
Collagen 66 µg/ml	Latency	185	38	33.2	69	< 0.20
	OD	44.11	6.71	35.33	6.70	< 0.02

Table 4. Prostaglandin E1 and intra-vascular coagulation

	Number of segments examined at each time	Number of thrombosed venous segments before and after human serum injection (1.32 ml/kg)			
		Before	15 sec after	2 min after	5 min after
Controls (saline)	6	0	6	5	0
PGE1 5 μg/kg/min	5	0	4	3	0

Table 5. Heparin and experimental platelet thrombosis

	Number of injuries				Initial bleeding time			Initial embolisation time			Rebleeding
	Total	Inactive	Producing 1.5 WB	Producing > 5 WB	Mean	SD	P	Mean	SD	P	
10 controls	33	3	11	19	41.8	41.9		131.1	84.1		0
4 Heparin 1000 U/kg	10	0	8	2	88	86.7	<0.05	114	69	<0.60	0
2 Heparin 4000 U/kg	5	0	4	1	62	46.9	<0.40	94	61	<0.40	multiple
	White bodies looked small, instable with small fragments being detached often embolising before becoming occlusive										

Table 6. ADP (30 mg/kg i.v.) and platelet aggregation

Agent		Before injection		5-10 min after injection			30 min after injection		
		Mean	SD	Mean	SD	P	Mean	SD	P
ADP 0.16 µg/ml	OD	24.6	4.5	15.5	3.9	< 0.01	23	5.6	< 0.60
ADP 0.16 adrenaline 4.1	OD	44.8	6	29.8	3.7	< 0.05	28.1	4.4	< 0.02
Collagen 66 µg/ml	latency	85	26.1	170	50.5	< 0.01	130	50.5	< 0.30
	OD	46	5.5	29.7	7.1	< 0.02	34.3	7.9	< 0.20

Table 7. Ellagic acid experimental platelet thrombosis

	Number of injuries				Bleeding time			Initial embolisation time			Rebleedings
	Total	Inactive	Producing 1.5 WB	Producing >5 WB	Mean	SD	P	Mean	SD	P	
10 Controls	33	3	11	19	41.8	41.9		131.1	84.1		0
3 Ellagic 10 mg/kg											
< 20 min	8	1	7	0	41.6	38.5	> 0.90	315.7	208	< 0.001	0
> 20 min	6	0	2	4	53.3	64	< 0.40	96.5	42.2	< 0.40	0
2 Ellagic 22 mg/kg/min											
< 15 min	5	0	2	3	59	79	< 0.40	172	71.5	< 0.40	1
> 15 min	8	1	7	0	63.1	56.9	< 0.30	321.4	202	< 0.001	2

Table 8. Ellagic acid (10 mg/kg) and platelet aggregation

		Before injection		5 min after injection		
		Mean	SD	Mean	SD	P
ADP 0.16 µg/ml	OD	32.5	4.03	25.5	4.68	< 0.10
ADP 0.16 adrenaline 4.1 µg/ml	OD	56.4	3.7	35.5	4.9	< 0.001
Collagen 66 µg/ml	latency	78	14.3	90	14.1	< 0.50
	OD	54.5	5.1	34.2	4.4	< 0.001
Platelet count		366000	17.900	329000	14.791	< 0.02

Table 9. Ellagic acid and intravascular coagulation

	Number of segments examined at each time	Number of thrombosed venous segments before and after injection			
		Before	15 s after	2 min after	5 min after
Human serum 1.32 ml/kg	6	0	6	5	0
Ellagic acid 5 mg/kg	15	0	10	14	8

References

Born, G.V.R.: Quantitative investigation into aggregation of blood platelets. J. Physiol. (Lond.) 162, 67 (1962)

Born, G.V.R., Honour, A.J., Mitchell, J.R.A.: Inhibition by adenosine and by 2 chloroadenosine of the formation and embolisation of platelet thrombi. Nature (Lond.) 202, 761 (1964)

Dyken, M.L., Kolar, O.J., Haven Jones, F.: Differences in the occurrence of carotid transient ischemic attacks associated with antiplatelet aggregation therapy. Stroke 4, 732 (1973)

Emmons, P.R., Hampton, J.R., Harrison, M.J.G., Honour, A.J., Mitchell, J.R.A.: Effect of Prostaglandin E_1 on platelet behaviour in vitro and in vivo. Brit. med. J. II, 468 (1967)

Harrison, M.J.G., Marshall, J., Meadow, J.C., Ross Russel, R.W.: Effects of aspirin in amarirosis fugax. Lancet 2, 743 (1971)

Honour, A.J., Ross Russel, R.W.: Experimental platelet embolism. Brit. J. exp. Path. 43 (4), 350 (1962)

Jørgensen, L., Hovig, T., Rowsell, H.C., Mustard, J.F.: ADP induced platelet aggregation and vascular injury in swine and rabbits. Amer. J. Path. 61, 161 (1970)

Kinlough, R.L., Packham, M.A., Mustard, J.F.: The effect of Prostaglandin E_1 on platelet function in vitro and in vivo. Brit. J. Haemat. 19, 559 (1970)

Kobayashi, I., Didisheim, P.: Systemic effects of ADP-induced platelet aggregation and their modification by aspirin and by pyridinolcarbamate. Thrombos. Diath. haemorrh. (Stuttg.) 30, 178 (1973)

Medical Research Council: Effect of aspirin on post-operative venous thrombosis. Lancet II, 441-444 (1972)

Mundahl, J., Quintero, P., von Kaulla, K.N., Harmon, R., Austin, J.: Transient monocular blindness and increased platelet aggregability treated with aspirin. A case report. Neurology 84, 525 (1972)

O'Brien, J.R.: Platelet aggregation. Some results of a new method of study. J. clin. Path. 15, 452 (1962)

Ratnoff, O.D., Crum, J.D.: Activation of Hageman factor by solutions of ellagic acid. J. Lab. clin. Med. 63, 359 (1964)

Sinakos, Z., Caen, J.P.: Platelet aggregation in Mammalians. Thromb. Diath. haemorrh. (Stuttg.) 17, 99 (1967)

Wessler, S.: Studies in intravascular coagulation. I. Coagulation changes in isolated venous segments. J. clin. Invest. 31, 1011 (1952)

Wessler, S.: Experimental intra vascular thrombosis induced by serum fractions containing serum prothrombin conversion accelerator. Fed. Proc. 12, 152 (1953)

Platelet Abnormalities in Cerebrovascular Diseases

C.A. Bouvier

If I try to remember my first experience as a young resident in neurology, I immediately think of the extraordinary logical precision of the neurologic semeiology. A correct knowledge of the CNS neuroanatomy made it possible to logically locate the lesion by systemic grouping of symptoms, with a precision unknown to other fields of internal medicine.

Already at this time, EEG, Gazeous encephalography, and radiology, as well as examination of liquor cerebri by puncture, helped to confirm the results of the semeiologic investigation.

In the past 20 years, neurologic investigation has made striking progress: at present, selective arteriography, cerebral scintigraphy, gamma-encephalography, cerebrography, etc. are of current use in every major hospital. In other terms, our diagnostic possibilities in cases of CNS pathology are practically unequalled in any other medical discipline.

Meanwhile, considerable therapeutic progress has been accomplished in diseases like Parkinson's, because of the important acquisitions in pharmacologic knowledge, or for cerebral tumours because of earlier detection and the growing skill of surgeons.

But for the largest group of CNS diseases, those of vascular origin, I feel that the therapeutic management has not changed at all since the days of my internship. This implies that while the neurologic diagnosis was becoming more and more exquisitely precise, no real progress in therapy has been achieved. The fear of hemorrhage and the fact that high blood pressure is frequently associated with cerebral accidents preclude the use of anticoagulant drugs like heparin and coumadin derivatives. There is nothing like a specific cerebral vasoactive drug, and systemic vasodilators imply a serious risk of systemic hypotension, diminution of cerebral flow, and something like a steal syndrome. So here we stay, with papaverine and carbogen like 20 years ago. Furthermore, there is not the smallest bit of doctrine concerning the prevention of cerebrovascular accidents.

This state of affairs is obviously related to our ignorance of the mechanisms of thrombogenesis in arteries, and at this point in time, it is worth heeding some new information on the early mechanisms in arterial thrombotic disease.

Some of those mechanisms are concerned so early in the thrombotic process that they might never achieve what a pathologist would call "thrombus," much less determine a complete vascular occlusion. But they are the responsible mechanism for incipient thrombogenesis, and may explain transient episodes of cerebral ischemia and of microembolic manifestations. Moreover, their knowledge may open new therapeutic approaches to the management and to the prevention of cerebrovascular accidents.

This knowledge is the fruit of a confrontation which should have taken place long ago, a confrontation between the people interested in *blood clotting* and those concerned with *blood vessels' pathology*.

At present it seems to be a nonsense that these two divisions of general pathology were kept so much apart of each other for such a long time. The beneficiary of this encounter has undoubtedly been the blood platelet, and its relationship with the vessels' intimal surface. Circulating platelet behavior in the immediate proximity of an intimal lesion has become an object of intensive study, and already in 1965 Emmons et al. (1965a) studied the formation of "white bodies" (platelet aggregates) at the level of a small experimental lesion of meningeal arteries in rabbits. They also demonstrated that those aggregates were able to further embolize in smaller branches, and that both white body formation and embolization could be prevented by administration of dipyridamole, the effect of which had already been demonstrated on platelet aggregation in vitro. This led many workers to study platelet reactivity with various in vitro systems, and prompted us to use a modification of Hellems' trapping device as a tool for widespread clinical investigation (Rosner et al., 1967). Some years ago, we presented evidence that platelet behavior is highly abnormal in myocardial infarction, and that administration of persantin during 10 days determines rapid normalization of this parameter, while untreated patients maintained an abnormal pattern. Many different methods, though sometimes difficult to compare with each other, have led to the same conclusion: many pathologic vascular conditions are accompanied by hyperreactivity of platelets in vitro and probably in vivo. This in turn stimulated experimental and pharmacologic research on other drugs able to induce a "controlled thrombasthenia" (Boneu et al., 1972) in order to prevent arterial thromboembolic disease, whereby formation of platelet aggregates at the level of an endothelial lesion is the first stage of the disease.

In this line of thought the most remarkable clinical application was proposed and evaluated by Harker and Slichter (1970) in patients with prosthetic heart valves. Such patients had a high rate of cerebral emboli and though the number of circulating platelets was usually normal, because their bonemarrow was able to produce a higher number in response, an extremely short half-life (or rapid turnover) of autologous isotope-labeled platelets was observed. In this clinical trial both turnover and half-life could be normalized by the administration of 400 mg of persantin daily, or 3 g of aspirin, or even better by the combination of both drugs in smaller amounts. Moreover, the incidence of cere-

Table 1. Abnormal platelet trapping (PT) in patients with cerebral accidents of vascular origins

Patients	No.	PT > 50%	% of No.
	144	102	64%

Mean PT: 57.6% (20-92); N = 42% ± 8

bral emboli was definitely lessened as confirmed by other studies (Sullivan et al., 1968; Matlof et al., 1969; Kummer et al., 1974; Wyss et al., 1974).

In a recent investigation Franceschetti et al. (1974) were able to demonstrate a striking improvement and recuperation of vision after arterial and venous retinal thrombosis when heparin and the combination of aspirin and persantin were administered as soon as possible. Moreover, the onset of secondary glaucoma was definitely reduced. On admission, 74% of those patients had pathologically increased platelet trapping in our system and in most cases correction was obtained and maintained after 10 days treatment.

During the years 1973 and 1974 we have investigated platelet behavior in more than 200 patients admitted to the Neurology Division of Geneva University Hospital. Most patients were controlled several times. After eliminating nonvascular neurologic affections (discal hernia, cerebral tumor, peripheral neuritis, multilocular sclerosis, etc.) 144 patients remain with cerebral accidents of vascular origin. There are 102 (70%) who have been found to have a definitely abnormal platelet trapping (more than 50% trapped platelets). The overall mean is 56.7% trapped platelets, which differs significantly from our normogram established for healthy donors with the same mean age (42% ± 8). If we distribute our patients into age groups, 18 patients over 70 years divide into 11 with pathological increment of platelet reactivity and 7 with normal (Table 2).

In all aged cases the CNS involvement was definite and permanent and related with important modifications of cerebral arteries when investigated. In the group 51-70 (82 patients), only 25 had a normal platelet test while 57 (69%) were pathologic (Table 3). Only a small group (12) had transient episodes.

If we turn to the younger age group (less than 50 years) of 42 patients examined (Table 4), 27 (64%) had a pathologic test at least on two successive occasions after admission and before treatment. In this group 18 had a transient episode with complete restitution. They have all been treated with the combined therapy schedule, and Figures 1 and 2 show the results of repeated controls of platelet behavior in two cases. It can be seen that with our eldest case, started with persantin alone, aspirin test has twice shown a good response, and that the combination of the two drugs seems to have controlled platelet reactivity on a very long term. The second case has been followed for only a few months. She is very young (36) and I would like to stress that those par-

Table 2. Patients over 70 years

No.	PT > 50%	PT < 50%
18	11	7

All CNS symptoms permanent

Table 3. Patients 51-70 years

No.	PT > 50%	PT < 50%
82	57 (69%)	25

Only 12 transient CNS symptoms

Table 4. Patients under 50 years

No.	PT > 50%	PT < 50%
42	27 (64%)	15

18 with transient CNS symptoms (reversible)

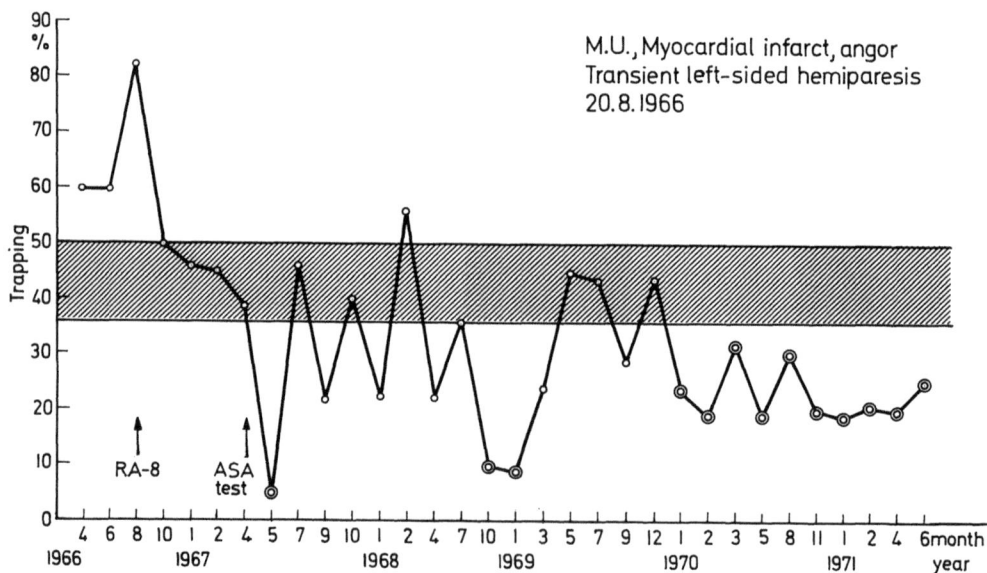

Fig. 1

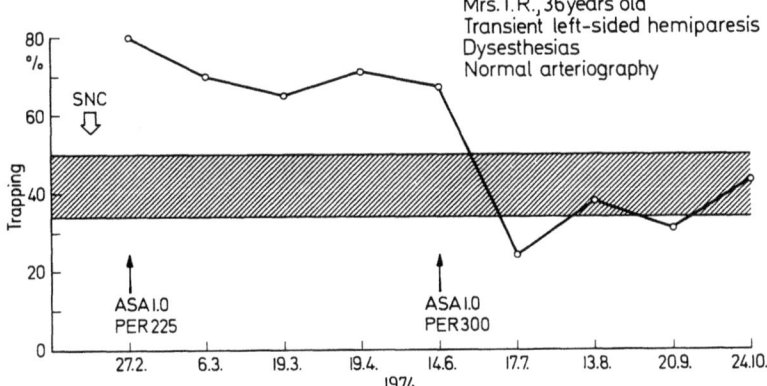

Fig. 2

ticular cases whereby a transient, early episode has been a serious warning should deserve all our attention in terms of long-term prevention. This is especially so when, as in this last case, no gross arterial pathology can be detected on arteriography. Obviously a correlation must be searched with abnormal lipid blood levels and latent diabetes, as will be dealt with in another publication. Meanwhile, many more patients should be followed on long term, both for their neurologic evolution and platelet abnormal behavior before we can obtain a proper evaluation of the efficiency of our drug combination.

References

Boneu, B., Guiraud, B., Fernet, P.: Traitement antiagrégant par l'aspirine. Nouv. Presse méd. 1, 863 (1972)

Emmons, P.R., Harrison, M.J.G., Honour, A.J., Mitchell, J.R.A.: Effects of a Pyrimido-pyrimidine Derivative on Thrombus formation in the rabbit. Nature (Lond.) 208, 255-257 (1965a)

Emmons, P.R., Harrison, M.J.G., Honour, A.J., Mitchell, J.R.A.: Effects of Dipyridamole on human platelet behaviour. Lancet II, 603-606 (1965b)

Franceschetti, A.Th., Bouvier, C.A., Scheppens, J.M.: Troubles du Comportement plaquettaire en pathologie vasculaire rétinienne. Ophtalmologica 172, 259-263 (1976)

Harker, L.A., Slichter, S.J.: Studies on platelet and fibrinogen kinetics in patients with prosthetic heart valves. New Engl. J. Med. 282, 1302 (1970)

Kummer, H., Hunziker, H.R., Althaus, U.: Medikamentöse Beeinflussung des Thrombozytenumsatzes bei Patienten mit künstlichen Herzklappen. Schweiz. med. Wschr. 104, 142-144 (1974)

Matlof, J.M., Collins, J.J., Sullivan, J.M., Gorlin, R., Harken, D.E.: Control of thromboembolism from prosthetic heart valves. Ann. Thorac. Surg. 8, 133-145 (1969)

Rosner, P., Bouvier, C.A., Berthoud, S.: Mesure de l'adhésivité plaquettaire au verre (modification de la méthode de Hellem). Nouv. Rev. franç. Hematol. 7, 185-194 (1967)

Sullivan, J.M., Harken, D.E., Gorlin, R.: Pharmacologic control of thromboembolic complications of cardiac valve replacement. New Engl. J. Med. 279, 576 (1968)

Wyss, M., Diez, C., Marron, H., Bouvier, C.A.: Traitement par les inhibiteurs de l'adhésivité plaquettaire chez des enfants porteurs de valves cardiaques artificielles. (1974, in press)

Inhibition of Platelet Aggregation by Synthetic Organic Acids: Quantitative Relationships Between Chemical Structures and Biological Activities

V. Čepelák, M. Kuchař, B. Brůnová, J. Murotová, and Z. Roubal

In the abstracts of the present communication, the antiaggregation effects of various drug groups were compared quantitatively in the light of some of our findings published earlier (Cepelák, 1971, 1974; Cepelák et al., 1971, 1972a,b) our present paper, on the other hand, is limited to a single large group of compounds.

From the chemical point of view, the compounds to be discussed are organic acids, embracing a majority of nonsteroidal antiinflammatory agents presently known. Our choice of this group to be lectured about is based on three reasons, as follows.

1. We wish to emphasize, and to illustrate in more detail from the pharmacologic point of view, the close relationship between the pathogenetic mechanisms of thrombosis and of inflammation proclaimed by Nachman in the introductory section.

2. Many of the compounds concerned have been clinically used for long-term treatment of rheumatic disease; a distinct effect on platelet functions doubtlessly is exerted at plasmatic concentrations of the agents well available with clinical dosages.

3. In a selected series of compounds belonging to the group of organic acids, we quite recently succeeded in revealing quantitative relationships between the chemical structures and inhibitory effects on platelet aggregation as well as other biological activities.

Such biological activities include, besides the antiinflammatory activity, also the capacity of a direct induction of plasmatic fibrinolysis, as tested in the hanging clot test according to von Kaulla. Recently, Kluft and Brakman (1974) identified the mechanism of the above effect as blockade of the C1-inactivator, an important natural inhibitor of the plasmatic fibrinolytic system. Besides activation of fibrinolysis, the compounds in question exhibit many additional activities in vitro, as follows: stabilization of erythrocytary membrane against hypotonic osmotic hemolysis (Kalbhe et al., 1970), inhibition of erythrocyte sedimentation and aggregation induced by various factors (Kovács and Görög, 1972), and stabilization of plasmatic albumin against thermal denaturation (Mizushima, 1964).

Table 1

Compound	Inhibition of platelet aggregation plasma conc. (mM)				Hanging clot lowest conc. (mM) inducing complete lysis
	ADP 50% max. deflection	Thrombin	Adrenaline 2nd wave absence	Collagen 50% slope	
Aspirin	10,4	2,0	0,2	1,0	-
Indomethacin	3,3	2,5	1,4	1,6	6
Flufenamate	1,8	2,7	0,5	1,2	2
Mefenamate	3,3	6,2	1,0	2,3	2
Ibufenac	11,7	3,0	0,5	0,5	2
Phenylbutazone	3,5	1,6	0,5	0,8	8
Kebuzone	6,7	5,4	0,5	2,0	40
Trimethazone	5,2	3,0	0,3	0,3	3
Benzopyrazone	2,6	1,9	1,4	1,2	4
VUFB 4671	1,3	1,2	1,3	1,3	2
FGY 161	0,5	0,6	1,7	0,8	-
FGY 215	0,2	0,3	1,5	0,6	-

Table 1 shows the equieffective concentrations of compounds inhibiting platelet aggregation mediated by various mediators (ADP, thrombin, adrenaline, and collagen). Besides well-known nonsteroidal antiinflammatory drugs, the table also presents some of pyrazolidine derivatives synthesized within our research program. The most effective compounds inhibit platelet aggregation in concentrations lower than 1 mM, attainable with clinical dosages of the drugs concerned. At the concentrations mentioned, inhibitory effects are only detectable in collagen-induced aggregation and in the so-called secondary aggregation, induced by adrenaline in our experiments. Synthetic modifications of the molecule in the pyrazolidine series also led to compounds with intensified inhibitory activities against ADP-induced primary aggregation, such as the compound VUFB 4671 - omega-furyl derivative of kebuzone; FGY 215 - ethyl ester of di-(pyrazolidinyl)aliphatic acid. The endeavor for syntheses of maximally effective compounds in the pyrazolidine series, unfortunately, has also led to less soluble and sometimes even more toxic derivatives.

Nevertheless, from the new pyrazolidine derivatives there were successfully selected compounds with marked antiinflammatory activities and low toxicities, documented also by clinical trials. Of such compounds, the most remarkable one is the trimethyl derivative of kebuzone, named trimethazone, that not only in vivo but also in vitro belongs to the strongest activators of fibrinolysis and inhibitors of collagen-induced aggregation. In the hu-

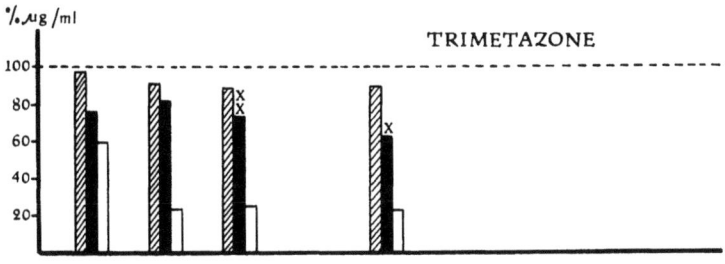
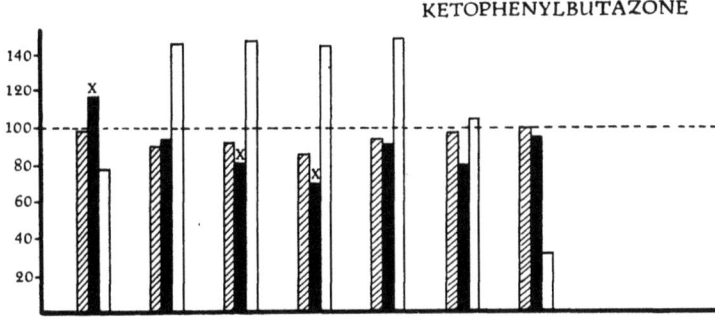
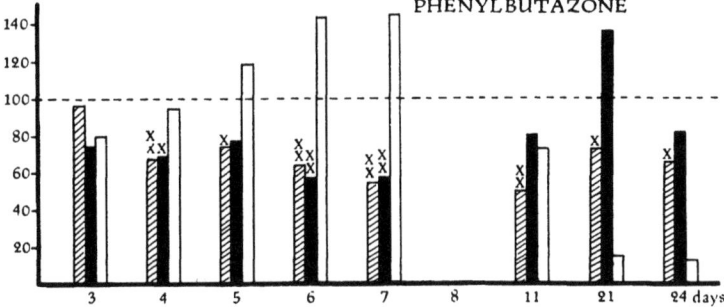

Fig. 1. Collagen-induced aggregation in a group of five volunteers after a week's oral application of trimethazone (3 × 250 mg), ketophenylbutazone (3 × 250 mg), and phenylbutazone (3 × 100 mg daily). First column (▨) indicates relative changes of maximal deflection, second column (■) maximal slope of aggregation curve. Significant differences from pretreatment values were tested on 5% (x) and 1% (^x_x) level. Third column (☐) indicates serum level of the drugs

man organism, a significant inhibition of collagen-induced aggregation takes place even with relatively low levels of trimethazone in the blood serum (Fig. 1).

Trimethazone has also shown a significant preventive activity in experimental arterial thrombosis in laboratory animals. In the first type of experimental thrombosis (Table 2), a platelet thrombus forms on platinum electrodes introduced into rabbit carotid artery; the size of the thrombus is estimated by means of conductivity measurement (Muratová et al., 1972). In rats,

Table 2. Experimental arterial thrombosis in rabbits

Site of thrombosis	Carotid artery
Principle	Formation of white thrombi on platinum electrodes
Drugs applied	1. Pyrazolidine derivatives 2. Derivatives of β-arylaliphatic acids } 40 mg/kg 3. Dipyridamol 10 and 40 mg/kg 4. Physiological saline
Mode of application	i.v., 10 min before production of thrombosis
Evaluation	Decrease in conductivity between electrodes

Table 3. Experimental arterial thrombosis in rats

Site of thrombosis	Carotid artery
Principle	Damage of arterial wall by electrical current (3 mA, 1.5 min)
Drugs applied	1. Pyrazolidine derivatives 2. Derivatives of β-arylaliphatic acids } 40 mg/kg 3. Dipyridamol 10 and 40 mg/kg 4. Physiological saline
Mode of application	i.v., 10 min before production of thrombosis
Evaluation	Decrease in temperature of the vessel

carotid arterial thrombosis (Table 3) is elicited by electrocoagulation, and the decrease in the temperature of the artery is recorded distally from the occlusion site (Hladovec, 1971). The effective drugs reduce the extent of thrombosis on the electrodes with resulting changes in conductivity in the first type (Fig. 2), and retard the occlusion formation and the resulting temperature decrease in the second type (Fig. 3). Apart from pyrazolidine derivatives, other antiaggregation agents have also been tested in both models (Table 4), especially the organic acids studied by us at present.

Our choice of the organic acids for more profound studies was motivated by their better solubility in comparison with pyrazolidines as well as by their better susceptibility to synthetic modifications of their molecules with a view to studies of the structure-activity relationships. In the series of derivatives of cinnamic and beta-arylaliphatic acids, we determined their equief-

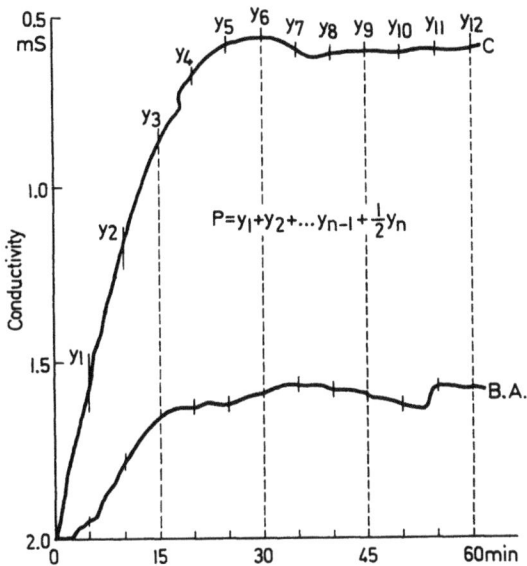

Fig. 2. Record from conductometric examinations performed in a control rabbit C and in another premedicated with one of the organic acids tested B.A.

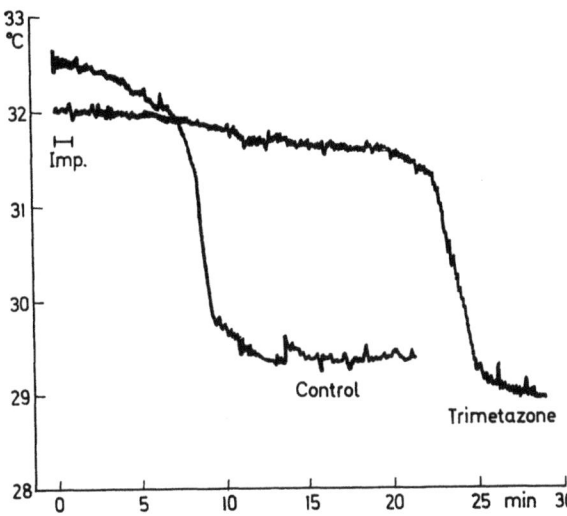

Fig. 3. Record from thermometric examinations performed in carotis of a control rat and of another premedicated with trimethazone (imp. - duration of electrical impulse)

fective concentrations for inhibition of individual types of platelet aggregation, fibrinolytic effect, inhibition of osmotic hypotonic hemolysis, and stabilization of serum albumin. The appropriate experimental findings were correlated with physicochemical parameters characterizing the lipophilic (π), electronic (σ, σ^*), and steric (E_s) effects of variable substituents of the acids studied. The quantitative relationships between the chemical structures and the several biological activities were expressed by multiparametric regression equations according to Hansch (1971).

Table 4. Prophylactic effect of drugs on arterial thrombosis

	Rabbits Change in conductivity (%)	Rats Time of temperature decrease (min)
Control (physiol. saline)	100	8
Phenylbutazone	88	14
Trimethazone	74	23
Derivatives of β-arylaliphatic acids	32 - 78	8 - 20
Dipyridamole 10 mg/kg	78	9
40 mg/kg	80	6

Values of conductivity below 80% and times of temperature decrease over 14 min were significantly different from those obtained in control animals

Table 5. Series of β-aryl aliphatic acids

$$X\text{—}\underset{R}{\underset{|}{C_6H_4\text{—}CH}}\text{—}CH_2\text{—}COOH$$

Type	Substituent R	Substituent X
β-aryl-n-butyric	CH_3	H, p-CH_3, p-C_2H_5, p-iC_3H_7, p-iC_4H_9, p-Cl, p-Br, m-Br, m-CF_3, p-CH_3O
β-aryl-n-valeric	C_2H_5	H, p-CH_3, p-i-C_4H_9, p-Cl
β-aryl-n-caproic	n-C_3H_7	H, p-CH_3, p-Cl
β-aryl-i-caproic	i-C_3H_7	H, p-CH_3, p-i-C_4H_9

Our studies were done in a series of beta-arylbutyric and other beta-arylaliphatic acids (Table 5). These compounds exhibit antiinflammatory activities in animal experiments and are chemically related to the clinically used drug ibuprofen. The correlations between their inhibitory effects on collagen-induced aggregation and the structural parameters are expressed by regression equations (Table 6); this table also shows the criteria used for appraising the statistical significance of individual equations.

Table 6. Correlation of inhibition of collagen-induced platelet aggregation

β-aryl-n-butyric acids	n	s	r	F
(1) $\log(1/C_n) = 0{,}4121\,\pi + 4{,}9377$	10	0,1842	0,8025	14,48
(2) $\log(1/C_n) = 0{,}4170\,\pi - 0{,}1261\,\sigma + 4{,}9396$	10	0,1935	0,8103	6,69

β-arylaliphatic acids	n	s	r	F
(3) $\log(1/C_n) = 0{,}2248\,\varepsilon\pi + 4{,}849$	20	0,2492	0,5329	7,14
(4) $\log(1/C_n) = 0{,}2513\,\varepsilon\pi + 0{,}3405\,\varepsilon\sigma + 4{,}8323$	20	0,2425	0,5997	4,77
(5) $\log(1/C_n) = 0{,}3518\,\varepsilon\pi + 1{,}0800\,E_s^\beta + 4{,}8229$	20	0,1641	0,8407	20,50

n = number of compounds; s = standard deviation; r = regression coefficient;
F = Fischer-Snedecor criterion

As demonstrated by the criteria, the correlations between the antiaggregation activities and the physicochemical parameters of beta-arylbutyric and beta-arylaliphatic acids are optimally expressed by the equations (1) and (5). Hence, it follows that the biological activity studied is dependent on the lipophilia (π) and on the steric effect (E_s) of each beta substituent. The decrease in the significance of the equations following the introduction of the parameter σ (electronic effects of substituents) indicates that the antiaggregation activity is not dependent on these effects.

In our studies of the capacity of inhibiting thrombin-induced aggregation we found a similar dependence on lipophilia, with a slightly more marked participation of the steric effect. In beta-arylaliphatic acids, of particular importance, in our opinion, is the revealed similarity of characters of the quantitative correlations between their chemical structures on the one hand and their fibrinolytic activities (Table 7) as well as their stabilizing effects on serum albumin (Table 8) on the other hand. In these instances, too, in the biological activity decisive roles are played by lipophilia, plus the steric effect in the case of beta substituents, whereas the electronic effects practically play no role.

Studies of the stabilizing effects on erythrocytes revealed a quantitative correlation between this activity and the chemical structure, different in comparison with the previous tests; besides lipophilia, the electronic effects of the substituents also play a significant role. In primary ADP-induced platelet aggregation, no quantitative evaluations have been done as yet because

Table 7. Correlation of fibrinolytic activity

β-aryl-n-butyric acids	n	s	r	F
(1) $\log (1/C_n) = 0{,}5506\,\pi + 3{,}7054$	12	0,2219	0,8707	31,35
(2) $\log (1/C_n) = 0{,}5541\,\pi + 0{,}0112\,\sigma + 3{,}6873$	12	0,2190	0,8952	18,16

β-arylaliphatic acids	n	s	r	F
(3) $\log (1/C_n) = 0{,}5879\,\varepsilon\pi + 3{,}3667$	21	0,2202	0,8668	57,38
(4) $\log (1/C_n) = 0{,}5942\,\varepsilon\pi + 0{,}3112\,\varepsilon\sigma + 3{,}3417$	21	0,2051	0,8919	35,00
(5) $\log (1/C_n) = 0{,}5916\,\varepsilon\pi + 0{,}7610\,E_s^\beta + 3{,}4446$	21	0,1787	0,9192	49,01

n = number of compounds; s = standard deviation; r = regression coefficient; F = Fischer-Snedecor criterion

Table 8. Correlation of inhibition of denaturation of serum albumin

β-aryl-n-butyric acids	n	s	r	F
(6) $\log (1/C_n) = 0{,}4693\,\pi$	13	0,1301	0,9419	86,52
(7) $\log (1/C_n) = 0{,}4646\,\pi + 0{,}0482$	13	0,1359	0,9424	39,67

β-arylaliphatic acids	n	s	r	F
(8) $\log (1/C_n) = 0{,}3505\,\varepsilon\pi + 0{,}7472$	23	0,2181	0,7665	29,92
(9) $\log (1/C_n) = 0{,}3975\,\varepsilon\pi + 0{,}7906\,E_s^\beta + 0{,}7713$	23	0,1569	0,8759	32,97

n = number of compounds; s = standard deviation; r = regression coefficient; F = Fischer-Snedecor criterion

the series of compounds studied influences this aggregation type at considerably high concentrations only. From a qualitative appraisal it ensues that the order of compounds arrayed by activities resembles those established in tests using collagen- and thrombin-induced aggregation. The relationship between the structure and the capacity of inhibiting secondary adrenaline-induced aggregation is slightly different even in the qualitative respect, and remains the subject of further study.

Conclusion

The results of studies of quantitative correlations between the antiaggregation activity and the structural parameters of beta-arylaliphatic acids, in our opinion, are also valuable for clinical medicine. The methodical approach used makes possible a mathematical prediction of a structure likely to possess a maximal activity, even in advance of its actual synthesis. As confirmed by the present study of ours, a fairly close correlation exists between the antiaggregation activity calculated from a suitable regression equation and the activity established experimentally (Fig. 4).

From the pathophysiologic point of view, we consider meaningful the finding that in a series of antiinflammatory compounds, in particular, that of beta-arylaliphatic acids, very similar quantitative correlations are valid for the dependence of either the antiaggregation activity or the fibrinolytic activity and the stabilizing effect on serum albumin on the chemical structure (Table 9). This finding, presently confirmed in the series of cinnamic acids as well, substantiates the concept that the biological activities discussed are governed by related mechanisms.

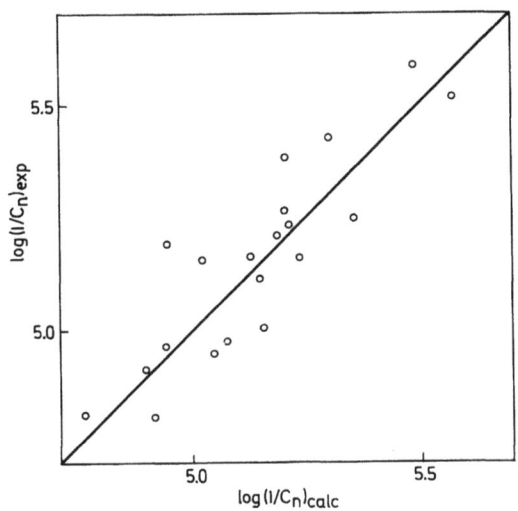

Fig. 4. Relationship between experimental (log $(1/C_n)$ exp.) and calculated (log $(1/C_n)$ calc.) inhibitory activities against collagen-induced platelet aggregation

Table 9. Biological activity of β-arylaliphatic acids

Effect	Inhibition of platelet aggregation	Activation of fibrinolysis	Binding of serum albumin
Lipophilic	+ +	+ +	+ +
Electronic	o	> o	o
Steric	+ +	+	+

References

Čepelák, V., Roubal, Z., Čepeláková, H., Němeček, O.: Chemical induction of fibrinolysis and inhibition of platelet aggregation (with special respect to effect of non-steroid antiinflammatory drugs. Acta Univ. Carol. Med. Monogr. (Praha) LII, 41 (1972a)

Čepelák, V., Čepeláková, H., Roubal, Z., Němeček, O.: Effect of drugs on platelet aggregation: attempt at quantitative comparison of inhibitory effects. Acta Univ. Carol. Med. Monogr. (Praha) LIII-IV, 221 (1972)

Čepelák, V., Roubal, Z., Čepeláková, H.: Agregace krevních destiček in vitro a její inhibice deriváty kumarinu. Čas. Lék. čes. 110, 863 (1971)

Čepelák, V.: Inhibice destičkové agregace a možnosti jejího klinického uplatnění. Čas. Lék. čes. 113, 51 (1974)

Čepelák, V.: Effect of Dipyridamol on platelet aggregation induced by various mediators. First Meeting Europ. Div. Int. Soc. Haemat. Milano, Sept. 10-12, 1971, Abstr. p. 158

Hansch, C.: Drug Design, Vol. 1, p. 271. E.J. Ariëns (ed.). London: Academic Press 1971

Hladovec, J., Svobodová, J., Rossmann, P.: Influence of citrate on platelet adhesion to the endothelial surface. Physiol. bohemoslov. 19, 421 (1970)

Hladovec, J.: Experimental arterial thrombosis in rats with continuous registration. Thromb. Diath. haemorrh. (Stuttg.) 24, 407 (1971)

Kalbhen, D.A., Gelderblom, P., Domenjoz, R.: Effect of antirheumatic drugs on human erythrocyte membranes. Pharmacology 3, 353 (1970)

Kaulla, K.N., von: A simple test tube arrangement for screening fibrinolytic activity of synthetic organic compounds. J. med. Chem. 8, 164 (1965)

Kaulla, K.N., von: Inhibitors of platelet aggregation with dual action. The example of fibrinolytic organic anions. Thromb. Diath. haemorrh. (Stuttg.), Suppl. 45, 83 (1971)

Kluft, C., Brakman, P.: The effect of flufenamate on euglobulin fibrinolysis: involvement of Cl$^-$-inactivator. Second Int. Conference on Synthetic Fibrinolytic-Thrombolytic Agents, Paris, Sept. 30 - Oct. 1, 1974

Kovács, I.B., Görög, P.: The effect of anti-inflammatory drugs on the aggregation and adhesiveness of platelets, red cells and leukocytes. Acta Univ. Carol. Med. Monogr. (Praha) LII, 69 (1972)

Mizushima, Y.: Inhibition of protein denaturation by antirheumatic or antiphlogistic agents. Arch. int. Pharmacodyn. 149, 1 (1964)

Muratová, J., Roubal, Z., Trčka, V.: In vivo effects of dicoumarol derivatives on platelet behaviour. Acta Univ. Carol. Med. Monogr. (Praha) LIII-IV, 269 (1972)

Nachman, R.L.: The platelet as an inflammatory cell. This volume

Some Observations of Platelet Changes in Atherosclerosis and Some Observations in the Platelet Alterations Before and After Antiaggregant Drugs in Normal and Atherosclerotics

O. N. Ulutin and S. B. Ulutin

Recognition of platelets' role in atherosclerotics and thromboembolic incidencts, insufficiency of anticoagulant treatment especially in arterial thrombosis, and increasing observations concerning the beneficial effects of antiaggregating drugs brought about extensive studies of platelets. We shall summarize our observations of platelet changes in the cases of atherosclerosis and our findings concerning the effects of aspirin and dipyridamol on platelets (Ulutin and Ulutin, 1969a).

We shall not discuss the hypercoagulability, hypofibrinolysis, and lipid changes which are observed in atherosclerotics here; we want to limit our topic to the changes observed in platelets.

In our studies on atherosclerotics (coronary sclerosis and cerebrosclerosis) and on transient ischemic attacks, in accordance with the previous literature findings, the presence of a significant increase in platelet retention in respect to normal is demonstrated (Table 1). In a group of subjects, platelet retention was found to be $43.73 \pm 6.7\%$ in normal control groups, when it was $74.33 \pm 5.6\%$ in patients. The difference was highly significant ($P < 0.001$). In these cases the presence of an excessive sensitivity in response to ADP was demonstrated. The ADP concentrations, which induce only a small primary aggregation in normal controls, cause a high primary wave and a secondary aggregation in patients (Fig. 1).

In these cases, together with an excessive sensitivity to ADP, a hyperaggregation with adrenalin and collagen is observed.

This excessive sensitivity toward ADP becomes more significant during ischemic attacks and apart from the attacks this sensitivity turns to normal or decrease. On the other hand, in coronary sclerosis this sensitivity to ADP increases progressively and goes as far as causing infarction. We have observed that in certain cases during an ischemic attack, spontaneous aggregation occurs in citrated PRP.

It is demonstrated by various methods that in these cases platelet antiplasmin increases significantly in respect to normal (Table 1).

It was shown that in atherosclerotic cases platelets undergo a secretion more rapidly and in higher proportion compared to nor-

Table 1. Some laboratory findings related with platelets in normal and in atherosclerosis

		Normal	Atherosclerosis
Secondary aggregation curves with ADP (γ/ml)	0.1	None	Frequently
	0.25	None	Frequently
	0.5	Frequently	Frequently
ADP contents of platelets (mg/3 × 10¹⁰ platelets)	Platelet pellet	0.720 ± 0.39	0.362 ± 0.38
	Supernatant PPP	1.470 ± 0.36	1.509 ± 0.39
	Release in per cent	67%	82%
Aggregation time with DW		40"–50"	25"–35"
Plasma fibrinogen level		240 ± 18	360 ± 26.5
Platelet F 4 (s)	Supernatant PPP after aggregation	58.6 ± 3.2	46.1 ± 4.5
Thrombin time (s)		17 ± 2.1	13 ± 1.9
Platelet glass adhesion (Salzman)		41.9 ± 1.7	77.5 ± 2.1
Platelet antiplasmin	In units	2.4 ± 0.8	10.1 ± 1.9
	Inhibition in per cent	3.5%	12%

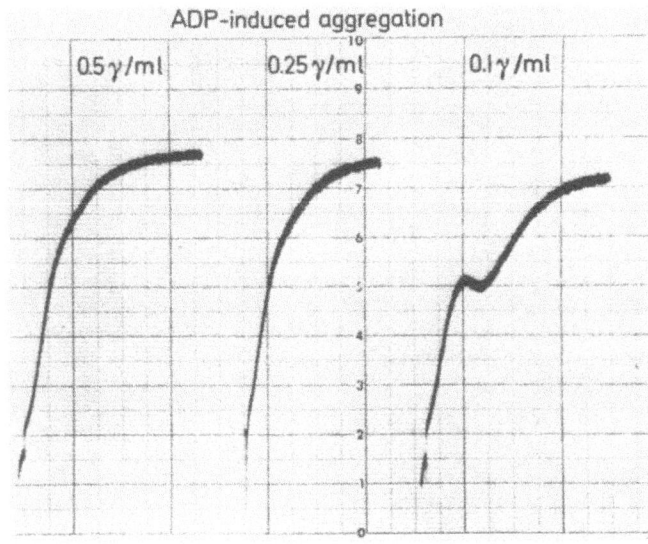

Fig. 1. ADP hypersensitivity in a case of transient ischemic attacks with high fibrinogen level

mals (Akman, 1971). As it can be seen in Table 1, in response to the same release inducer, more PF4 and adenine nucleotides are released.

I would like to point out that in these cases fibrinogen levels are higher compared to normal. In our series of cases, average fibrinogen levels in the same age group is found to be 240 ± 8.1 mg% in normals, when it was 360 ± 15.6 mg% in cases of chronic atherosclerotic cerebrovascular disease. These findings are in accordance with the findings of coronary sclerosis cases of the same age group.

In transient ischemic attacks or in ischemic heart disease, high fibrinogen levels are observed.

An interesting observation in these cases is the similarity of the laboratory findings with the preceding stage of DIC (Ulutin and Ulutin, 1973, 1974b). In this preceding stage of chronic DIC, findings such as high fibrinogen levels, hyperaggregation, excessive sensitivity to ADP, and a higher proportion of release are among the observations which need more attention.

In our cases, when the differential count of the adhesion of platelets to the formvar membrane was done in comparison to the normal, the presence of a higher proportion of spread form was observed (Ulutin and Ulutin, 1974a). Our findings show a parallelism with the findings of Walsh et al. (1974). Like these authors we have also observed that in acute attacks the amount of intermediate and spread form increases (Table 2).

Barnhart and coworkers (1970) in one of their studies showed that aggregates increase and this increase reaches the highest level in acute stroke and transient ischemic attacks follow that. Both in their studies and in ours, Rebuck and coworkers' (1960) procedure and classification are taken as a basis.

Table 2. Differential count of platelets on formvar membrane

	Round	Dendritic	Intermediate	Spread
Control	4	64	27	16
Patient	3	25	27	15

Table 3. Differential count of platelets on formvar membrane before and after aspirin and dipyridamol

	Round	Dendritic	Intermediate	Spread
Control	3	54	17	26
Dipyridamol	17	48	5	30
Aspirin	32	38	9	21

Now I shall try to summarize our observations concerning the effects of aspirin and dipyridamol on platelets in vitro and in vivo conditions.

Aspirin and dipyridamol causes a decrease in platelet glass retention, as we observed by means of Salzman test. Bölükbaşı (1971), by means of celite and glass bead methods demonstrated in dogs that the platelet glass retention decreases significantly, and he states that this effect comes out most with dipyridamol.

Aspirin and dipyridamol modifies the differential count of the platelets in formvar membrane (Table 3). Both aspirin and dipyridamol cause a decrease in intermediate spread form and cause an increase in the round form. The most distinct change occurs in the consecutive order aspirin, pyridinol carbamate, dipyridamol (Ulutin and Ulutin, 1974a). Furthermore, aggregate proportion and size in the formvar membrane decreases.

The excessive sensitivity to ADP in these cases decreases with these drugs. Dipyridamol has a decreasing effect on this sensitivity (Fig. 2). The effect of dipyridamol is time-dependent. Both aspirin and dipyridamol inhibit the platelet secretion. This inhibition can either be detected by the absence of secondary curves in aggregation curves, or by the inhibition of ADP, FP4, platelet fibrinogen, and antiplasmin releases (Table 4). Both aspirin and dipyridamol in vivo and in vitro conditions inhibit the glucose ultilization and lactate formation (Table IV). In our experiments, aspirin inhibited average glycose utilization 85% and lactate formation 42%, dipyridamol inhibited glycose utilization 36% and lactate formation 39%.

After examining the effects of these drugs on platelet ultrastructure in vivo and in vitro experimental conditions, the findings can be summarized as follows. Aspirin causes shape changes in granulae as well as membrane alterations. In the ultrastructural

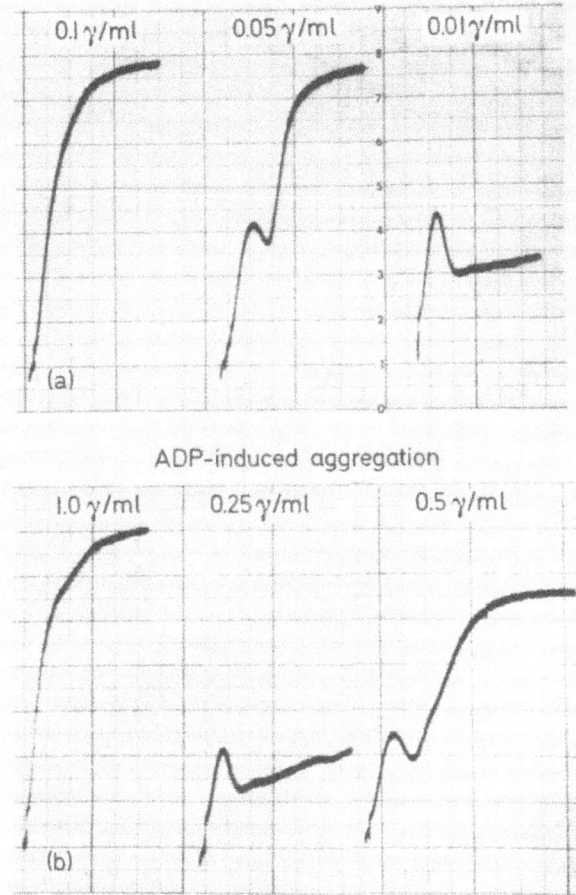

Fig. 2. ADP hypersensitivity in a case of cerebrosclerosis with high fibrinogen level. a) Before treatment; b) During dipyridamol therapy

examinations, after the release inducers exert their effect, an enlargement in the endoplasmic microcanalicular system and even cysterna-like vesicular shapes are noted. Granulae retention is among the observations.

With dipyridamol on the other hand, endomembrane formation and a thickening in mitochondria critae are noted.

In summary, acetyl salicylic acid decreases the platelet glass retention, inhibits the secondary curve, and also inhibits the PF3, PF4, adenine nucleotides, platelet fibrinogen, and platelet antiplasmin release. Moreover, glycose utilization and lactate formation decreases.

Dipyridamol inhibits the ADP-induced aggregation, lessens the platelets' sensitivity to ADP, and inhibits the secretion. Also glycose uptake and lactate formation decreases.

Table 4. The effect of antiaggregating drugs on the platelet

		Aspirin		Dipyridamole	
		Before	After	Before	After
Secondary aggregation curve with ADP (µg/ml)	0.25 g/ml	Exist	None	Exist	None
	0.5 g/ml	Exist	None	Exist	None
	1.0 g/ml	Exist	None	Exist	Exist
ADP release ($mg/3 \times 10^{10}$ platelets)	Supernatant	1.509 ± 0.39	0.326 ± 0.29		
	Pellet	0.813 ± 0,25	1.386 ± 0,31		
Platelet fibrinogen ($\mu g/10^9$ platelets)	Intact	119 ± 5.6		119 ± 5.6	
	After aggregating agents	46.2 ± 4.9	125.3 ± 10.3	46.2 ± 4.9	102.4 ± 4.9
Glucose utilization ($mg/h/3 \times 10^{10}$ platelets)		15.6 ± 3.4	2.4 ± 1.9	18.3 ± 5.1	11 ± 2.65
Lactate formation ($mg/h/3 \times 10^{10}$ platelets)		6.6 ± 0.6	3.8 ± 0.2	6.1 ± 1.2	2.1 ± 0.58
Ultrastructural alterations		Alterations in membranes and granulae		Endomembrane transformation thickening of mitochondrial cristae	

There are increasing clinical observations concerning the effects of using antiaggregating drugs on decreasing the transient ischemic attacks. However, it is too early yet to arrive at a conclusion on the prophylactic and to treating values of these antiaggregation agents clinically.

The matter we want to point out here is that laboratory findings and clinical findings do not always show a parallelism. In our series, together with cases showing severe clinical pictures but normal laboratory findings, there are cases with excessive hypercoagulability and hyperaggregation but not clinical symptoms. I would like to point out that in the same manner the effect of the drugs on laboratory tests and clinical data do not always show a parellelism.

Antiaggregating drugs gain an increasing importance in the prevention and treatment of this group of diseases. However, more studies and long clinical evaluations are necessary to reach a final conclusion.

References

Akman, N.: Platelet factor 4 activity and release of platelet factor 4 in patients with atherosclerosis. Proc. Second Mediterranean Congr. on Thromboembolism. Istanbul, Oct. 11-14, 1971
Aktulga, A., Ulutin, O.N.: Platelet fibrinogen and its release under different conditions. Int. Symp. on Blood Platelets. Istanbul, Aug. 24-27, 1974a
Aktulga, A., Ulutin, O.N.: Platelet fibrinogen and its release under different conditions. Int. Symp. on Blood Platelets. Istanbul, Aug. 24-27, 1974b
Barnhart, M.I., Gilroy, J., Meyer, J.S.: Dextran 40 in cerebrovascular thrombosis. Thromb. Diath. haemorrh. $\underline{42}$, 321 (1970)
Bölükbaşı, F.: On the effects of aspirin and persantin in platelet adhesiveness in dogs. Second Mediterranean Congr. on Thromboembolism. Istanbul, Oct. 11-14, 1971
Karaca, M., Kabakçı, T., Kocabaş, A.: Lipids and platelet stickiness in patients with coronary heart disease. Proc. IVth Congr. Asian-Pacific Soc. Hemat. New Dehlhi, Nov. 14-17, 1969, p. 234-242
Rebuck, J.W., Riddle, J.M., Johnson, S.A., Monto, R.W., Sturrock, R.W.: Contributions of electron microscopy to the study of platelets. Henry Ford Hosp. med. Bull. $\underline{8}$, 273 (1960)
Ulutin, O.N.: Introduction to Blood Platelets: Physiology, pathology and clinical application. Springfield (Ill.): Charles C. Thomas (in preparation)
Ulutin, Ş.B., Ulutin, O.N.: A study on the platelet antiplasmin activity in atherosclerotics and normal people. Vth Congr. Asian-Pacific Soc. Hemat. Istanbul, Sept. 1-6, 1969a
Ulutin, Ş.B., Ulutin, O.N.: Las cambios de la actividad fibrinolitica en los aterosclerosis Y en los condiciones thromboembolicas. Proc. I Congr. Mediterraneo Sobre Thromboembolicas. Bilbao, Oct. 22-25, 37-40, 1969b
Ulutin, O.N., Ulutin, S.B.: Some observations related with disseminated intravascular coagulation. New Istanbul Contr. clin. Sci. $\underline{10}$, 252 (1973)
Ulutin, Ş.B., Ulutin, O.N.: Some observations on the effect of antiaggregant drugs on platelets. Symp. on Blood Platelets. Istanbul, Aug. 24-27, 1974a
Ulutin, O.N., Ulutin, Ş.B.: Acquired storage pool deficiency in chronic disseminated intravascular coagulation. Int. Symp. on Blood Platelets. Istanbul, Aug. 24-27, 1974b
Ulutin, Ş.B., Aktuğlu, G., Ulutin, O.N.: Antiadeziv ilaçların trombositlere etkileri. II. Dipyridamol'un trombosit glikoz utilizasyonuna etkisi. Cerrahpaşa Tıp Fak. Dergisi $\underline{2}$, 344 (1971)

Ulutin, Ş.B., Yaramancı, T.E., Ulutin, O.N.: The effects of antiadhesive drugs on platelet function and metabolism. Symposium on the role of platelets in Haemostasis and Thrombosis. Prague, June 14-16, 1971. Acta Univ. Carol. (Med. Monogr.)(Praha) L<u>III-LIV</u>, 1972a

Ulutin, O.N., Ulutin, Ş.B., Yaramanci, T.E.: The effect of antiaggregant agents on platelet functions. IIIrd Congr. Int. Soc. on Thrombosis and Haemostasis. Washington, D.C., Aug. 22-26, 1972b

Walsh, R.T., Bower, R.B., Barnhart, M.I.: Platelet function in transient ischaemia and cerebrovascular disease: effects of aspirin and contrast media. Int. Symp. on Blood Platelets. Istanbul, Aug. 24-27, 1974

Yaramancı, T.E., Ulutin, Ş.B., Ulutin, O.N.: The effect of dipyridamole on the ultrastructure of platelets. First Meeting European Div. Int. Soc. Haemat., Milano, Sept. 10-12, 1971. Aggregazione Piastrinica, 25-33, 1973

Pharmacology and Clinical Pharmacology

Chairmen: C. A. Bouvier, W. S. Fields, Orhan N. Ulutin,
A. Rascol

Inhibition of Platelet Thrombus Formation by Pharmacological Agents

R. Kadatz

The main physiologic function of platelets is to protect blood vessels from injury. This might either be minor damage of the endothelium, as in an early stage of atherosclerosis, which is covered by a layer of platelets, or a disruption, where platelets adhere to the severed vessel wall and to each other to form a plug thus stopping bleeding from a cut vessel.

In organisms with a circulatory system, aggregation of thrombocytes is a fundamental property, which in the course of evolution appeared earlier than plasma coagulation. It must be a very delicate equilibrium, which keeps platelets apart in the flowing blood under normal conditions, which enables platelets to stick to the vessel wall in case of a defect and - on the other hand - prevents them from an overshooting reaction with formation of an occluding thrombus in case of a minor defect. We are far from knowing exactly how this equilibrium is achieved.

Drug therapy of thromboembolic diseases is based on the idea of diminishing platelet reactivity; however, we have to be aware of the fact that we do not exactly know whether an enhanced platelet function is the cause of the illness or how far factors in the vessel wall or changes of the blood constituents might play a role.

During the past 10 years, a great many compounds were synthetized and have been shown to inhibit platelet function in vitro, and even some older well-known drugs have shown to have antiplatelet activity.

Many of them are listed in Table 1 which, however, is still incomplete. Aspirin is the most advanced example of several nonsteroidal antiphlogistics, which depress platelet function by inhibiting its release reaction. Persantine and its congeners form a group on their own, being more active in the intact organism than in in vitro tests. The next two groups are said to influence platelet cAMP either by stimulating membrane adenyl cyclase or by depressing phosphodiesterase. Papaverine also belongs to group 4.

Only some of these substances have been tested in vivo in animals and still fewer have been studied in man. I will discuss in detail only those substances which have shown some activity in man in experimental or clinical trials.

Table 1. Substances with antiplatelet activity

1. Nonsteroidal antiinflammatory agents:

   Acetylsalicylic acid, Indomethacin, Phenylbutazone, Sulfinpyrazone, Sudoxicam, Flurbiprofene, Ditazol, Benzidamine

2. Pyrimido-pyrimidine-derivatives:

   Persantine, R-A 233, 433, V-K 744, 774

3. Substances supposed to stimulate adenyl cyclase:

   Adenosine, 2-Chloradenosine, PGE_1, Isoprenaline, S-H 869, M-H 220

4. Methylxanthines:

   Caffeine, Aminophylline, Theobromine

5. Tricyclic antidepressants:

   Chlorpromazine, Imipramine, Aminophylline

6. Sympathetic blocking agents:

   Phentolamine, Dibenamine, Dibenzyline, Propranolol

7. Histamine- and serotonin-antagonists:

   Reserpine, Methysergide, Promethazine, Cyproheptadine, Diphenhydramine, Pyrilamine

8. Miscellaneous drugs:

   Dextran, Glycerylguaiacolate, Nialamide, Penicilline, Atromid, Localanaesthetics, Clofibrate

Table 2 shows the inhibitory activity of such compounds on platelets in ADP-induced aggregation. The experiments were performed with Born's turbidimetric method in platelet-rich plasma, and demonstrate only primary aggregation (first phase). Here and in the following tables I have compiled the results of our own experiments and of others from the literature.

Although the comparison is somewhat hampered because results with blood of different species are put together, it can easily be seen that aspirin and other antiphlogistics are either weak or ineffective in this test. Persantine and its congeners are also relatively weak compared to their effectiveness in the intact animal. Most effective in low concentrations are prostaglandin E_1, adenosine, and three of our synthetic compounds. All of them are, however, potent vasodilators and the hypotensive effect may restrict their clinical use as antithrombotics. This group probably acts by increasing the level of platelet cAMP via stimulation of adenyl cyclase.

S-H 862 and M-H 220 are the most effective inhibitors of primary ADP-aggregation I know, and it is an interesting pharmacologic

Table 2. Effects on primary ADP-induced aggregation, PRP, Born's method

Substance	Concentration in PRP	Effect % inhib.	Species	Source
Aspirin	2×10^{-3} M	++	Rabbit	Caprino et al.[b]
Sudoxicam	10^{-4} M	-	Man	Constantine and Purcell[c]
Ditazol	3×10^{-4} M	++	Rabbit	Caprino et al.[b]
Persantine	3×10^{-4} M	20	Man	Own experiments
R-A 233	5×10^{-5} M	50	Man	Own experiments
V-K 744	5×10^{-5} M	54	Man	Own experiments
PGE_1	4×10^{-8} M	50	Man	Own experiments
	15 ng/ml	+++	Rat	Muirhead et al.[d]
Adenosine	2×10^{-5} M	+++	Man	Born et al.[a]
S-H 869	1.9×10^{-6} M	50	Man	Own experiments
M-H 220	5×10^{-7} M	69	Man	Own experiments
S-H 862	1×10^{-8} M	53	Man	Own experiments

[a] Born, C.V.R., Haslam, F.J., Goldman, M., Lowe, R.D.: Nature (Lond.) 205, 678 (1965)

[b] Caprino, L., Borrelli, F., Falchetti, R.: Arzneimittel-Forsch. 23, 1277 (1973)

[c] Constantine, J.W., Purcell, J.M.: J. Pharmacol. exp. Ther. 187, 653 (1973)

[d] Muirhead, C.R.: Thromb. Diath. haemorrh. (Stuttg.) 30, 138 (1973)

problem to elucidate further the relation between antiplatelet activity and smooth muscle relaxing effect in the wall of blood vessels.

Collagen-induced platelet aggregation starts only after a short delay and is caused by the release of the platelets own ADP. This so-called release reaction is inhibited by aspirin and other anti-inflammatory drugs. In vitro, sudoxicam and benzidamine are most active with 1-2 µM/ml. The effect of aspirin on human platelets was similar at a 20-fold greater concentration. This concentration of aspirin, which inhibits the release reaction in vitro, however, is of the same order of magnitude as that found in plasma in ordinary clinical use. Persantine and related substances are somewhat weaker in this test and their effect apparently comes about by an inhibition of the released ADP and not so much by a specific inhibition of the release process (Table 3).

Although these tests described so far are extremely useful in assessing the effect of drugs on platelets in quantitative terms, they are inherently unphysiological and caution should be exercised in extrapolating from the results to any effect that a drug may have on the physiological properties of platelets in the organism under normal conditions or during thrombus formation.

Table 3. Effect on collagen-induced platelet aggregation, PRP, Born's method

Substance	Concentration in PRP	Effect % inhib.	Species	Source
Sudoxicam	1 µM	++	Man/dog Rabbit	Constantine and Purcell[b]
Benzidamine	2 µM	++	2	Fortunato et al.[c]
Aspirin	20 µM	++	Rabbit	Caprino et al.[a]
	50 µM	+++	Man	Weiss et al.[e]
	40 µM	50	Man	Own experiments
Flurbiprofen	?	+++		Nishizawa et al.[d]
	10 µM	87	Man	Own experiments
Dibazol	3 µM	++	Rabbit	Caprino et al.[a]
Phenylbutazon	0.3 - 0.6 mM	++	Man	Zucker and Peterson[f]
Persantine	5×10^{-5}	50	Man	Own experiments
R-A 233	2.7×10^{-5}	50	Man	Own experiments
V-K 744	30 µM	72	Man	Own experiments
M-H 220	0.1 µM	68	Man	Own experiments
S-H 862	0.1 µM	69	Man	Own experiments
S-H 869	3 µM	48	Man	Own experiments

[a] Caprino, L., Borrelli, F., Falchetti, R.: Arzneimittel-Forsch. 23, 1277 (1973)

[b] Constantine, J.W., Purcell, J.M.: J. Pharmacol. exp. Ther. 187, 653 (1973)

[c] Fortunato, G., Belisario, A., Coccheri, S.: IV. Int. Congr. Thromb. Haemost., Wien, 1973

[d] Nishizawa, E.E., Wynalda, D.J., Suydam, D.E.: Platelets, Thrombosis and Inhibitors, East-West Conference Center Honolulu, 1973

[e] Weiss, H.J., Aledort, L.M., Kochwa, S.: J. clin. Invest. 47, 2169 (1968)

[f] Zucker, M.B., Peterson, J.: J. Lab. clin. Med. 76, 66 (1970)

Therefore, the oral application to laboratory animals and the observation of their effect on thrombus formation is the next step in the evaluation of new substances. In our hands, the measurement of the bleeding time in mice proved to be a reliable method. The formation of a hemostatic plug at the end of a cut vessel is a similar process to that involved in platelet thrombogenesis. We give the substances orally and cut the tip of the tail one or several hours later. The droplets of blood are absorbed carefully with a filter paper every 30 seconds and the normal bleeding time is exactly 4 minutes.

As can be seen from Table 4, persantine and congeners are quite effective in this test. The antiphlogistics need higher doses to be equieffective with the exception of flurbiprofen, which accord-

Table 4. Effect on bleeding time in mice after oral application

Substance	Dose mg/kg p.o.	% increase 1 h	3 h	5 h	Source
Persantine	10	63	113	65	Own experiments
R-A 233	10	93			Own experiments
V-K 744	10	181	21		Own experiments
S-H 862	0.5	88			Own experiments
S-H 869	10	138	100	58	Own experiments
	0.5	55			Own experiments
M-H 220	10	100	58		Own experiments
Aspirin	200	55	55	55	Own experiments
Sudoxicam	10	63	44	20	Own experiments
Sulfinpyrazone	10	13			Own experiments
Indomethazin	10	15			Own experiments
Ditazol	100	35			Caprino et al.[a]
Phenylbutazon	100	55			Caprino et al.[a]
Flurbiprofen	1	138 (rats)			Nishizawa et al.[b]
	10	22	39	22	Own experiments

[a] Caprino, L., Borrelli, F., Falchetti, R.: Arzneimittel-Forsch. 23, 1277 (1973)

[b] Nishizawa, E.E., Wynalda, D.J., Suydam, D.E.: Platelets, Thrombosis and Inhibitors, East-West Conference Center Honolulu, 1973

ing to Nishizawa, prologs bleeding time in rats ater 1 mg/kg by 138%. Measuring of the bleeding time allows one to estimate the duration of effects, which last more than 48 hours with aspirin. In man, bleeding time was also prolonged after oral ingestion of 1 g of aspirin according to several authors.

Table 5 shows another group of in vivo tests. The substances were given to laboratory animals and their effects were studied on experimental thrombosis induced by electrical or chemical irritation of the vessel wall or on platelet function in blood samples taken some time later. Again, comparability is limited because techniques and species are not uniform. I have compiled these results because they might give an impression at least of the existence and order of in vivo effectiveness.

Both groups, persantine and congeners, and nonsteroidal antiphlogistics have proven to be effective by these methods. PGE_1, in contrast to its high in vitro activity, had only a weak and very short-lasting effect.

Table 5. Effect in vivo in animal experiments

Substance	Dose	Species	Effect	Source
Persantine	10 mg/kg p.o.	Rat	44% inhib. electr. ind. thrombi	Own experiments
R-A 233	10 mg/kg p.o.	Rat	32% inhib. electr. ind. thrombi	Own experiments
R-A 744	10 mg/kg p.o.	Rat	62% inhib. electr. ind. thrombi	Own experiments
S-H 869	5 mg/kg p.o.	Rat	97% inhib. electr. ind. thrombi	Own experiments
PGE_1	0.25 mg/kg i.v.	Rabbit	ADP-ind. aggr. inhib. for 5 min	Kinlough et al.[d]
Aspirin	0.6 g/3 days	Dog	Inhib. of thrombi in injured art.	Danese et al.[b]
	0.6 g/4 days	Dog	Protect. myocard. necrosis by epinephrine	Haft et al.[c]
Sudoxicam	1 mg/kg i.v.	Dog	Everted artery	Constantine and Purcell[a]
	1 mg/kg p.o.	Dog	Inhib. coll. ind. aggr. 4-5 days	Constantine and Purcell[a]
Flurbiprofen	1 mg/kg p.o.	Rat	Inhib. coll. ind. aggr. 48 h	Nishizawa et al.[e]

[a] Constantine, J.W., Purcell, J.M.: J. Pharmacol. exp. Ther. 187, 653 (1975)
[b] Danese, C.A., Voleti, C.D., Weiss, H.J.: Thromb. Diath. haemorrh. (Stuttg.) 25, 288 (1971)
[c] Haft, J.I., Gershengorn, K., Kranz, P., Albert, F., Oestreicher, R., Fani, K.: Amer. J. Cardiol. 30, 838 (1972)
[d] Kinlough, R.L., Packham, M.A., Mustard, J.F.: Brit. J. Haemat. 19, 559 (1970)
[e] Nishizawa, E.E., Wynalda, D.J., Suydam, D.E.: Platelets, Thrombosis and Inhibitors, East West Conference Center Honolulu, 1973

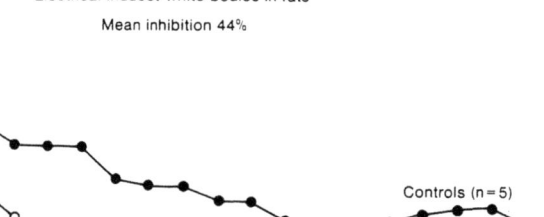
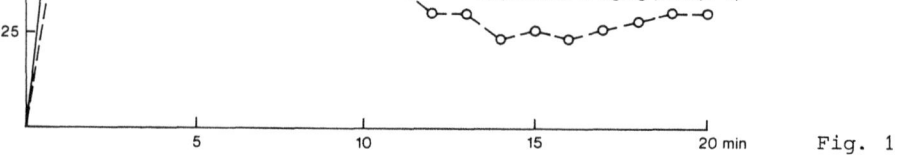

Fig. 1

In the experiments of Danese and Weiss, 0.6 g aspirin daily for 3 days reduced thrombi in the arteries of dogs after mechanical injury or local application of sulfuric acid; in these investigations persantine was ineffective at 0.2 g/day. Aspirin as well as persantine protected dogs in Hafts paper after 4 days pretreatment against myocardial necrosis by intracoronary infusion of epinephrine. Both substances have been reported by Mustard to decrease the amount of thrombotic deposits which form in extracorporeal shunts in pigs. Sudoxicam and flurbiprofen are also effective in vivo in low doses. Aspirin and sudoxicam have a very long-lasting effect which could be shown for 4-5 days after a single oral dose.

As an example of these animal experiments, Figure 1 shows a white body produced by electrical stimulation of mesenteric vessels of rats. We observe size and time course of such an experimental thrombus. In Figure 2, the percent occlusion of the vessel during a 20-min observation period is shown in a control group and after pretreatment with 10 mg/kg persantine orally.

This method permits the operator to observe *continuously* the mural thrombus formation, the extent of occlusion of the vessel and embolization rate rather than only *once*, as is the case with methods in which the vessel is removed at the end of the experiment and size and weight of the thrombus are recorded. One needs only observe these rapidly changing events in the microcirculation to become convinced of the dynamic nature of thrombogenesis, thrombolysis, and embolization, and of the inadequacy of methods restricted to pathologic observation.

Another informative study was published by Harker and Slichter. They measured the capacity of several agents to prevent consumption of circulating platelets in baboons. Platelet consumption was increased by chronic implantation of arterial catheters or

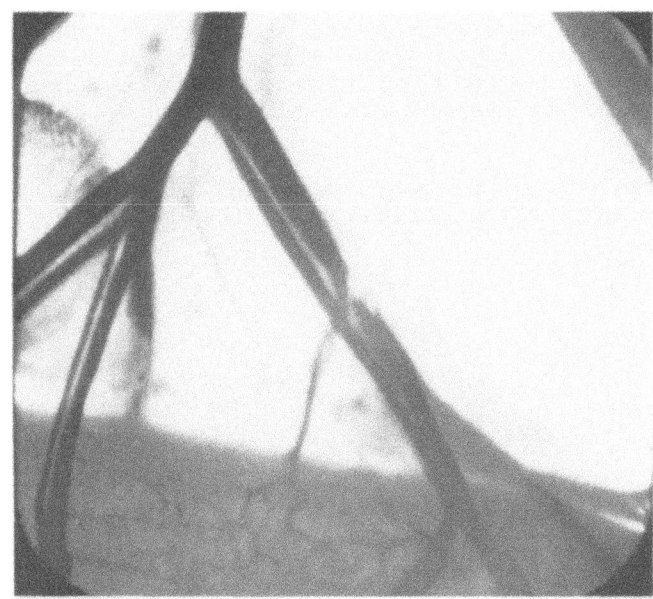

Fig. 2

Table 6

Substance	Dosage normalizing shortened platelet survival time in baboons	
Persantine	10 mg/kg	
R-A 233	4 mg/kg	
V-K 744	4 mg/kg	
S-H 869	2 mg/kg	
Sudoxicam	2 mg/kg	
Phenylbutazone	50 mg/kg	partially effective
Sulfinpyrazone	100 mg/kg	
Aspirin	50 mg/kg	no effect

Harker, L.A., Slichter, S.J.: IV. Intern. Congr. on Thrombosis and Haemostasis, Vienna 1973

aterio venous silastic canulas in these animals, and platelet survival time was taken as indicator of enhanced consumption.

Minimal daily oral dosage that prevents platelet consumption is shown in Table 6.

In this report, I have limited myself to the experimental effects of antithrombotics. Clinical evaluation of a few of these substances is being carried out in many places. I think that it is still too early to give definitive statements about their therapeutic effect and the suitable indications. For the pharmacologist it is a big problem to know which of his tests corresponds best with the therapeutic effectiveness of a substance in the patient.

Only reliable clinical results with such compounds will make possible the development of new and better drugs in the laboratory.

A Long-Term Clinical Trial with Antiplatelet Agents in Cerebrovascular Ischemia: Biological and Methodological Aspects

B. Guiraud, B. Boneu, J. David, G. Geraud, R. Bierme, and A. Rascol

Several months ago, we started a long-term clinical trial, with antiplatelet and vasoactive drugs, in the prevention of ischemic stroke.

We shall present three points:

1. The biological basis of the trial
2. The methodology
3. The preliminary results.

I. The Biological Basis of the Trial

In 1972, we published some biological results which are the basis of this clinical trial. At that time (Bonea et al., 1972b), we pointed out a platelet hyperaggregability phenomenon in patients with ischemic cerebrovascular diseases. Born's method was used with ADP at the final concentration of 1.2; 0.6; 0.3; and 0.15 μM. There were 15 patients and 200 normal subjects studied at that time (Fig. 1). The lowest concentration of ADP able to produce an irreversible platelet aggregation curve was determined. We have found that the two populations (patients and normal subjects) are different.

In ordinate is plotted the percentage of normal subjects or patients who present an irreversible platelet aggregation curve, and in abcisse the different concentrations of ADP. At the concentration of 0.3 μM, 75% of the patients present an irreversible aggregation curve but only 25% of normal subjects.

Thus, platelet hyperaggregability is defined as the property of the platelets to aggregate irreversibly in presence of low concentration of ADP. This phenomenon is observed more frequently in patients than in normal subjects, but a patient may have platelets without a hyperaggregable behaviour.

In another work (Bonea et al., 1972a), we studied some characteristics of the inhibitory effect of aspirin on blood platelets release reaction.

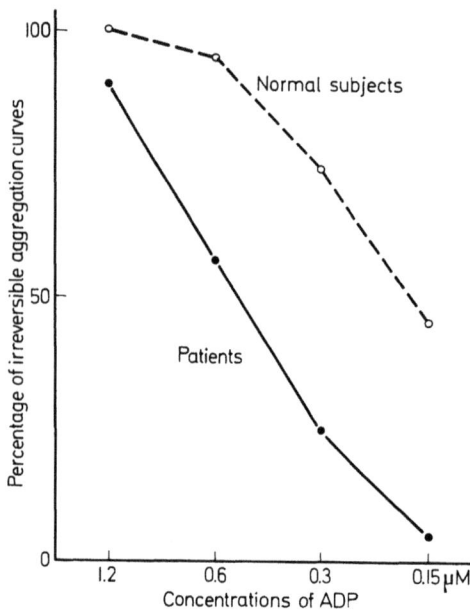

Fig. 1. Platelet aggregation studies with ADP: Normal subjects and patients (see text)

Thirty patients took a single dose of aspirin (2 g of lysin acetyl salicylate). The mean value and confidence interval was calculated with a computer. The inhibitory effect of aspirin on release reaction was still present on the fifth day (Fig. 2), but the dispersion of the values increased with the time. This fact suggests that the length of the inhibitory effect of aspirin varies with the different subjects.

The variability of the length of the inhibitory effect of one single dose of aspirin was studied in relation with hyperaggregability. Hyperaggregability was defined as the minimal concentration of ADP able to produce irreversible aggregation. In hyperaggregable subjects the duration of the inhibitory effect of aspirin is very short: less than 24 h. In normal subjects this inhibitory effect varies from 2 to 5 days (Fig. 3).

Thus it is necessary to fractionate the daily aspirin intake in order to obtain a constant inhibitory effect of the release reaction.

In order to determine if the biological effect of antiaggregant agent is followed by some clinical effect in cerebrovascular disease, we decided to undergo a clinical trial which compares the evolution of patients with transient neurologic deficit under vasoactive or antiaggregant drugs.

II. Methodology of the Trial

We organized a long-term clinical trial. The first group was treated with Hydergine, which is the headline of vasoactive drugs in France for cerebrovascular disease. We studied platelet aggregation to see if it is modified by Hydergine.

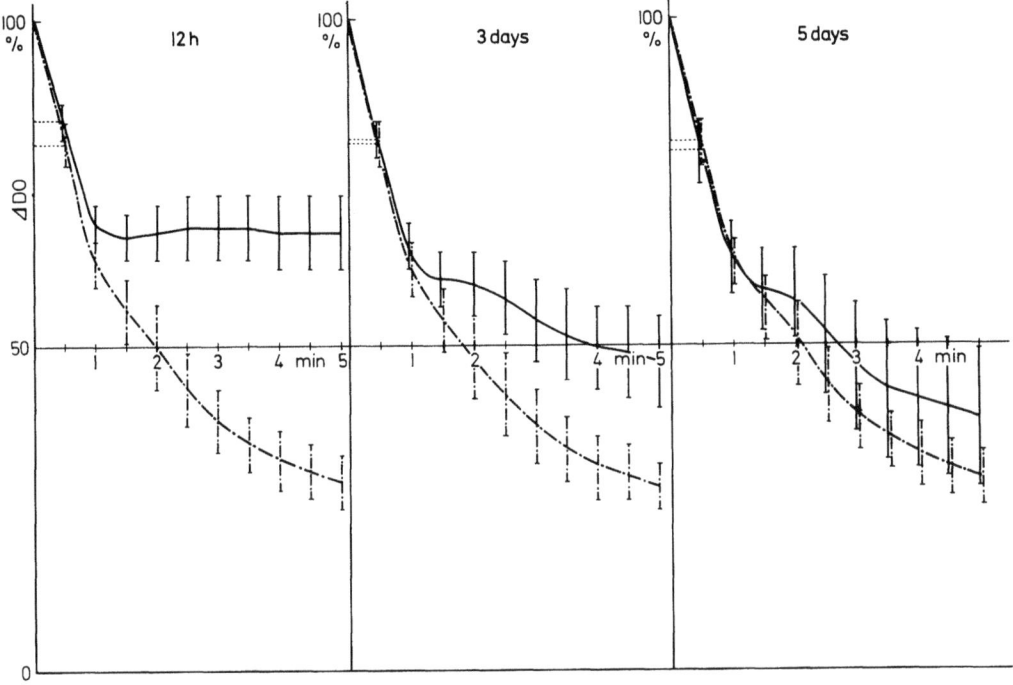

Fig. 2. Studies on the inhibitory effect of aspirin on platelet release reaction. In 30 patients platelet aggregation tests were performed at various times after one single injection of 2 g of lysin acetyl salicylate. Each curve represents the mean value and confidence interval calculated by a computer. The dotted line is the curve obtained before the injection of aspirin

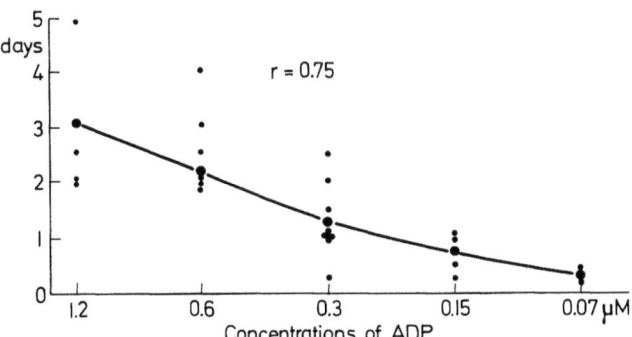

Fig. 3. Length of the inhibitory effect of aspirin and hyperaggregability. The length of inhibitory effect of one intake of aspirin on release reaction was studied in 24 patients or normal subjects. The platelet aggregation test with adrenaline was performed each day. The duration of inhibition of the second wave is expressed as days (ordinate). The degree of hyperaggregability (abcisse) is defined as the minimal concentration of ADP able to produce irreversible aggregation (see Fig. 1). Hyperaggregable subjects are plotted on the right of the diagram

Table 1

Group I	Group II	Group III
Hydergine 45 mg/day	Hydergine 45 mg/day	Hydergine 45 mg/day
	Aspirin 1 g/day	Aspirin 1 g/day
		Dipyridamole 150 mg/day

Each patient is randomized in one of the three groups

In the condition of our trial we did not find any modification of platelet aggregation by Born.

The second group was treated with the association of Hydergine and aspirin. The daily dose of aspirin was 1 g. This dose was fragmented in three intakes. We chose that dose because our preceding work showed that in some hyperaggregable patients a smaller dose does not inhibit the release reaction during 24 h.

The third group was treated by the association of three drugs. We chose the dose of 150 mg/day for dipyridamole because, for Harker, the clinical action on embolism from artificial aortic valve needs 450 mg of dipyridamole, when this drug is used alone, but only 150 mg when dipyridamole is associated with aspirin.

All our patients were treated because we could not take the responsibility of not treating a patient for at least three years; patients of each group received the same vasoactive drugs, we assume that the only difference between the three groups was under the control of antiaggregation drugs.

Each group must have at least 100 patients. Some patients could not be followed as we wanted. When we planned our trial, we estimated that we should have 50% loss. So we decided to include 150 patients to a group. The first patient was included in June, 1973. We thought 18 months were necessary to obtain the formation of our three groups. The first patient will be followed 4 1/2 years and the last one at least 3 years. In October 1974, we were a little late for inclusions, but our loss was not 50%, only 10%. So only 350 patients were necessary and sufficient in the formation of our groups.

Our trial was not multicentric. Only four physicians decided if a patient was to be included in the trial. They did not decide the treatment, that is decided at random. After randomization, the physician knowns which treatment is given to the patient. So, our study differed from American and Canadian cooperative studies of aspirin in transient ischemic attacks by two fundamental points:

1. Our study was not a double-blind study

2. We judged not only the effects of aspirin but also an association of two antiaggregant agents: aspirin plus dipyridamole.

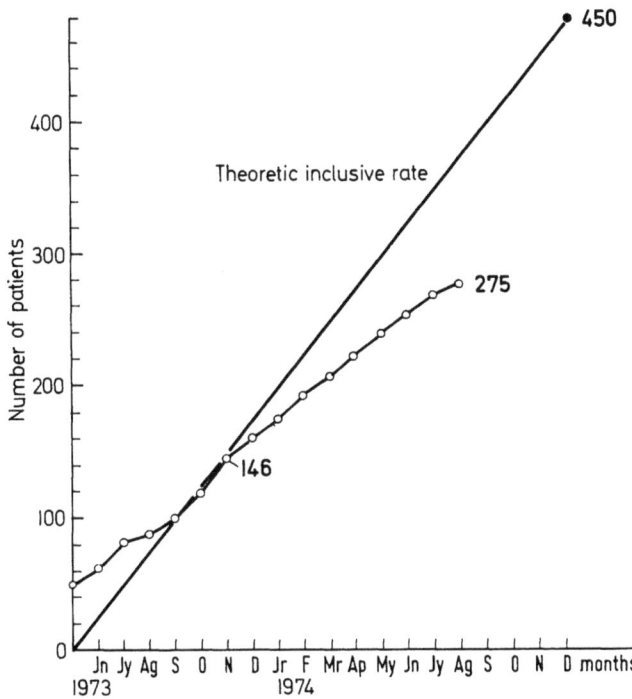

Fig. 4. Inclusive rate of patients in the trial

Table 2. Clinical trial: methodology inclusion criteria

TIA or regressive stroke	
EEG, EKG, lumbar punction	} Obligatory
Biological examination	
Angiography	Facultative
Red cells and platelet counts	
Lipids, cholesterol, triglycerides	} Once a year
Blood level of glucose, uric acid, creatinin	
Platelet aggregation test	Three times a year

Some patients present a clinical story which corresponds to inclusion's criteria, but the clinical state does not fit with a long-term therapy by antiaggregating drugs (cardiac and renal failure, rhumatismal valvulopathies, chronic intake of anticoagulant or antiplatelet drugs).

We call patients for clinical reexamination and survey of treatment three times a year. At each visit we study platelet aggregation to make sure that patient has taken aspirin when he has to. Each year we seek again the level of principal blood components.

Table 3. Preliminary results

	Hydergine	Hydergine + aspirin	Hydergine + aspirin + dipyridamole
Number	84	86	85
TIA	40	28	35
Regressive stroke	44	58	50
Hypertension	38%	38%	41%
Diabetes	14%	10%	10%
Hyperuricemia	25%	19%	27%
Hyperlipemia	9%	18%	14%
Stroke under therapy	4	1	1

III. Preliminary Results

Our preliminary results 14 months into the study concern 255 patients. We fed some questions to the computer (Table 3). Each group seems comparable as far as hypertension, diabetes, hyperuricemia, and hyperlipemia is concerned. Of the patients 46% have been submitted to arteriography.

The repartition in each group of TIA and total regressive stroke is approximately the same. We must point out that the group with Hydergine and aspirin contains twice as many regressive strokes than TIA, as the group with Hydergine alone contains approximately equal percentages in the two types of ischemic accidents.

It is too early to state positively a difference between the three types of treatment.

In conclusion, we want to point out some elements of discussion:

1. Our trial is homogeneous by the team who examines patients but not homogeneous by pathogenetic mechanisms of clinical events. We think that the following of more than 300 patients during three years is a very heavy enterprise. It was impossible for us to undergo a bigger job. It is sure that a multicentric trial may permit a bigger population or a better division in subgroups.

2. We have some individual cases where the effect of antiaggregating drugs seems really efficient, but it is impossible to assert that the amelioration of those patients is only due to aspirin or dipyridamole. It seems to us that in 1974 the only way to predict a real prevention of stroke by vasoactive or antiaggregating drug consists in a randomized study.

3. Our trial is randomized, but it is not a double-blind study. We hope that a 3-year survey will permit us to determine if patients with cerebrovascular disease may benefit in a long-lasting treatment by antiaggregant agents. Our study cannot give a definitive answer before January 1978. Before this date, any preliminary results must be considered as partial and inconclusive. After 1978, if there is not any statistically significant difference between our three groups, we will say that our job was to define better the natural history of ischemic stroke.

References

Boneu, B., Bierme, R., Boneu, A., Guiraud, B., Rascol, A.: Renouvellement plaquettaire et durée d'inhibition de la deuxième vague d'aggrégation après prise unique d'Aspirine. Path. Biol. 20, 71-75 (1972a)

Boneu, B., Guiraud, B., Fernet, P.: Traitement anti-agrégant plaquettaire par l'Aspirine. Bases biologiques. Nouv. Presse méd. 1, 863 (1972b)

Fields, W.S., Callen, P.W.: Aspirin in Cerebral Ischemia. In: Platelet Aggregation in the Pathogenesis of Cerebrovascular disorders. Round Table conference, Rome, Oct. 30-31. Abstract book p. 101, 1974

Guiraud, B., Boneu, B., Boneu, A., David, J.: Abnormal platelet behaviour in atheromatous disease of cerebral vessels. Management by agents inhibiting platelet aggregation. Cerebral vascular disease. 6th International Conference Salzburg. Meyer, J.S., Lechner, H., Reivich, M., Eischorn, O. (eds.)., 31-40, 1972

Simard, D.: The canadian cooperative studies of the effect of Platelet suppressing drugs. In: Platelet Aggregation in the Pathogenesis of Cerebrovascular disorders. Round Table conference, Rome, Oct. 30-31, Abstract book p. 104, 1974

Effect of Aspirin and Dipyridamole on Platelet Function and on Neurologic Evaluation of Patients Affected by Stroke or Transient Ischemic Attacks

E. E. Polli, M. Cortellaro, L. Frattola, A. Randazzo, L. Candelise, E. Pogliani, A. Politi, S. Bassi, G. Scotti, and S. Santambrogio

While the role of platelets in atherogenesis is still discussed, it is well known that at sites of altered, nonendothelialized arterial vessel wall, platelets play a fundamental role in the evolution of an arterial thrombus (Harker and Slichter, 1974; Hellem, 1970; Harrison et al., 1971; McNicol et al., 1974).

In particular, considerable evidence suggests that platelets play a significant role in the occlusive variety of transient ischemic attacks (TIA), and in the completed or in evolution strokes (Fields and Hass, 1970; Harrison et al., 1971; Ross Russel, 1968). These considerations suggest a rationale for the use of antiplatelet agents in these diseases. However, unconclusive results have been reported in the literature because only a few of the studied patients fit in a well-defined experimental design.

The same considerations can be drawn from the trials performed in order to evaluate the effect of platelet-suppressive agents on the natural history of these diseases since clinical evidence of such an effect is still conflicting.

For these reasons, a cooperative prospective study on platelet function and on effect of antiplatelet aggregation drugs is now running in patients affected by transient ischemic attacks (TIA) and acute completed stroke. In order to exclude that the investigation could be performed in a heterogeneous group, the patients were selected and properly assessed and the criteria for admission of the patients with TIA or acute completed stroke to the study were established as it is shown in Table 1.

In order to evaluate the results in those patients affected by stroke in which the anterior cerebral circulation was involved, only those cases with a positive carotidography were considered. A total of 40 patients had the provisional diagnosis of completed stroke and 15 of TIA, but only 27 and 11 patients, respectively, answered all the criteria quoted above. Age and sex distribution of each group are listed in Table 2.

As far as TIA is concerned, the frequency, length, and extent of the last attacks of each patient was recorded. The time from the attack to admission ranged from a few hours to 3 days.

In the patients with acute completed stroke, the neurologic evaluation was performed after a short period (hours to 3 days) from

Table 1. Transient ischemic attacks (TIAs)[a] and acute completed stroke (ACS)[b] criteria for inclusion of patients in the study

1. Age (70 years)
2. Diastolic arterial pressure (110 mm Hg)
3. Carotid arterial system was visualized angiographically
4. No overt evidence of liver-renal-hematologic alterations - no cardiac arrhythmia - peptic ulcer - hypoglycemia - epilepsy
5. No meningeal signs
6. No bloody cerebrospinal fluid
7. No posterior circulation insufficiencies
8. No other illnesses which will require other drugs that will alter blood clotting mechanism

[a] Transient hypofunction less than 24 h
[b] Neurologic deficit without improvement of 24 h duration

Table 2. Case material

Total cases admitted to the study	38
Acute completed strokes (CS)	27
Age	25-70
Male	20
Female	7
Transient ischemic attacks (TIAs)	11
Age	25-65
Male	8
Female	3

The first patient was admitted to the study in March 1974

Table 3. Acute completed stroke: parameters and score of neurologic evaluation[a]

Coma	0 - 15
Invalidity	0 - 15
Motor function	0 - 8
Sensation	0 - 6
Visual field	0 - 6
Speech	0 - 6
Maximal severity of loss of functions	56

[a] Periodically performed before and after treatment

Table 4. Laboratory tests for platelet function[a]

Platelet aggregation (Born and Cross) by:	ADP (0.2 µM–0.4 µM–0.8 µM)[b]
	Adrenaline (90 µM–1.8 µM)[b]
	Collagen (STAGO) 1:2–1:8–1:16[b]
Platelet retention test (Hellem II)	
Bleeding time (template system modified)	
"PF$_4$ assay": heparin-thrombin time (H-TT) (Harada and Zucker, 1971)	

[a] All tests are performed periodically before and after treatment
[b] Final concentration

cerebrovascular accident by utilizing an opposite chart. Each parameter reported in Table 3 has been evaluated by a conventional functional scale. The score for the maximal severity of loss of functions is 56.

The neurologic score was repeated at the 20th day, after a period free from substances affecting platelet function; then at monthly intervals after treatment.

There are three main groups of drugs which interfere with platelet function; antihistamines-antiserotonins substances, pirimido-pirimidine compounds, and nonsteroidal antiinflammatory drugs (Zucker and Peterson, 1970). We have selected for this trial acetylsalicilic acid (ASA) and dipyridamole, because they rarely cause severe side effects and they are effective in oral administration, which is an important task in a long-term therapy.

Each patient of TIA and completed stroke groups, respectively, were included in the trial by randomization using 500 mg of aspirin twice a day or 150 mg of dipyridamole 3 times a day. For ethical reasons we recognized the opportunity of eliminating the placebo test.

The first patient was included in trial on March 1, 1974; up to that day 19 completed stroke patients and 11 TIA patients were included in the survey.

Each patient has a base-line platelet function determination in early phase (1-3 days from the acute phase) and in late phase (20th day); the study, repeated at monthly intervals after the beginning of ASA or persantin, was performed also on 30 healthy individuals of the same age which, during the last two weeks before testing, did not take drugs affecting platelet function.

Laboratory investigation (Table 4) include platelet aggregation (Born and Cross, 1973) by ADP, adrenaline, and collagen; platelet retention test (Hellem, 1970), platelet factor four (PF$_4$) assay (Harada and Zucker, 1971), and bleeding time (Praga et al., 1972). To emphasize eventual differences between normal subjects and patients, aggregating agents were used at different final concentration.

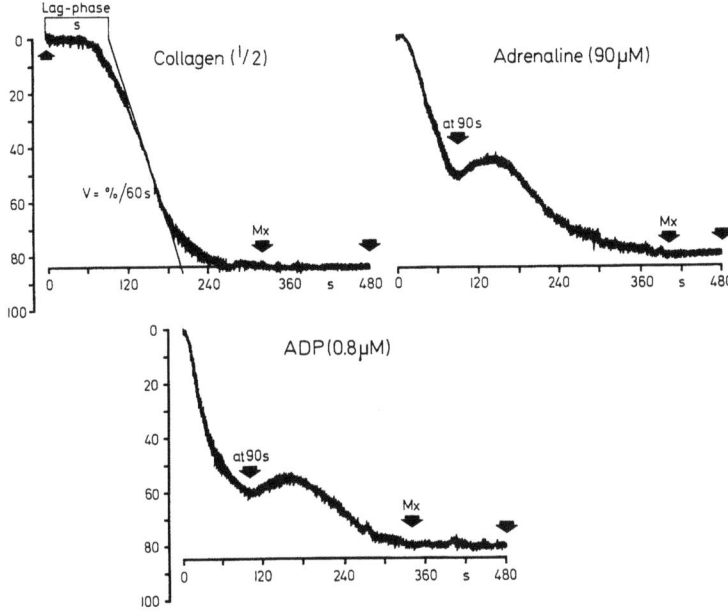

Fig. 1. Standard curves obtained in 30 normal subjects at highest concentration of each aggregating agent

Figure 1 shows the standard curves obtained in 30 normal subjects at the highest concentration of each aggregating agent. Various parameters of each curve had been evaluated; in ADP and adrenaline curves the percent of aggregation at 90 s (initial aggregation) and Mx aggregation have been evaluated; the curve was followed up to 8 min to substantiate eventual disaggregation. In collagen curve the lag-phase, the slope, and Mx aggregation were recorded. These parameters have also been tested at different concentration of the aggregating agent.

The mean, standard deviation, and standard error for each parameter were computed for all groups (normal, TIA, completed stroke); an analysis was done first to test the significance of platelet functional parameters of normal versus completed stroke and normal versus TIA; then, the results of parameters before and after treatment were tested for significance.

Let us now consider the data on normal subjects and patients before treatment. The comparison between normal subjects and acute completed stroke showed no significant difference in adrenaline platelet aggregation at the different concentrations used. The platelet aggregation by collagen at high concentrations does not show any significant difference between the two groups; on

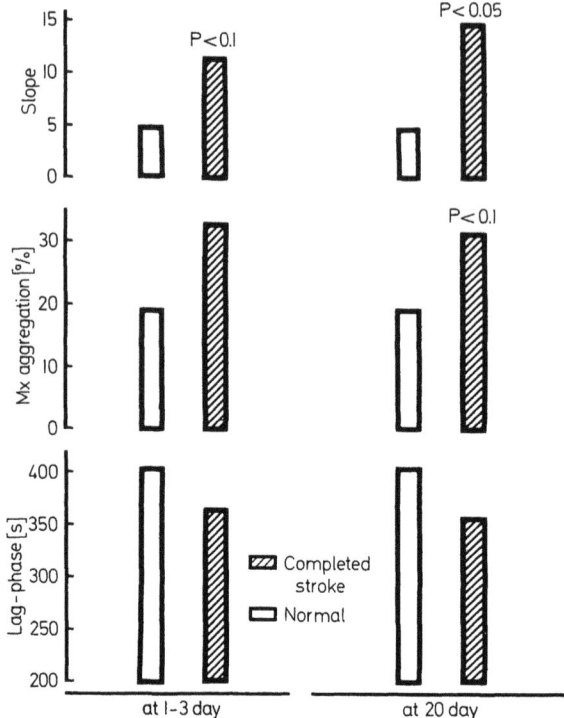

Fig. 2. Mean values of collagen-induced (1/16) platelet aggregation at day 1 and 20

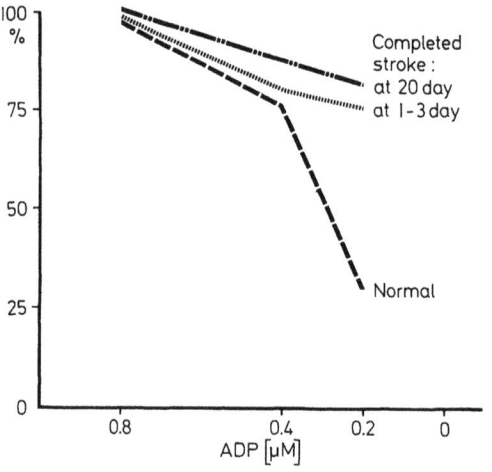

Fig. 3. Percent of irreversible platelet aggregation by decreasing concentration of ADP

the contrary, at highest dilution of collagen (1/16) the slope at 1-3 days, the slope and Mx aggregation at 20th day of the completed stroke differ from normal subjects significantly as shown in Figure 2.

The results of the aggregation by ADP are summarized in Figure 3 where on the ordinate line are the percentage of subjects who present a stable and irreversible curve, and on the abscissa, the final concentration of ADP.

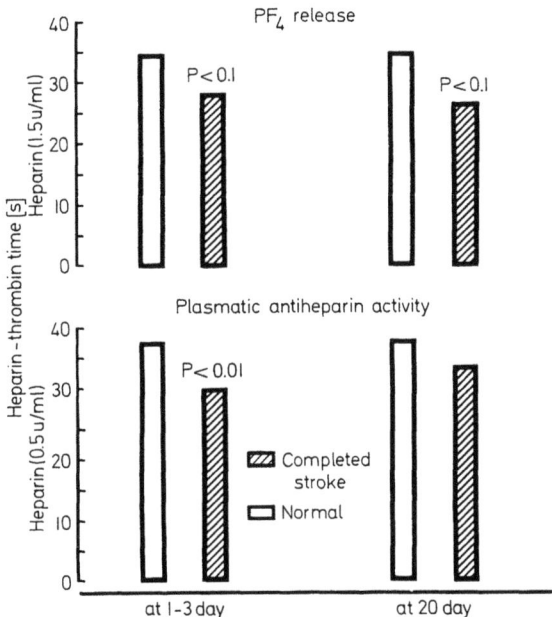

Fig. 4. Mean values of plasmatic antiheparin activity and PF_4 release by collagen

Generally in normal subjects disaggregation occurs within 8 min, particularly if ADP is at the minor concentration. On the contrary, as far as completed stroke patients are concerned, we have found that this group presents a stable curve, more frequently than normal volunteer group, and this difference is higher for minor concentration of ADP (0.2 M). These data are in agreement with the results of Danta (1973) and Boneu et al. (1972).

The statistical analysis of the results of the TIA group has not shown significant differences from normal group, except for the antiheparin plasmatic activity (AHA), which in TIA group is significantly higher than control group.

In completed stroke we could show this plasmatic AHA and the PF_4 release significantly higher than controls, both at 1-3 days and at day 20 (Fig. 4). This result acquires relevance, if we consider that the PF_4 release values of completed stroke are significantly correlated with the seriousness of the neurologic score (Fig. 5).

In conclusion, these studies, according to other authors (Boneu et al., 1972; Born and Cross, 1973; Danta, 1973) suggest a hyperreactivity of the platelets in complete stroke; such an irritability can be shown by using very low concentrations of aggregating agents.

Moreover, the increase of PF_4 activity in TIA and completed stroke, which is considered a reliable parameter of release reaction, may be involved in further thromboembolic complications by its paracoagulability activity (Niewiaroski et al., 1968; O'Brien et al., 1974).

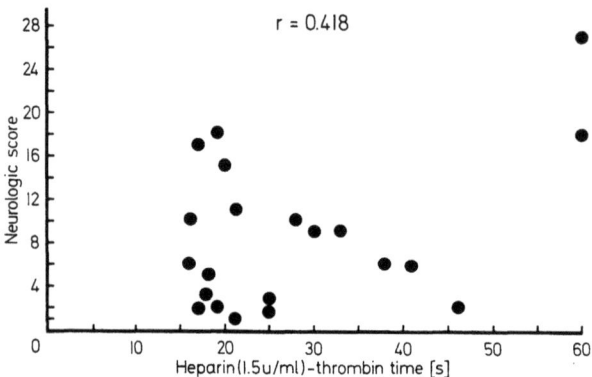

Fig. 5. Completed stroke: correlation between PF$_4$ release by collagen and clinical score

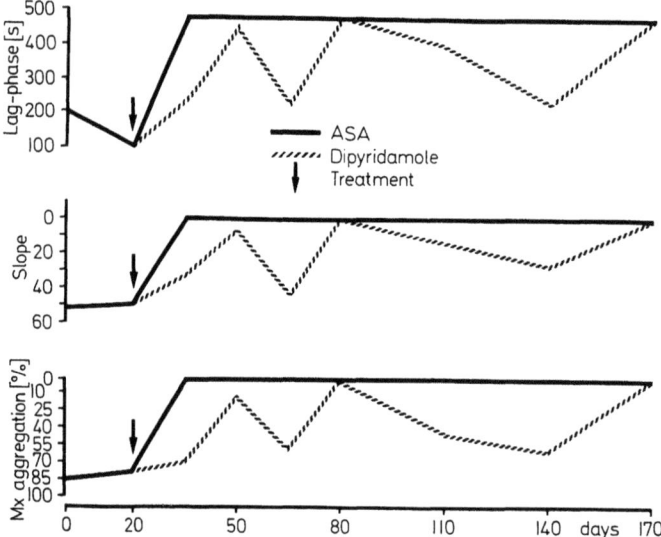

Fig. 6. Completed stroke: mean changes of collagen-induced platelet aggregation with ASA (——) and dipyridamole (····) treatment (↓)

When the effect of the antiaggregating agents was tested, comparable quantitative data were obtained concerning the inhibitory action of ASA and dipyridamole on platelet aggregation. As already known, the ASA inhibiting the release reaction, shows its activity, particularly in collagen aggregation curve (Fig. 6). In fact, ASA increases the lag-phase, inhibiting aggregation within 8 min; dipyridamole, also, at the dose of 450 mg/day increases the lag-phase and reduces the other constants; however, its effect is time-dependent.

The effect of ASA on release reaction is confirmed by the inhibitory effect of PF$_4$ release, as it is shown in Figure 7.

In the ADP aggregation curve, dipyridamole reduces the various constants more than aspirin (Fig. 8). In adrenaline aggregation, on the contrary, the ASA influences, particularly the parameters as Mx aggregation at 480 s, depending on release reaction, while

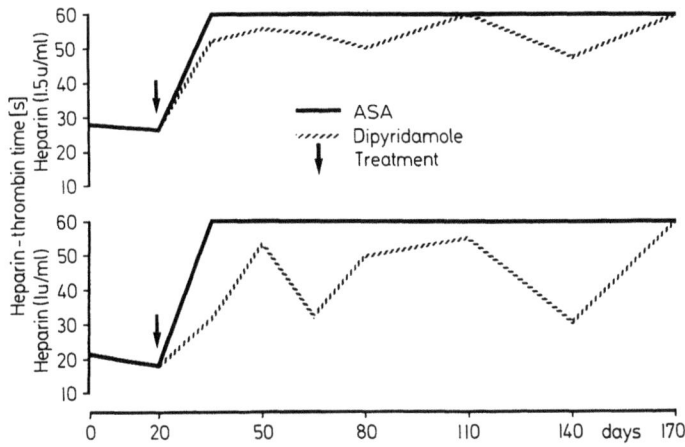

Fig. 7. Completed stroke: mean changes of PF$_4$ release by collagen with ASA (———) and dipyridamole (····) treatment (↓)

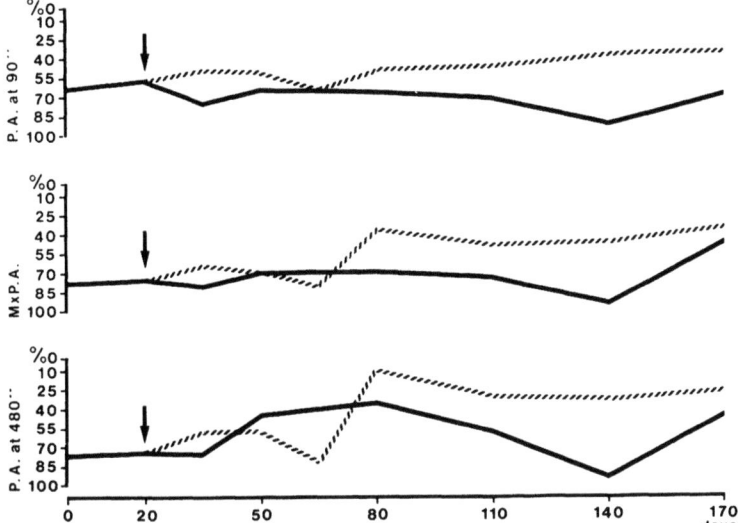

Fig. 8. Completed stroke: mean changes of adrenaline-induced (0.8 μM) platelet aggregation (PA) with ASA (———) and dipyridamole (····) treatment (↓)

in this reaction dipyridamole is not effective (Fig. 9). As far as the effect of drugs in glass-bead columns is concerned, we could confirm that ASA (1 g/day) does not reduce platelet retention of the patients, while dipyridamole reduces it a little (Fig. 10).

The bleeding time, which reflects the effects of drugs on thrombosis, is increased after 3 months of treatment (Fig. 10).

The aim of antiaggregant treatment in TIA is to reduce the number of ischemic attacks and to prevent stroke; in completed stroke

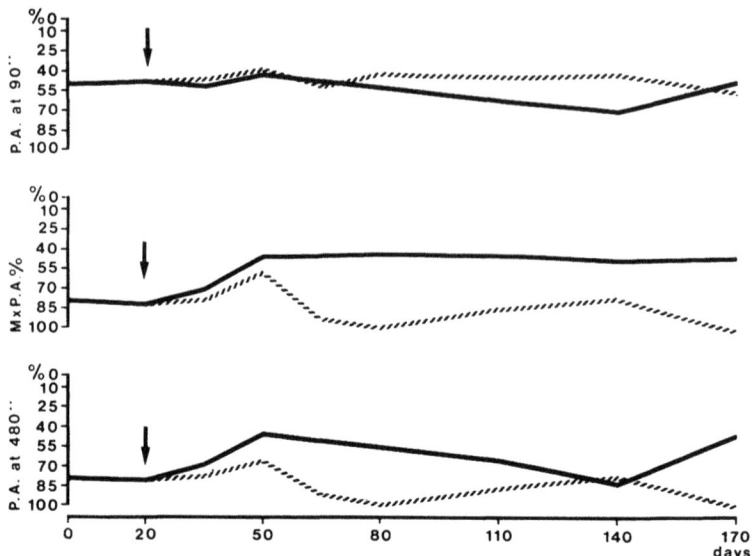

Fig. 9. Completed stroke: mean changes of adrenaline-induced (90 µM) platelet aggregation (PA) with ASA (——) and dipyridamole (····) treatment (↓)

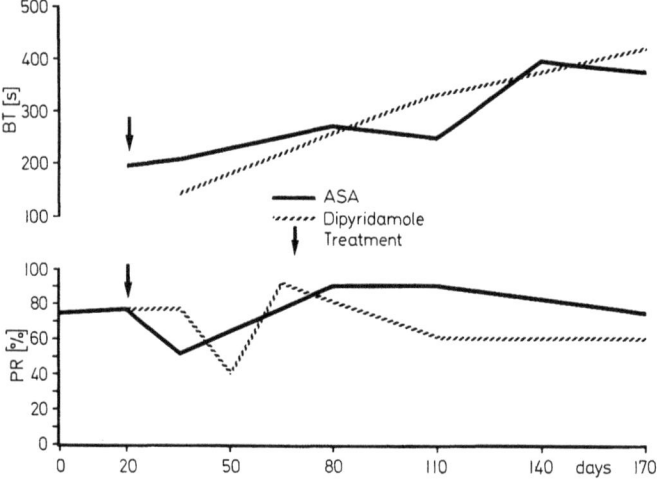

Fig. 10. Completed stroke: effect of ASA and dipyridamole treatment (↓) on mean values of platelet retention (PR) and bleeding time (BT)

the aim is to prevent recurrent thrombosis or embolism in all parts of the body, including the brain, and to increase life expectancy. Obviously these conclusive clinical results require a case history of hundreds of patients and a few years of observation. It has to be emphasized that the thrombotic process is an intravital event and the result of various interacting fac-

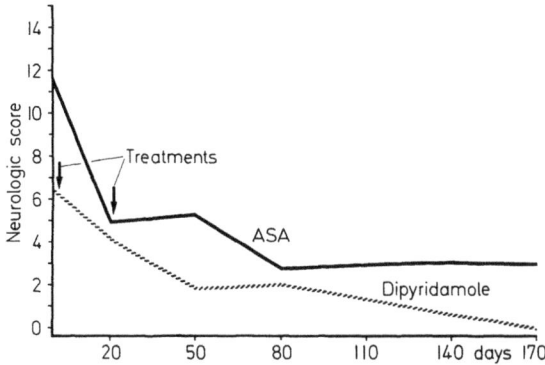

Fig. 11. Mean average of neurologic score before and several times after ASA (———) and dipyridamole (·····) treatment (↓)

tors, and the demonstration that some drugs inhibit the function of one of these factors, even as important as platelets, is not enough to define such drugs as antithrombotic. This will only be obtained by the results of polycentric trials.

The short time during which the present trial has been carried out allows us to state that the mean average neurologic score is similar in patients treated respectively with ASA and dipyridamole (Fig. 11).

References

Boneu, B., Guiraud, B., Fernet, P.: Traitement anti-agrégant plaquettaire par l'aspirine. Nouv. Presse méd. 1, 863 (1972)
Born, G.V.R., Cross, M.J.: The aggregation of blood platelets. J. Physiol. (Lond.) 168, 178 (1973)
Danta, G.: Platelet aggregation in patients with cerebral vascular disease and in control subjects. Thromb. Diath. haemorrh. (Stuttg.) 29, 730 (1973)
Fields, W.S., Hass, W.K. (eds.): Aspirin, Platelets and Stroke. St. Louis, Missouri: Warren H. Green, Inc. 1970
Harada, K., Zucker, M.B.: Simultaneous development of platelet factor 4 activity and release of ^{14}C serotonin. Thromb. Diath. haemorrh. (Stuttg.) 25, 41 (1971)
Harker, L.A., Slichter, S.J.: Arterial and venous thromboembolism: kinetic characterization and evaluation of therapy. Thromb. Diath. haemorrh. (Stuttg.) 31, 188 (1974)
Hellem, A.J.: Platelet adhesiveness in von Willebrand's disease. A study with a new modification of the glass bead filter method. Scand. J. Haemat. 7, 374 (1970)
Harrison, M.J.G., Marshall, J., Meadows, J.C., Ross Russel, R.W.: Effect of aspirin in amaurosis fugax. Lancet 2, 743 (1971)
Mc Nicol, G.P., Mitchell, J.R.A., Reuter, H., van de Loo, J.: Platelets in thrombosis. Their clinical significance and the evaluation of potential drugs. Thromb. Diath. haemorrh. (Stuttg.) 31, 379 (1974)

Niewiaroski, S., Lipinski, B., Farbiszewski, R., Poplawski, A.: The release of platelet factor 4 during platelet aggregation and the possible significance of this reaction in Hemostasis. Experientia (Basel) $\underline{24}$, 343 (1968)

O'Brien, J.R., Etherington, M., Jamieson, S., Lawford, P.: Blood changes in atherosclerosis and long after myocardial infarction and venous thrombosis. Int. Soc. Hematol. Meeting Jerusalem, 1974

Praga, C., Cortellaro, M., Pogliani, E.: Standardized bleeding time in the study of drugs interfering with platelet function. In: Platelet Function and Thrombosis. Mannucci, P.N., Gorini, S. (eds.). New York: Plenum Press 1972, p. 149

Ross Russel, R.W.R.: The source of retinal emboli. Lancet $\underline{ii}$, 789 (1968)

Zucker, M.B., Peterson, J.: Effect of acetylsalycilic acid, other non steroidal anti-inflammatory agents and ipyridamole on human blood platelets. J. Lab. clin. Med. $\underline{76}$, 66 (1970)

The Pharmacologic Control of the Enhanced Platelet Aggregation in Preventive Neurology

F. Federici, S. Biagini, R. Eggér, F. Marchionni, and G. Penchini. In Collaboration with A. Ferroni, E. Signorini, C. Tardioli, and F. Bazzanella

It is well known that the coaction of hereditary, metabolic, and environmental factors play an important role in the pathogenesis of cerebral vasculopathy. Each one of these three factors may be hypothesized and analyzed with extreme severity. This extreme severity must, however, be measured according to how rough the diagrams of clinical integration are: "better," "the same," "worse" are the final references to the man before us and not to the integrated model we have analyzed.

Just what is it that neurologic clinics must analyze in reference to the central nervous system? Bloodstream variations, and anoxia decrease are certainly noteworthy data but aside from acute episodes they do not inform us about the state of the brain with the language peculiar to it.

We feel, therefore, that the answer to external stimuli, their memorization, and behavioural organization, all furnish certain reliable parameters for evaluating the evolution of a noxa or hypothesizing on the risk of disease.

Two other research sections have considered the social and environmental context of patients and the parameters concerning hematologic, biochemical, and fat organization profiles.

This study technique on the superior nervous activity has utilized: the analysis of visual-motor and acoustic-motor time responses, various tests, and behaviour when confronted with hospitalization.

Before presenting the apparatus used for generating and recording time responses it would be opportune to set down certain points that we feel furnish a basis for neurofunctional analysis.

In neurologic semeiotics, the anatomical articulation of optical and acoustical pathways allows us to hypothesize with great likelihood as to: prevalent monohemispheric involvement, increase or decrease of signals, simultaneous or program-phased stimulation.

Examples of some of these experimental conditions can be seen in the first variation level in Figure 1 with the sound frequency that is used as a stimulus. Figure 2 shows the biauricularity and the possibility of a 30-70% distribution or a substantially equivalent bitemporal involvement. The third level, constituted

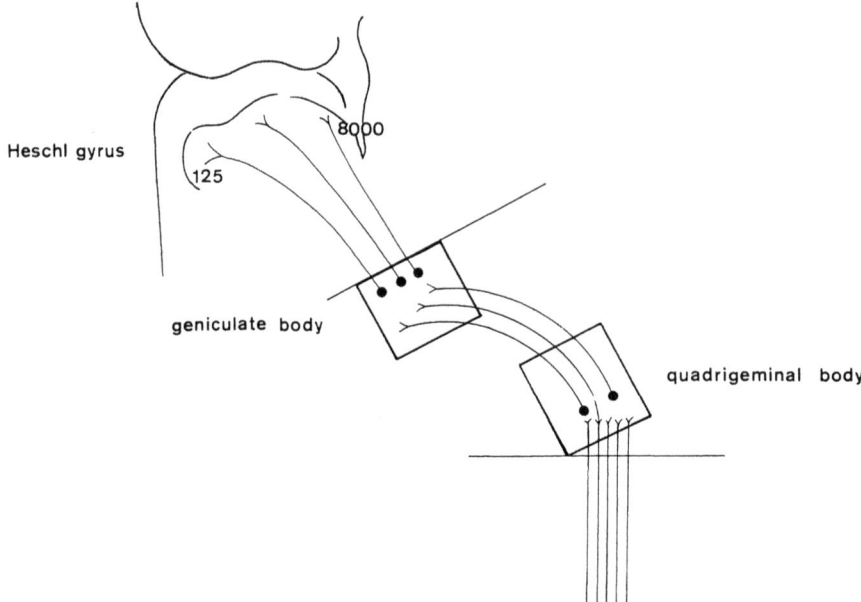

Fig. 1. Sound frequency used as stimulus

by the motor response which is expressed by the hand used by the hemisphere concerned in the stimulation.

The optic pathway in Figure 3 exemplifies the enrichment in the visual field which can be induced by simple mono- or binocularity.

Figures 4 and 5 are reminders of the ideal condition, i.e., "lumière dirigée" for a prevalently monohemispheric occipital stimulation, and in this case the particular stress that the motor response assumes when expressed by the homo- or contralateral hemisphere. All this makes more sense if we consider that the visual and acoustical motor response times are remarkably influenced by the intensity of the stimulus.

Figure 6 represents the apparatus used in the neurological semeiotics of the superior nervous activity laboratory. The patient is placed in the correct position in front of a field (range) meter and is trained beforehand to respond by pressing a button leading to a light stimulation with either the right or the left hand. This is the beginning of the test to consider the density of stimulation, the hemi-fields in which it is placed, the use of the right eye and that of the left one, of both eyes, of the response with the right hand, and with the left one. For each of these conditions 10 time tests are recorded. The same criteria subtend the acoustical stimulation tests that can be deduced from the stimulation possibilities represented by the identical formalities used for collecting the responses.

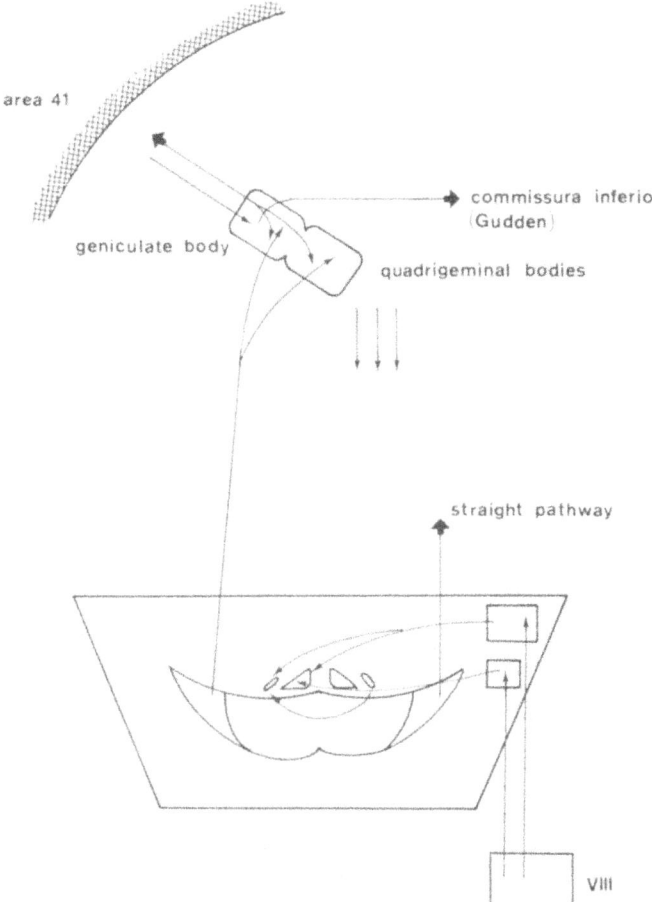

Fig. 2. Biauricularity and possibility of 30-70% distribution or substantially equivalent bitemporal involvement

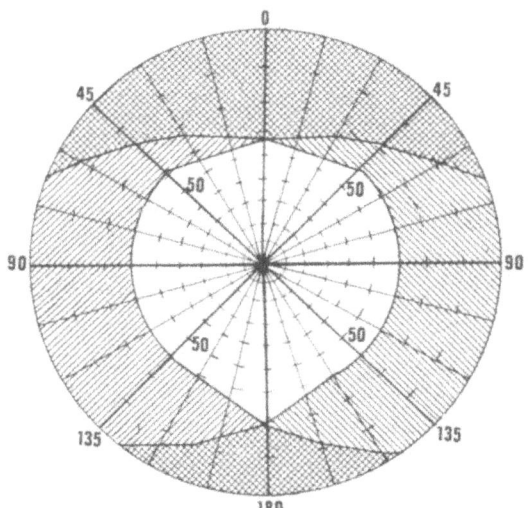

Fig. 3. Visual field with enrichment induced by binocularity

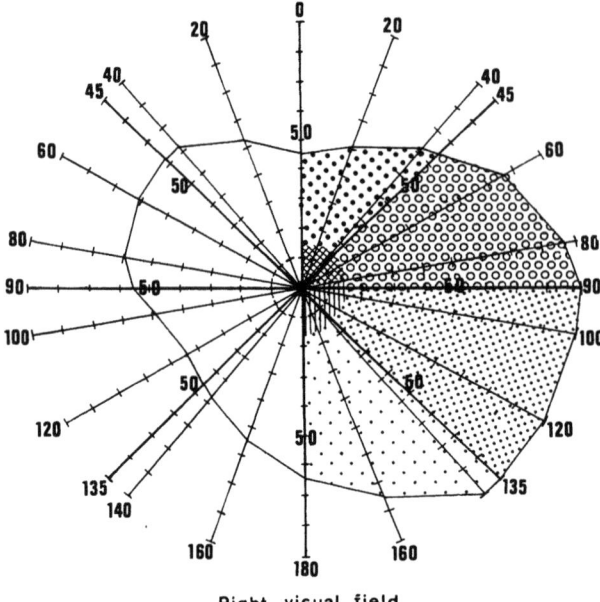

Right visual field   Fig. 4

Fig. 5

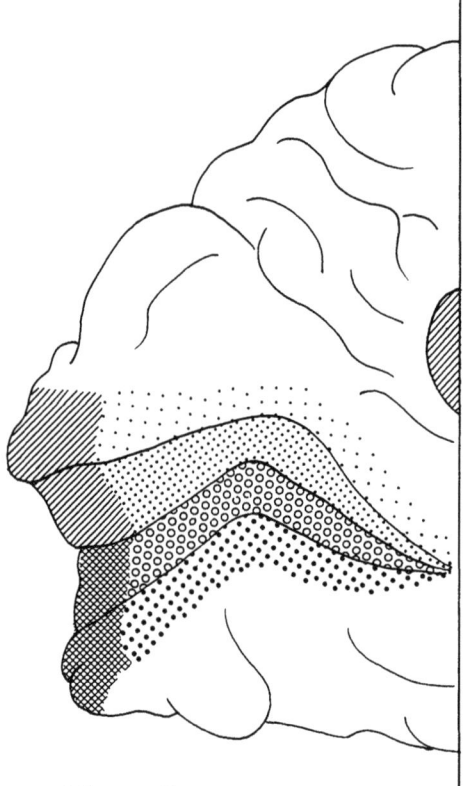

left area 17

Figs. 4 and 5. Stress assumed by motor response when expressed by homo- or contralateral hemisphere

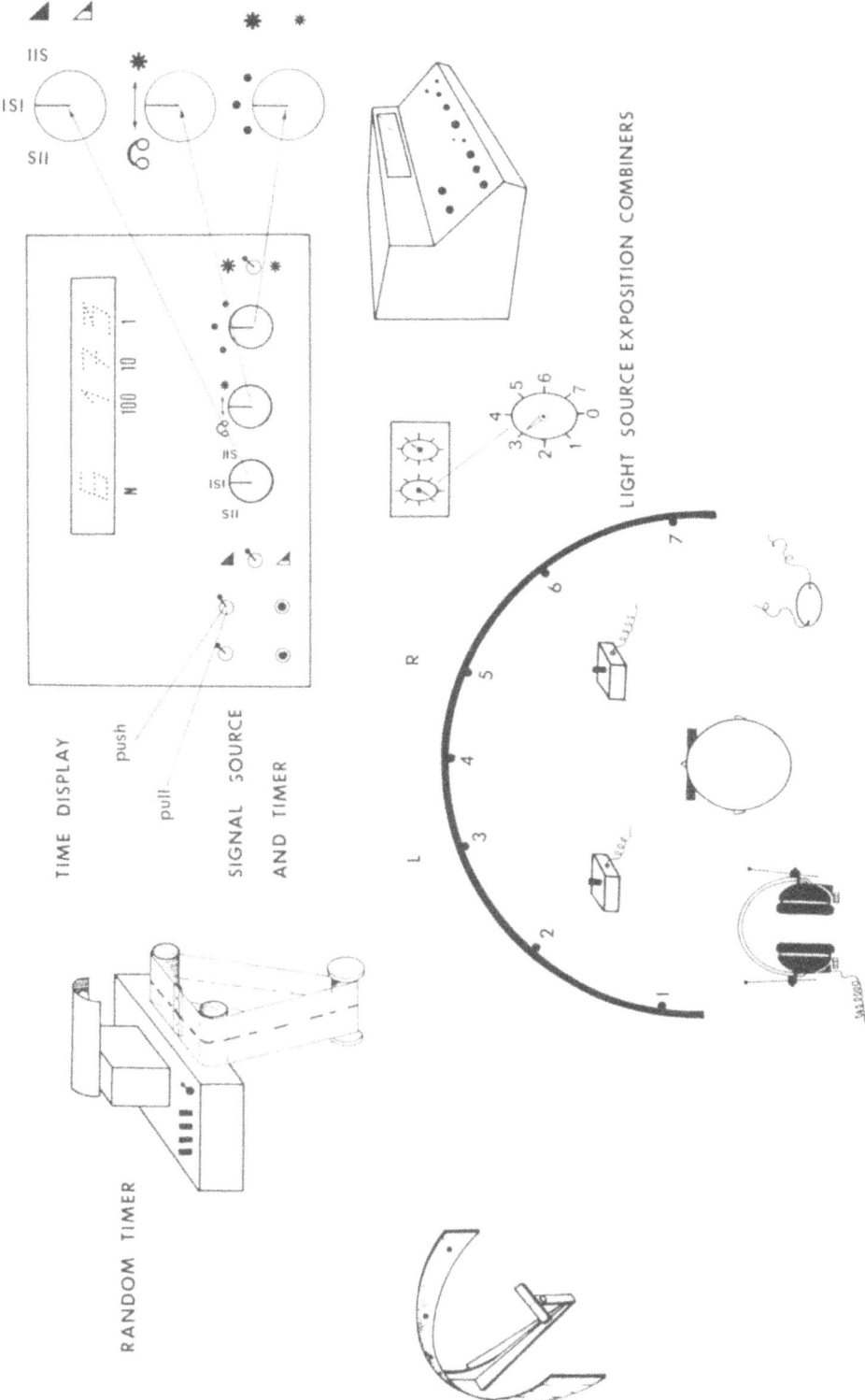

Fig. 6. Laboratory apparatus

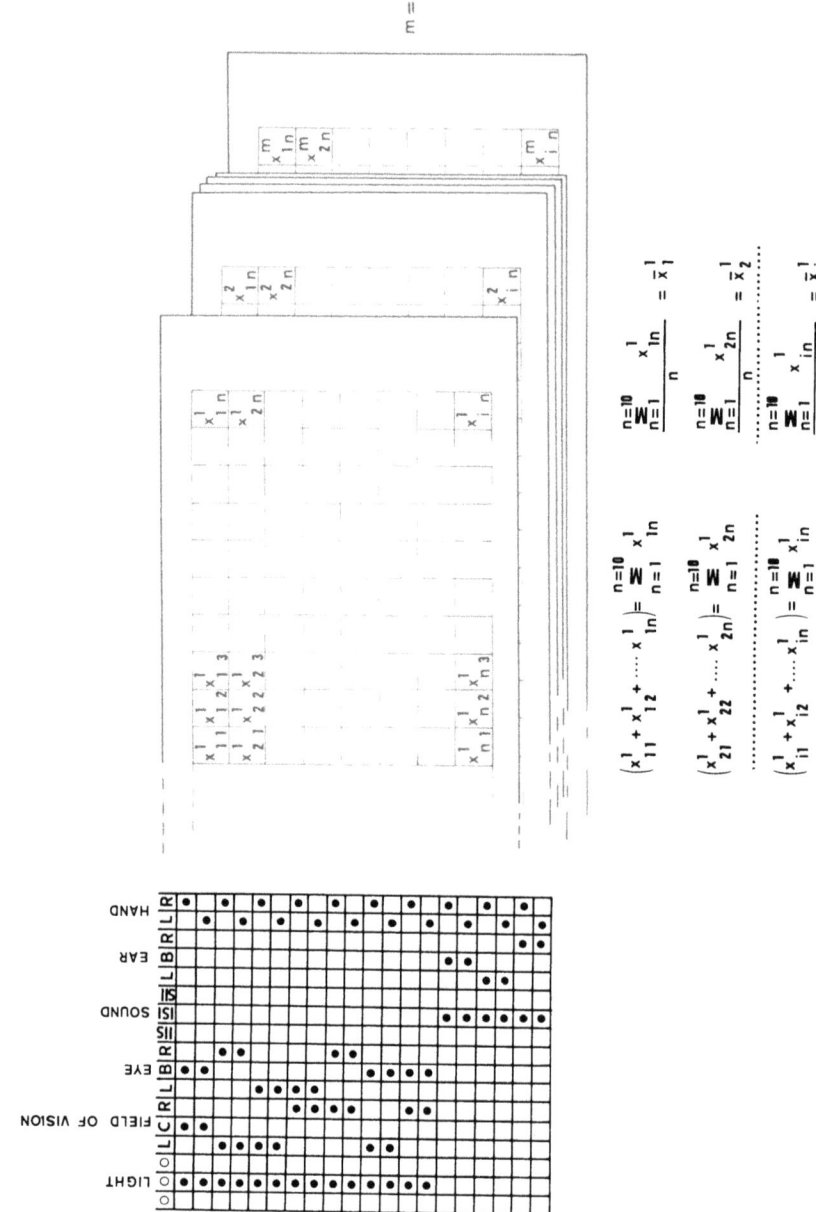

Fig. 7. Characteristics of index synthesizing experimental data

Let us briefly look at the mathematical elaboration techniques that led to the preparation of our grills. Figure 7 represents the characteristics of the index synthesizing the combination of the various experimental conditions and the data obtained. After determining the mean times for each test on each individual, the table seen in Figure 8 is constructed by calculating the mean averages of the whole and the fiduciary limits. Figure 9 shows the ratio between each mean value with succeeding ones calculated in each subject. With this information we obtain the mean ratio for all the subjects examined and these determine the fiduciary limits. This elaboration which has tended to furnish the possibility of integrating the data and corrections of casual variations in the evaluation comparisons, allows for faster reading regarding the averages and an articulated comparison obtainable regarding the ratios.

Following an initial evaluation of superior nervous activity centered on the stimulus-response times, there is an investigation on the memorization capacity with a technique using the possibility of remembering standardized drills for both short and long periods of time.

Figure 10 shows the examining criteria of the tests, retesting occurs at various times for sampling.

The behavioural response as quantifiable observation data is very simply constituted by rehospitalization while the subject is under treatment. This already indicates that we may hypothesize that the evolution pattern is getting worse.

One technique must be verified and this has made it necessary to examine paradigmatical cases of sure clinical checks. In Table 1 various clinical parameters are examined for their percentage distribution within the group studied; it is to be noted that this grouping also comprises cerebral neoplasy and primitive atrophy. In these cases our aim was to evaluate semeilogic localization values of the techniques used with specific tests and to understand just how the various lesional levels influenced the various tests.

In Table 2 it is interesting to note the percentages of positivity of the various diagnostic techniques in relation to the functional diagnostics we considered. Finally, Table 3 presents the specific aspect under examination: we consider 18 months of observation carried out during two different periods, on two groups of patients, all in their 50s with subjective evaluation of deficit in their superior nervous activity.

All observations took place at the University Neurological Clinic. It must be specified that no noteworthy behaviour changes characterize the two periods, if not the onset of using antiaggreganting substances in the sequels to vascular brain incidents and in chronic circulatory insufficiency after dismissal.

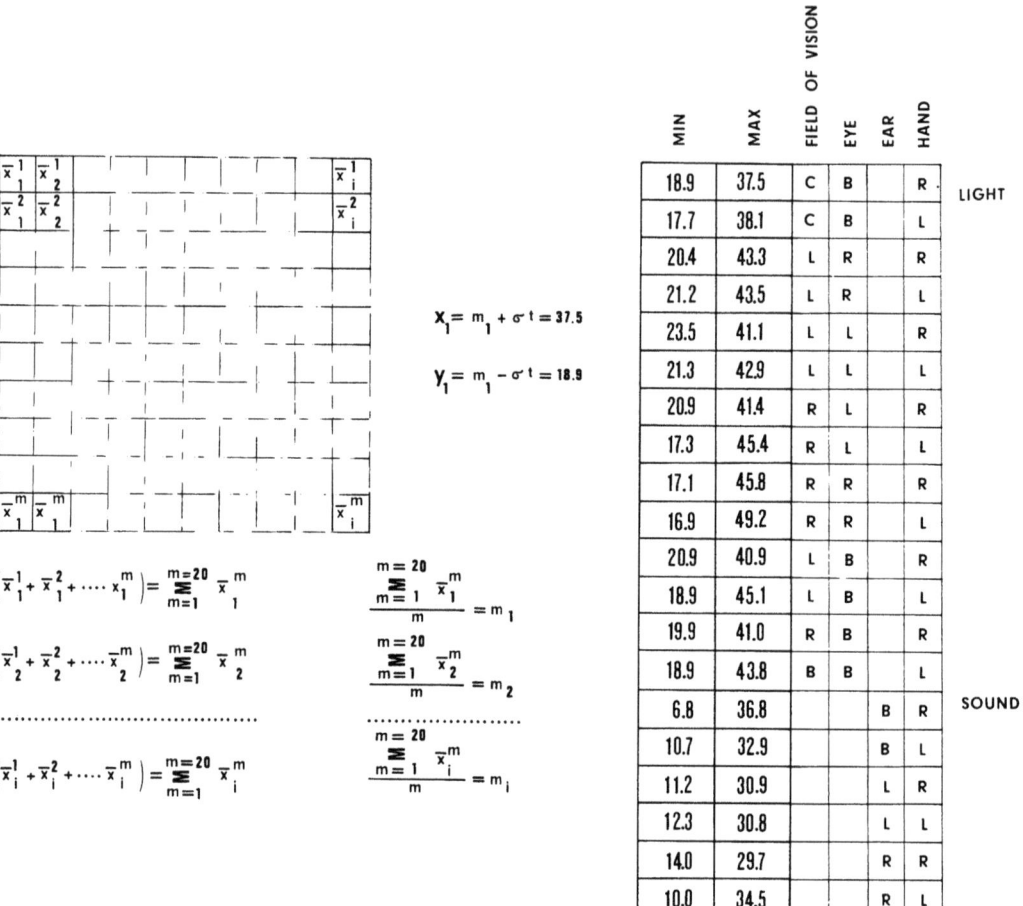

Fig. 8. Mean averages of whole and fiduciary limits

The criteria of administration were clinical and any contraindications to treatment were constituted by the hypothesis of improved hemorrhagic pathology.

The limited number of observations in the result of having to stress the qualitative aspect of analysis and the sureness of being able to protract the observations of subjects for a reasonably long time.

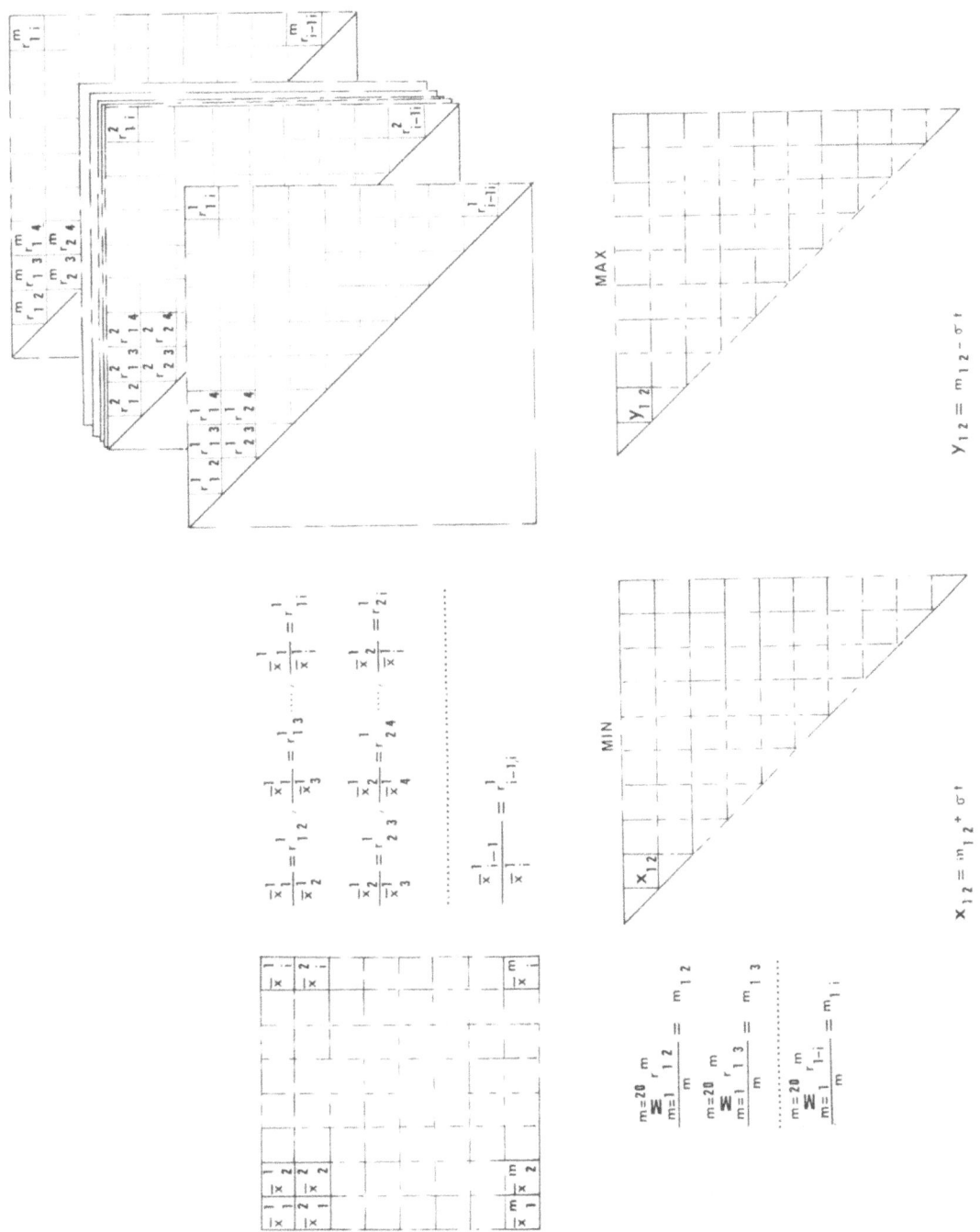

Fig. 9. Ratio between mean values with succeeding ones in each subject

Items	Psychosensory performances Not verbal symbolization	Motor and psychomotor	Psychosensory performances Verbal symbolization	Short term memory	Long term memory
Motor Performance		●			
Stereognosis	●				
Visual Perception	●				
Color Agnosia	●				
R. 15	●				
B.G.T.V.M.	●				
House, pot, tree				▲	
8, 3, 26				▲	
Kohs blok designs	●				
Stylus Maze	●	●		▲	■
Praxic memory	●	●		▲	
Memory of numbers	●		●	▲	
Memory of significance	●		●	▲	
Not structured profile	●			▲	
Correction of normal illusion	●			▲	
Reversible Figures	●			▲	
Number – Object	●		●	▲	
Babcoock's Law			●	▲	
Verbal symbolic net			●	▲	■
Time Scansion	●			▲	
Self valuation for short term memory					
Self valuation for long term memory					■

Fig. 10. Examining criteria of tests

Table 1

Reason for admission	By family 12.6	His own 32.2	By family doctor 49.3	Specific 27.2	Acute illness 7.5	Chronic illness 4.4	Other 1.8
Family pathologic history	Normal 25.1	Metabolism disease 12.9	Vascular disease 51.9	Psychiatric illness 11.4	Other 32.8		
Deficit ANS subjective evaluation	Normal 3.1	"I do not remember" 17.7	"I am different" 25.3	"I am more troubled" 24	"I work worse" 16.4	Other 20.8	
Deficit ANS neighborings' evaluation	Normal 83.9	"He is different" 13.7	"He is less efficient" 6.8	"He is less sociable" 3	Other 5.3		
Trouble's beginning	+ 15 years 7.5	+ 10 years 4.4	+ 5 years 11.3	+ 1 years 32.2	+ 6 months 12.6	− 6 months 31.6	
Former treatments	Antiasthenic drugs 1.2	"Vascular" drugs 10.7	Psychiatric drugs 18.9	Pyramidal + cerebellar + sensorial 2.1	Physiotherapy 3.1	Other 50.6	None 21.5
Other concomitant diseases	Yes 36.7	No 61.3					
Neurologic symptoms	R 15.8	L 12.4	Pyramidal 6.2	Pyramidal + cerebellar + sensorial 2.1	Neuro-muscular 1.3	Extra-pyramidal 4.8	Other 2	None 54.4
Cranial nerves' diseases	None 82.7	1 15.1	+ 1 2				
Epilepsy	Absent 83.4	Partial 6.8	Diffuse 9.6				
Visual field	Not performed 91	Normal 30.7	Pathological 69.2				
Cochleovestibular examination	Not performed 84.8	Normal 36.3	1 Pathological sign 36.3	Pathological 69.2	+ 1 Pathological sign 31.8		

Continuation of Table 1

Hematologic profile	Normal 80.1	1 Pathological sign 16	+1 Pathological sign 4.5				
Blood biochemical profile	Normal 75.5	1 Pathological sign 18.3	+1 Pathological sign 5.3				
Electroencephalogram	Normal 15.9	Bitemporal theta activity 25	Right abnormalities 9	Left abnormalities 18	Diffuse abnormalities 31.9	Not performed 25.3	
Skull x-Ray	Normal 71.5	Calcifications 12.5	Erosions 10.2	Convolutional markings 0	Pineal dislocation 0	Several signs 3.4	Other signs 5.6	Not performed 39.3
Scintigraphy	Normal 54.5	Right abnormalities 22.7	Left abnormalities 18.1	Other signs 4.5	Not performed 84.8			
Pneumoencephalography	Normal 26.9	Right space occupying lesion 11.5	Left space occupying lesion 3.8	Right atrophy 3.8	Left atrophy 3.8	Diffuse lesion 41.6	Other signs 3.8	Not performed 82
Angiography	Normal 38.2	Not performed 67.5	Right space occupying lesion 17	Left space occupying lesion 19.1	Right vascular abnormalities 2.1	Left vascular abnormalities 10.6	Diffuse lesion 8.5	Other signs 4.2
Illness' evolution in hospital	Improvement 52.6	Aggravation 2.2	Changeless course 45					

Table 2

Electroencephalogram	Normal	Bitemporal theta activity 25	Right abnormality 9	Left abnormality	Diffuse abnormality 31	Not performed 25		
Skull x-ray	Normal 71	Calcification 12	Erosions 10	Convoluted markings 0	Pineal dislocation 0	Several signs 3	Other signs 5	Not performed 39
Scintigraphy	Normal 54	Right abnormality 22	Left abnormality 18	Other signs 4	Not performed 84			
Angiography	Normal 38	Right space-occupying lesion 17	Left space-occupying lesion 19	Right vascular abnormality 2	Left vascular abnormality 10	Diffuse lesion 8	Other signs 4	Not performed 67
Pneumoencephalography	Normal 26	Right space-occupying lesion 11	Left space-occupying lesion 3	Right atrophy 3	Left atrophy 3	Diffuse lesion 46	Other signs 3	Not performed 82
ANS evaluation 0-4	Deficit ANS 0-1 65	Deficit ANS 2-3 27	Deficit ANS 4 7	Not performed 0				

From Nov., 1969 to May, 1971 - Patients: no. 56   For 16 Patients: For 6 Patients:
                                                     2 admissions   3 admissions
From Oct., 1971 to Dec., 1972 - Patients: no. 47   For 9 patients:
                                                     2 admissions

Group A: "Vascular" drugs + antiedema drugs + oxygen (1964-1968)
Group B: Antiedema drugs + oxygen + blood pressure measuring + hypotension management (1969-1970)
Group C: Idem + antiaggregation treatment soon after vascular accident + antiaggregation treatment
         after discharge (1971-1972)

ANS evaluation: Group B: 1   = 51%
                                               2-3 = 39%
                                               4   = 10%

                                   Group C: 1   = 56%
                                               2-3 = 36%
                                               4   = 8%

Family doctor's ANS evaluation after 3-4 months: Group B: ± 65%
                                                              - 27%
                                                              + 13%

                                                   Group C: ± 71%
                                                              - 13%
                                                              + 16%

Second test after 6 months with above-mentioned methodology performed on: 29% B and 31% C with these results:

                                   Group B: + 13%
                                              - 8%
                                              ± 79%

                                   Group C: + 28%
                                              - 6%
                                              ± 66%

References

Aaronson, Doris: Temporal factors in perception and shortterm memory. Psychol. Bull. 67, 130 (1967)

Averbach, E., Sperling, G.: Short term storage of information in vision. In: Information Theory. Cherry, C. (ed.). London: Butterworths 1961, pp. 196-211

Brown, J.: Some tests of the decay theory of immediate memory. Quart. J. exp. Psychol. 10, 12 (1958)

De Nicola, P.: First meeting of the european division of international society of haematology. Ed.: Boehringer, Milano 1971

De Renzi, E.: Deficit gnosici, prassici mnesici e intellettivi nelle lesioni emisferiche unilaterali. Atti del XVI Cong. Naz. di Neurol. 371, 430 Ed. "Il pensiero Scientifico", Roma 1967

De Renzi, E.: Nonverbal memory and hemispheric side of lesion. Neuropsychologia 6, 1968

De Renzi, E., Spinnler, H.: The influence of verbal and nonverbal defects on visual memory tasks. Cortex 2, 322 (1966)

Emmons, P.R., Harrison, M.Y.G., Honour, A.J., Mitchell, J.R.A.: Effect of a dipyridamole on human platelet behaviour. Lancet II, 603 (1965)

Federici, F., Quattrini, A.: Un test di sensopercezione per lo studio della traccia mnesica in soggetti con gravi disturbi della memoria. Rivista di Neurobiologia XIV, 1 (1968)

Luria, A.R., Sokolov, G.N., Klimkovski, M.: Towards a neurodynamic analysis of memory disturbances with lesions of the left temporal lobe. Neuropsychologia 5, 1 (1967)

Melton, A.W.: Implications of short term memory for a general theory of memory. J. verb. Learn. verb. Behav. 2, 1 (1963)

Milner, Brenda: Impairment of visual recognition and recall after right temporal lobectomy in man. Paper read at 1st Ann. Meeting, Psychonomic Society, Chicago, Sept. 1960

Milner, Brenda: Laterality effects in audition. In: Interhemispheric Relations and Cerebral Dominance. Mountcastle, V.B. (ed.). Baltimore: The Johns Hopkins Press 1962a, pp. 177-195

Milner, Brenda: Les trubles de la mémoire accompagnant des lésions hippocampiques bilatérales. In: Physiologie de l'Hyppocampe. Colloques Internationaux No. 107. Paris: C.N.R.S. 1962b, pp. 257-272

Milner, Brenda, Taylor, L.B., Corkin, Suzanne: Tactual pattern recognition after different unilateral cortical excision. Paper read at 38th Ann. Meeting, Eastern Psychol. Ass., Boston, April 6-8, 1967

Rey, A.: L'examen psychologique dans les cas d'encêphalopathie traumatique. Arch. Psychol. 28 (No. 112) (1942)

Teuber, H.-L., Battersby, W.S., Bender, M.B.: Performance of complex visual tasks after cerebral lesions. J. nerv. ment. Dis. 114, 413 (1951)

Wechsler, D.: A standardized memory scale for clinical use. J. Psychol. 19, 87 (1945)

Zangwill, O.L.: Clinical tests of memory impairment. Proc. Roy. Soc. Med. 36, 576 (1943)

Aspirin in Cerebral Ischemia

W. S. Fields, P. W. Callen, and M. M. Preslock

It has become increasingly apparent that aspirin induces a marked delay in the aggregation properties of platelets. It causes an inhibition of collagen-induced platelet aggregation in association with and probably due to a decreased release of ADP. This occurs in vitro. The response of platelets to epinephrine-induced release of ADP in the second wave of aggregation is also blocked by aspirin. In addition, the drug inhibits both collagen and ADP-induced activation of platelet factor III (phospholipid-like factor) and the release of platelet factor IV in vitro and in vivo, but does not impair the response to thrombin. These effects are uniquely confined to aspirin, since sodium salicylate neither prolongs the bleeding time nor significantly inhibits platelet ADP release.

For the above reasons it was considered imperative to test the effectiveness of aspirin in vivo in order to determine whether it had a clinical benefit in the prevention of arterial thromboembolism. Support for this endeavor was received from the National Heart and Lung Institute.

The Study of Aspirin in Transient Ischemic Attacks is an interinstitutional cooperative study designed to determine the effectiveness of the oral administration of acetylsalicylic acid (aspirin) in the reduction or prevention of transient cerebral ischemic attacks of the hemispheric type and amaurosis fugax. There are 11 institutions involving 15 hospitals and clinics throughout the United States participating in the controlled, clinical drug trial. Only subjects having monocular blindness or the hemispheric type of transient ischemic attack which results in either full recovery or only minimal residual deficit are eligible for admission to the study. Previous extracranial arterial reconstructive surgery is not considered reason for exclusion of an individual patient if that individual has at the time of his work up a reason for inclusion such as hemispheric or monocular events of a transient nature.

Once an individual has met all the criteria for eligibility, he is then carefully questioned with respect to drug sensitivity or idiosyncrasy. All effort is made to determine whether there is a tendency to bleeding disorder or a history of peptic ulcer. Each patient has a base-line platelet aggregation determination and in addition to this the presence of bleeding or clotting disorders is studied from the stand point of various hematologic tests. Stool examinations for blood are performed at the time of follow-up examination in order to determine whether there is any suggestion of gastrointestinal bleeding. Thus far in the study there has been only an occasional report of occult blood in the feces of the patients both in the aspirin and placebo groups. This has not been reported in two consecutive monthly examinations. It was not expected to be a problem with the relatively small dose of aspirin (650 mg twice daily) being used. The average daily dose prescribed for arthritis and chronic pain syndromes is two to three times this amount.

All patients have arteriographic studies of a sufficient extent to demonstrate both carotid artery circulations, intracranially and extracranially as well as the basilar artery. This may be accomplished by any one of several techniques, depending upon the one usually employed in the particular participating institution. Following this, a clinical decision is made regarding surgical intervention, based on more than 50% stenosis of vascular lumen and this determines into which group the patient will be assigned, medical treatment or surgical treatment. After the clinical decision has been made with respect to the treatment, randomization takes place within each group independently and there is an assignment made to aspirin or placebo therapy.

The patients return to the clinic or hospital for follow-up examinations every 4 weeks for the first 6 months and then every 3 months for the duration of the study. Platelet aggregation tests are done on follow-up visits for the purpose of surveillance of adherence to treatment allocation. A patient is considered to have reached an absolute end point if any of the following events ensue:

1. Death

2. Cerebral hemorrhage

3. Cerebral infarction

4. Myocardial infarction

5. After first 7 days, the patient has hemispheric attack producing objective neurologic deficit persisting longer then 48 h.

6. Establishment of a serious complication resulting from aspirin or placebo therapy at any time after admission to the trial.

Patients are withdrawn from the study and only relevant data included in the analysis if one of the following occurs:

1. Failure to take medication for any reason for a period of 6 weeks.

2. Inadvertently taking aspirin-containing compounds (nearly 300 are available in pharmacies in the United States).

3. Refusal or inability to return for regular follow-up visits.

4. Development of a nonvascular intracranial tissue-destructive lesion.

5. Patient originally admitted to nonsurgery group undergoes arterial reconstructive surgery involving cerebral circulation.

6. Patient requires other anticoagulant or platelet-inhibiting drug during a period lasting longer than 6 weeks.

The first patient was admitted to the study in October, 1972. Between then and mid-September, 1974, 1070 patients had been screened for possible admission. Of these, 811 were excluded for one or more of the reasons cited above. Of the 259 included, 151 were medically treated and 108 had carotid artery surgery. More than half of the patients in the study have now been followed for a period of at least 6 months. It has been decided by the steering committee for the study not to break the double-blind code until an additional 5-6 months experience has been realized.

Primary analysis made only by our biostatisticians suggests to them that there is a trend in the reduction or cessation of transient ischemic attacks in favor of the aspirin-treated groups in both the surgical and medical patients. This, however, must not be construed at this time to be a recommendation for aspirin therapy in this disease.

There has been no other controlled study of platelet suppressant therapy in transient ischemic attacks thus far reported other than the small double-blind crossover study of sulfinpyrazone reported by Evans at the Princeton Conference on Cerebral Vascular Diseases in 1972. A large-scale trial comparable to the one reported here is being conducted in Canada. In that trial aspirin and sulfinpyrazone are being tested individually and together against placebo therapy. In both trials a determined effort is being made to avoid breaking the double-blind code until the point of statistical significance has been reached.

There have been numerous reports of anecdotal experience with aspirin and many patients in the United States are currently being treated by primary physicians and specialists on an empirical basis with aspirin for the control of both cerebral and myocardial ischemia. This has made it very difficult to recruit larger numbers of subjects into the ongoing studies. The participants in both studies feel that the ultimate goal must be to determine the effectiveness of these drugs in the prevention of cerebral and myocardial infarction and not merely the recurrent transient events which may preceed such infarctions.

References

Boston Collaborative Drug Surveillance Group: Regular aspirin intake and acute myorcardial infarction. Brit. med. J. **1**, 440-443 (1974)

Elwood, P.C., Cochrane, A.L., Burr, M.L., Sweetman, P.M., Williams, G., Welsby, E., Hughes, S.J., Renton, R.: A randomized controlled trial of acetyl salicylic acid in the secondary prevention of mortality from myocardial infarction. Brit. med. J. **1**, 436-440 (1974)

Evans, G., Packham, M.A., Mustard, J.F.: Chapter VII, Thromboembolism. In: Aspirin, Platelets and Stroke. Fields, W.S., Hass, W.K. (eds.). St. Louis Mo.: Warren H. Green, Inc. 1970

Evans, G.: Effect of platelet-suppressive agents on the incidence of amaurosis fugax and transient cerebral ischemia. In: Cerebral Vascular Diseases. 8th Conference, McDowell, F.H., Brennan, R.W. (eds.). New York: Grune and Stratton 1973

Stuart, R.K.: Platelet function studies in human beings receiving 300 mg of aspirin per day. J. Lab. clin. Med. **75**, 463-471 (1970)

Weiss, H.J., Danese, C.A., Voletti, C.D.: Prevention of experimentally induced arterial thrombosis by aspirin. Fed. Proc. **29**, 381 (1970)

Zucker, M.B., Peterson, J.: Effect of acetylsalicylic acid, other non-steroidal anti-inflammatory agents and dipyridamole on human blood platelets. J. Lab. clin. Med. **76**, 66-75 (1970)

The Canadian Cooperative Studies of the Effect of Platelet-Suppressing Drugs in Transient Cerebral Ischemic Attacks*

D. Simard

Twenty-five Canadian neurologic units are engaged in two cooperative studies to determine the possible usefulness of platelet-suppressing drugs in altering the clinical course of patients threatened with cerebral transient ischemic attacks.

The first one, the study of the effect of platelet-suppressing drugs, of recent recurrent presumed cerebral emboli, began in January, 1972. The second one is a multi-center randomized trial of sulfinpyrazone in patients following reconstructive cerebral vascular surgery and it began in January, 1974.

I. The Effect of Platelet-Suppressing Drugs on Recent Recurrent Presumed Cerebral Emboli

There is considerable evidence (Ashby et al., 1963; Denny-Brown, 1960; Fisher, 1959; Gunning et al., 1964; Honour and Ross Russel, 1962; Hughes and Tonks, 1962) that embolic material, arising from platelet thrombi formed at sites of irregularity and narrowing of extracranial cerebral arteries, is a cause of transient cerebral ischemia and stroke.

Platelets, in response to exposure to collagen, other subendothelial structures, and foreign substances such as atheromatous debris, change shape and become adherent.

The response to adherence is mainly the release of ADP, ATP, and serotonin. In the presence of calcium, ADP induces aggregation which occasions further adherence, release, and extended aggregation.

The possibility that pharmacologic agents were effective against this platelet reaction first arose in Toronto in 1965 with the report of Smythe et al. (1965). They demonstrated that collagen-induced aggregation was inhibited by sulfinpyrazone. Two years later Evans et al. (1967) observed that acetylsalicylic acid also inhibited platelet aggregation.

Besides numerous experimental models (Didesheim, 1969; Weiss et al., 1970; Dyken et al., 1973; Philp et al., 1972), the clinical

---

*On behalf of the committee of the Canadian cooperative study of the effect of platelet-suppressing drugs on recent recurrent presumed cerebral emboli

evidence respecting platelet-inhibiting drugs has come from a variety of conditions, most not connected with transient cerebral ischemic attacks.

One of the most impressive studies is the double-blind trial with sulfinpyrazone recently reported by Kaegi et al. (1974) in preventing thrombosis in arteriovenous shunts in patients submitted to renal dyalysis. There is also the report by Philp et al. (1972) on the protective action of pyrimido-pyrimidine compounds against thrombocytopenia of divers.

The first published controlled trial on transient ischemic attacks was that of Acheson et al. (1969) who reported on the effect of dipyridamole in 169 patients. There was no benefit of this drug in their study. The same authors also reported a similar negative trial with clofibrate in 95 patients (Acheson et al., 1972).

Evans (1972) carried out a double-blind study in 20 patients with amaurosis fugax, using sulfinpyrazone (200 mg) four times daily and reported a clinically significant improvement.

The aim of the study is to determine whether sulfinpyrazone, acetylsalicylic acid, or a combination of these drugs which suppress platelet adhesion are effective in significantly reducing the numbers of ischemic attacks, and if so, in preventing stroke or in increasing life expectancy. A 2-year pilot phase was to examine the first of these objectives and the expanded 5-year phase, the latter. We, however, decided recently to extend the first phase for an extra year.

The phenomenon of cerebral transient-ischemic attack is not due to a single cause. These attacks are usually of hemodynamic, cardiacembolic, or arterialembolic origin.

In our study, we are dealing strictly with the transient-ischemic episodes of arterialembolic origin because the available therapy to date, mainly anticoagulants and thromboendarterectomy, is still controversial.

A large number of agents have been shown to impair platelet aggregation and release. Of the many platelet function-suppressing drugs three are practical because they can be given orally and because they have minimal side effects, these are sulfinpyrazone, acetylsalicylic acid, and dipyridamole.

In this study it was decided to use sulfinpyrazone and acetyl salicylic acid. Sulfinpyrazone is combined with ASA in one group of patients in an attempt to augment the effects on platelet reactivity. This combination will be compared to each of the individual drugs and to the placebo group.

Patients are accepted if they had more than one cerebral transient ischemic attack within the past three months, and if they are judged to have recurrent emboli which are presumed to be of arterial origin.

Table 1. Stratification of patients according to the site of ischemia

Patient classification

1. Carotid - No residua
2. Carotid - With residua
3. Vetrebral-basilar - No residua
4. Vertebral-basilar - With residua
5. Carotid and vertebral-basilar - No residua
6. Carotid and vertebral-basilar - With residua

Table 2. Timetable for clinical and laboratory assessments

Assessment	Before treatment	End of week 1	2	4	13
Clinical assessment	*			*	*
Platelet studies	*	*	*	*	*
Blood studies	*			*	*
Urinalysis	*			*	*
Biochemistry	*			*	*
Stools for Blood				*	*
Blood levels				*	*

Assessments continue every 3 months

Patients fulfilling the criteria of entry are allocated to their respective treatment groups according to a prescribed randomized arrangement within each of six categories of patients (Table 1).

This stratification is according to the site of the ischemia, namely carotid and/or vertebral basilar, and the presence or absence of residue.

The exclusion criteria have been clearly specified and a monthly record is kept of patients who, after satisfying the inclusion criteria, have had to be excluded.

There is a central adjudicating committee that continually monitors the clinical evaluations to ensure that cases entered in the study satisfy the criteria laid down.

There is a regular time table for clinical and laboratory assessments (Table 2). The platelet studies include aggregation studies, using collagen and epinephrine, platelet adhesiveness test using glass bead columns, template bleeding times, and platelet survival studies.

The problem of drug-compliance is of major concern in long-term controlled clinical trials.

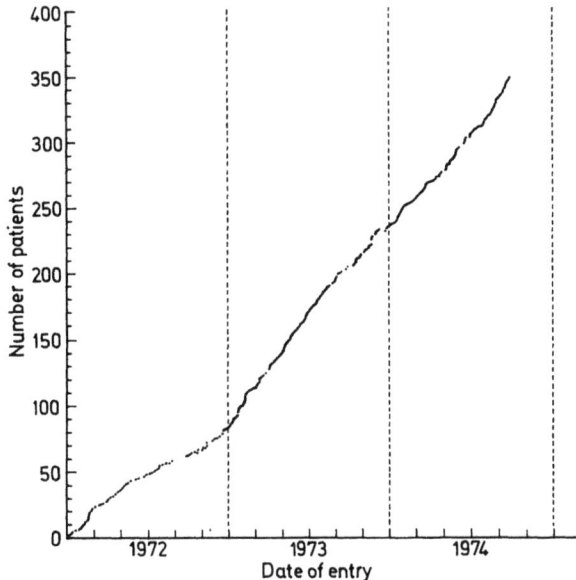

Fig. 1

Several measures of patient drug compliance are used in this study. Returned pills at follow-up assessments are counted; the serum uric acid which is lowered by sulfinpyrazone is measured; the epinephrine aggregation curve, the secondary wave of which is suppressed by aspirin, is assessed; blood-level determinations of the two drugs are made.

At the end of October, 1974, 351 patients had entered the trial (Fig. 1).

We have not yet broken the treatment code except for the patients who have been withdrawn from the trial. Considering only these withdrawals for end-points, it is reasonable to state that we are encouraged by them to continue this trial.

We should be able to determine without equivocation if these two platelet-inhibiting drugs are useful in preventing strokes.

We expect to be able to reach a total of 600 cases entered in the study and to report on the two end-points of stroke and death in 2 years.

II. The Canadian Multi-Center Randomized Trial of Sulfinpyrazone in Patients Following Reconstructive Cerebral Vascular Surgery

The main objective of this study is to determine if a platelet-suppressive drug can alter the long-term prognosis in patients who have had an appropriate surgical procedure involving the carotids, vertebrals, innominate, or subclavian arteries in order to improve their cerebral circulation.

When a thromboendarterectomy is performed the vessel is denuded of its intima and this lesion is a potential site for thrombosis in the pre-endothelization stage after operation. This is the rationale of our study which will be double-blind using, however, only sulfinpyrazone and a placebo.

This will be a five-year study and it began in January, 1974. At the end of October, 1974, 36 patients had been admitted to this study.

The clinical and laboratory assessments are similar to the medical study except for platelet studies which are not done. The patient stratification is also similar to the one used in the medical study.

References

Acheson, J., Danta, G., Hutchinson, E.C.: Controlled trial of Dipyridamole in cerebral vascular disease. Brit. med. J. 1, 614-615 (1969)
Acheson, J., Hutchinson, E.C.: Controlled Trial of Clofibrate in Cerebral Vascular Disease. Atherosclerosis 15, 177-183 (1972)
Ashby, M., Oakley, N., Lorentz, I., Scott, D.: Recurrent transient monocular blindness. Brit. med. J. 2, 894-897 (1963)
Denny-Brown, D.: Recurrent cerebrovascular episodes. Arch. Neurol. 2, 194-210 (1960)
Didisheim, P.: Inhibition by Dipyridamole of arterial thrombosis in rats. Thromb. Diath. haemorrh. (Stuttg.) 20, 257-266 (1968)
Dyken, M.L., Campbell, R.L., Muller, J., Feuer, H., Horner, T., King, R., Kolar, O., Solow, E., Jones, F.H.: Effect of aspirin on experimentally induced arterial thrombosis during the healing phase. Stroke 4, 387-389 (1973)
Evans, G.: Effect of drugs that suppress platelet surface interaction on incidence of amaurosis fugax and transient cerebral ischemia. Surg. Forum 23, 239-241 (1972)
Evans, G., Nishizawa, E.F., Packham, M.A., Mustard, J.F.: The effect of acetylsalicylic acid (aspirin) on platelet function. Blood J. Hematol. 30, 550 (1967)
Fisher, C.M.: Observations of the fundus occuli in transient monocular blindness. Neurology (Minneap.) 9, 333-347 (1959)
Gunning, A.J., Pickering, G.W., Robb-Smith, A.H.T., Russell, R.R.: Mural thrombosis of the internal carotid artery and subsequent embolism. Quart. J. Med. 33, 155-195 (1964)
Honour, A.J., Ross Russel, R.W.R.: Experimental platelet embolism. Brit. J. exp. Path. 43, 350-362 (1962)
Hughes, A., Tonks, R.S.: Intravascular platelet clumping in rabbits. J. Path. Bact. 84, 379-390 (1962)
Kaegi, A., Pineo, G.F., Shimizu, A., Trivedi, H., Hirsh, J., Gent, M.: Arteriovenous-shunt thrombosis-prevention by Sulfinpyrazone. New Engl. J. Med. 290, 304-306 (1974)
Philp, R.B., Inwood, M.J., Warren, B.A.: Interactions between gas bubbles and components of the blood: Implications in decompression sickness. Aerosp. Med. 43, No. 9, 946-953 (1972)
Philp, R.B., Inwood, M.J., Ackles, K.N., Radomski, M.W.: Changes in platelets and the hemostatic mechanism of human subjects decompressed from a hyperbaric environment an the effects of orally-administered dipyridamole. Presented at a symposium on Blood-Bubble Interactions at DCIEM, Downsview, Ont., Feb. 1973

Smythe, H.A., Ogryzle, M.A., Murphy, E.A., Mustard, J.F.: The effect of sulfinpyrazone (anturan) on platelet economy and blood coagulation in man. Canad. med. Ass. J. 92, 818-821 (1965)

Weiss, H.J., Danese, C.A., Voleti, C.D.: Prevention of experimentally induced arterial thrombosis by aspirin. Fed. Proc. 29, 381 (1970)

On the Relationships Between the Activation of the Complement System and the Platelet Aggregation

T. Di Perri, A. Auteri, A. Vittoria, and F. Laghi Pasini

Our interest in this field comes from our researches on inflammation. As is well known the inflammation process is a very complex phenomenon which is difficult to synthetically describe. According to the more accepted view, we consider as inflammatory all the processes which are characterized by morphologic and functional changes of the vessel's wall cells, thus leading to the exudation of some plasma protein, to the liberation of so-called chemical mediators, and finally to the mobilization of some fix and mobile cellular elements.

Inflammation is characterized by the participation of humoral and cellular factors, most of which are known (Fig. 1).

A schematic, but surely incomplete, dynamic view of the development of the inflammation reaction (Fig. 2) emphasizes cellular changes as a trigger mechanism, the possibility of an immunologic mediation being considered, and in a successive phase the activation, almost contemporaneous of several plasmatic proteic systems, which lead to the liberation of the chemical mediators, to the cellular chemotaxis with secondary liberation of its intracellular content, and to the final picture of the inflammatory process. In the last years we have particularly studied the role of the activation of the complement system in various experimental model and clinical situations, all related to the inflammatory reaction, and at the same time we have tried to identify a lot of substances active to inhibit the activation of one or more components of the complement system in order to standardize a model of blocking, in a precocious phase, the sequential reaction of inflammation.

In these researches we have observed that several drugs, known to actively inhibit the platelet aggregation, show also an inhibitory power on the activation of the complement system. These findings agreed with the results of Jobin and Gagnon (1970), and led us to consider the opportunity of a more profound study on the eventual interrelationship between complement activation and the platelet aggregation processes.

Here we present the first findings of these investigations, which we have schematized in three groups.

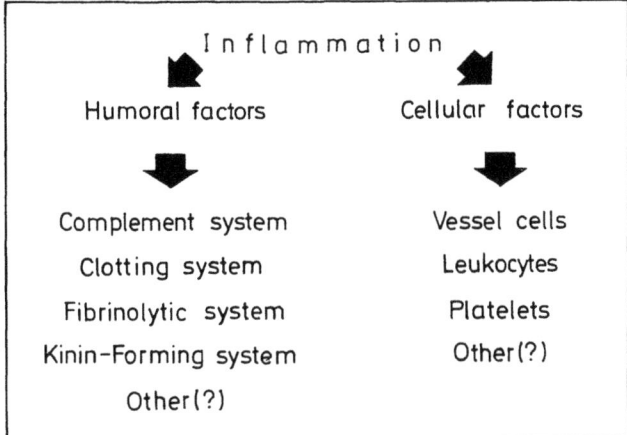

Fig. 1. Inflammatory factors

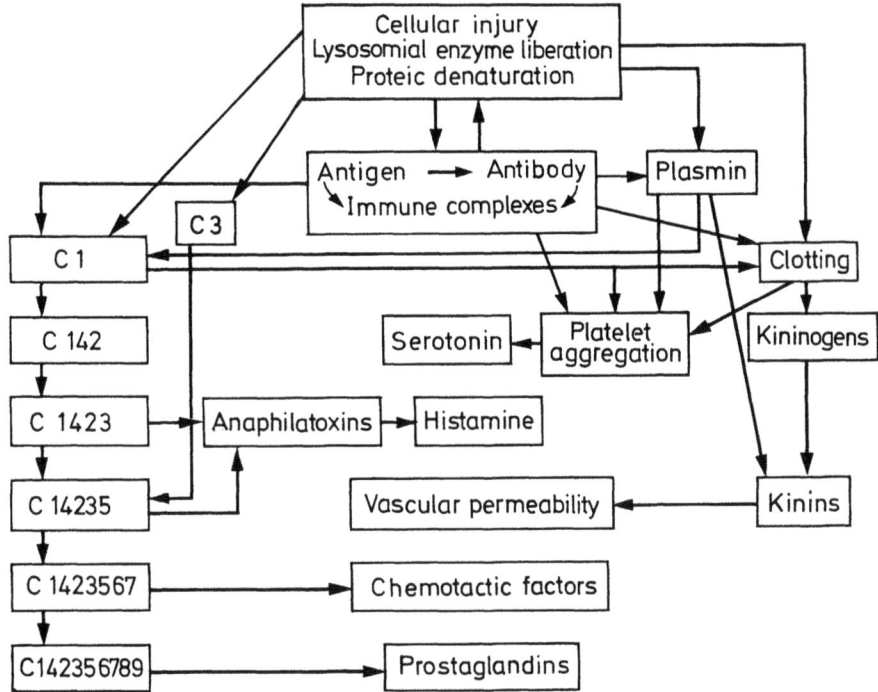

Fig. 2. Dynamics of acute inflammation (after Willoughby mod.)

Firstly we have tried to see if there is a participation of some component of the complement system in the platelet aggregation process. We have divided this problem in two parts:

1. In the first one we have tried to see if it is possible to find the presence of some complement component in the platelet aggregate. The complement components have been identified by

Table 1. Complement participation in platelet aggregation process. Immunofluorescent study on the presence of C_1, C_4 and C_3 in the platelet aggregate

Type of aggregation	Control (ALB.)	C_1	C_4	C_3
ADP	- - -	- - -	+ + +	+ + +
Epinephrine	- - -	+ + +	- - -	+ + +
Collagen	- - -	- - -	+ + +	+ + +

Table 2. Concentration of several complement components in human platelet-rich plasma before and after platelet aggregation mean and SE (n=5)

Type of aggregation	C_{1q} (mg%)		C_4 (mg%)		C_3 (mg%)	
	B	A	B	A	B	A
ADP	17,5±2,3	17,3±2,2	44,6±2,8	44,5±2,8	103±2,9	71,1±3[a]
	p = n.s.		p = n.s.			
Epinephrine	18,4±2,4	17,9±2,5	45,2±1,9	45,3±2	107±3,1	74,9±2,9[a]
	p = n.s.		p = n.s.			
Collagen	18 ±1,9	17,9±2,2	46,2±2,2	46,4±2,3	110±3,8	84,7±3,2[a]
	p = n.s.		p = n.s.			

[a] $p < 0.001$

means of the fluorescent revelation of the proteic fraction employing Coon's technique based on the treatment of the fixed tissue by specific antiserum. We have obtained the platelet aggregate with Born's technique, using as inducing agent either ADP, epinephrine, or collagen. The anti-C1, anti-C4, and anti-C3 antiserum, previously conjugated with fluoresceine isothiocianate, were used for the specific identification of the single component (C1, C4, and C3) of the complement system. In the ADP-induced platelet aggregate we have found a positive fluorescent staining indicating the presence of both C4 and C3; in the epinephrine-induced platelet aggregate a positive fluorescent finding for C1 and C4 was found, while in the collagen-induced aggregate the presence of C4 and C3 was obtained. These results are summarized in Table 1.

2. In the second part of this study we have dosed the concentration of the same complement components (C1, C4, C3) in the platelet-rich plasma (PRP) before and after the platelet aggregation induced either by ADP, epinephrine, or by collagen according to the above-mentioned technique. The concentration of the complement components was evaluated by the immunoprecipitation method employing the fluoronephelometric Technicon Autoanalyser and specific antiserums. The findings of this research are exposed in Table 2 and summarized in Figures 3, 4, 5: no changes of C1 and C4 concentration in the supernatant were found before and

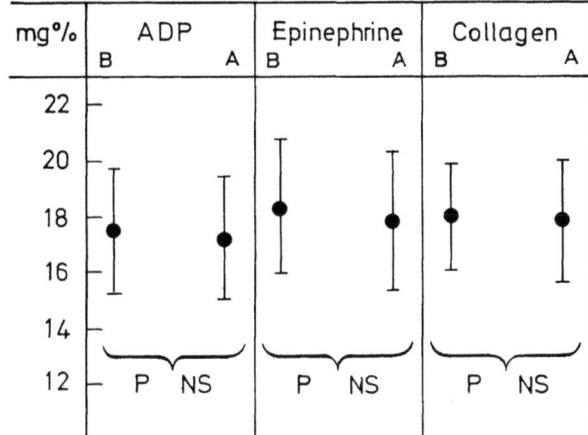

Fig. 3. Concentration of the first component of complement system (C1q) in human platelet-rich plasma before and after platelet aggregation (n=5)

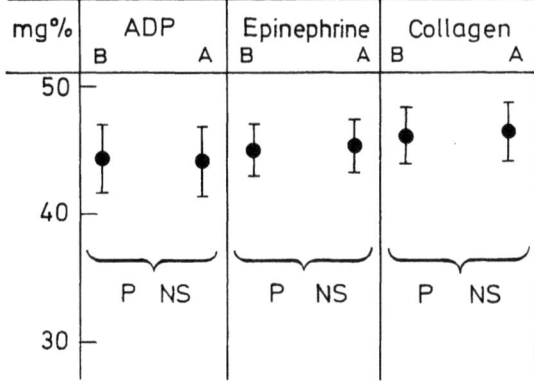

Fig. 4. Concentration of the fourth component of complement system in human platelet-rich plasma before and after platelet aggregation (n=5)

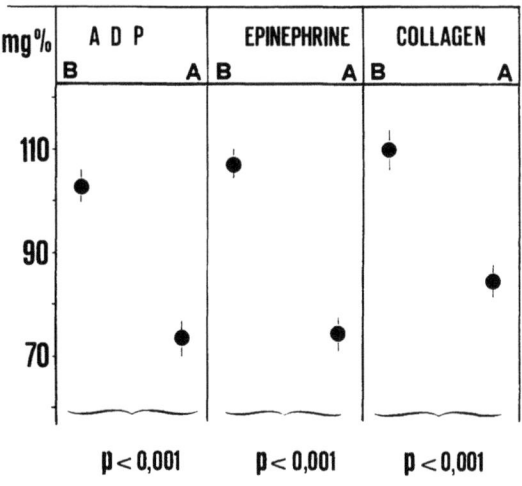

Fig. 5. Concentration of the third component of complement system in human platelet-rich plasma before and after platelet aggregation (n=5)

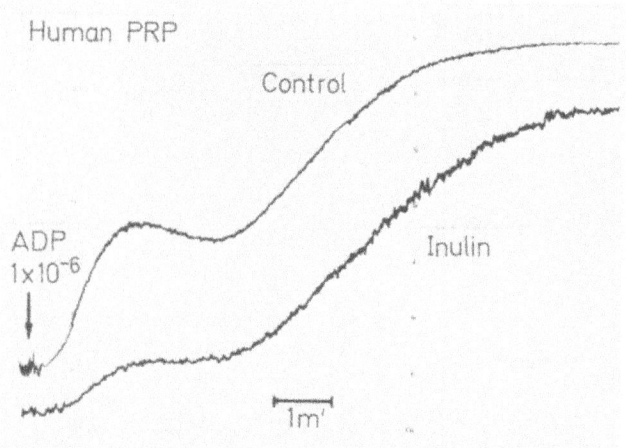

Fig. 6. ADP-induced platelet aggregation in human platelet-rich plasma previously incubated with inulin

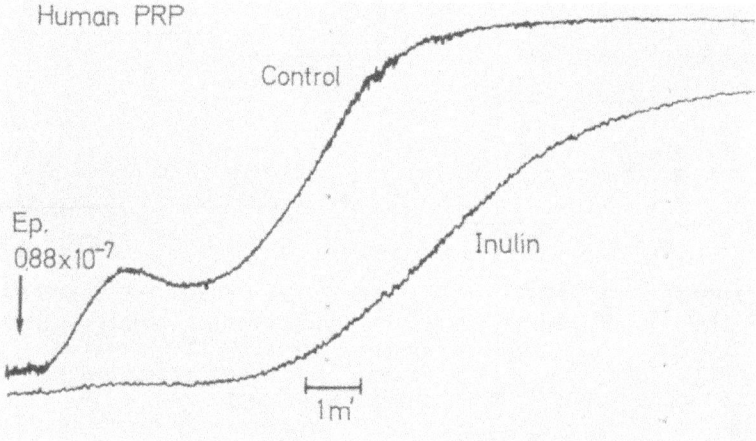

Fig. 7. Epinephrine-induced platelet aggregation in human platelet-rich plasma previously incubated with endotoxin

after the aggregation of the platelets in the ADP-, epinephrine-, and collagen-treated PRP, while in the same samples the concentration of C3 was significantly lowered after the platelet aggregation.

In conclusion, in these studies a participation of the first complement components to the platelet aggregation process, either by ADP, epinephrine, or by collagen-induced was found. The more involved factor seems to be the third component - C3 -, which, as it is known, play a very important role in the sequential activation of the complement system, as it is considered as the confluence point of both pathways of the activation process, either the classical or the alternate. On the basis of these findings and of these considerations we have also studied the influence of a preliminary depletion of C3 by the activation of the alternative pathway on the platelet aggregation. The alternate pathway of the complement activation has been activated by a preliminary incubation of the PRP with a solution of inulin,

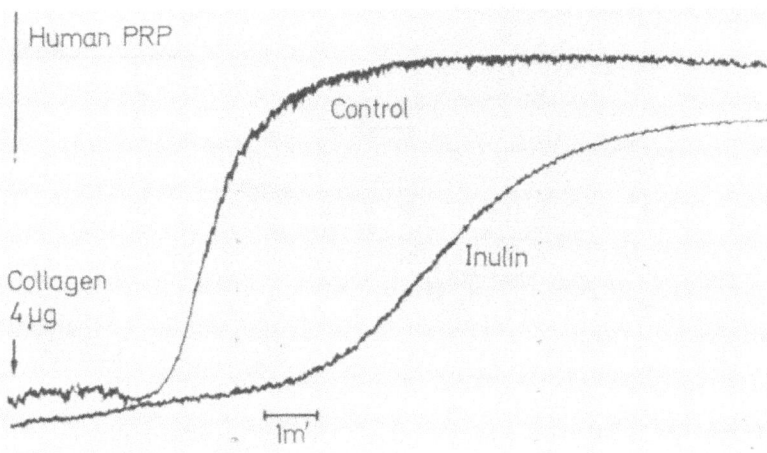

Fig. 8. Collagen-induced platelet aggregation in human platelet-rich plasma previously incubated with inulin

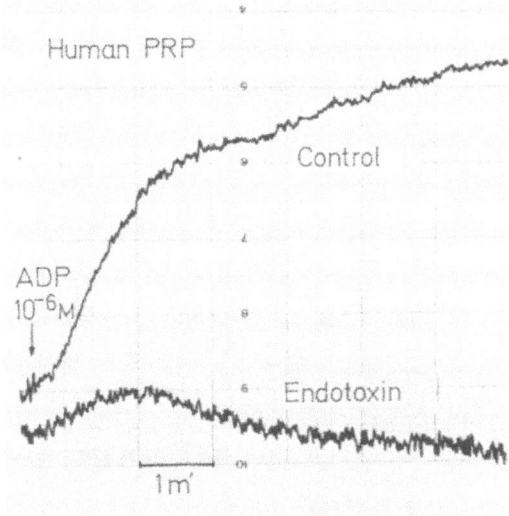

Fig. 9. ADP-induced platelet aggregation in human platelet-rich plasma previously incubated with endotoxin

which displays a specific action in this sense (Goetze and Mueller-Eberhardt, 1971), or by a preliminary incubation of PRP by a solution of bacterial endotoxin. The platelet aggregation in the preincubated PRP, either with inulin or with bacterial endotoxin, show a profound change in the curve after the addition of the inducing agent ADP, epinephrine, or collagen, thus indicating a decrease both of the speed and of the maximal level of the aggregating process (see Figs. 6-11).

In the last part of our paper we should like to present the actual situation of our studies on the pharmacologic experimental inhibition of both processes: the complement activation and the platelet aggregation.

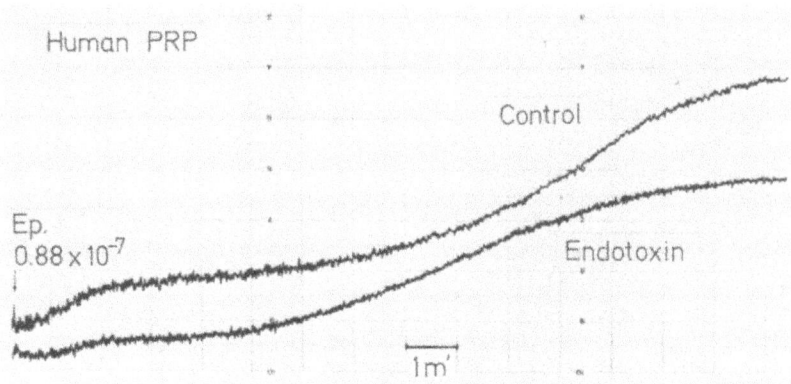

Fig. 10. Epinephrine-induced platelet aggregation in human platelet-rich plasma previously incubated with endotoxin

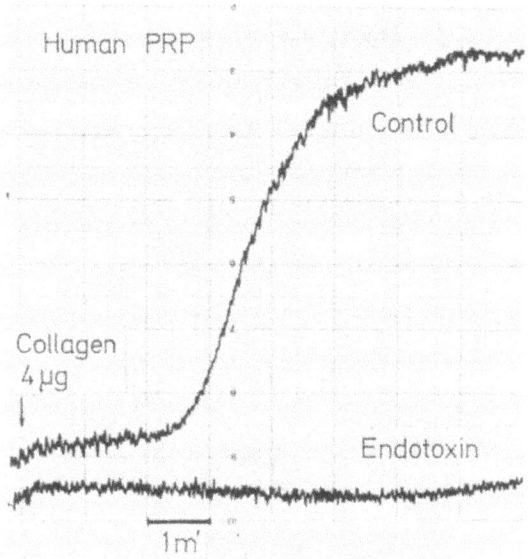

Fig. 11. Collagen-induced platelet aggregation in human platelet-rich plasma previously incubated with endotoxin

To date we have studied five groups of drugs, which are summarized in Table 3. They are:

1. Nonsteroidal antiinflammatory drugs
2. Beta-blocker agents
3. Pyrimidino-pyrimidinic compounds
4. Saluretic drugs
5. Drugs active on the clotting system.

The aggregation phenomenon was evaluated in PRP by an aggregometer with continuous registration of the transmitted light in order to obtain a curve for the measurement of the speed of the aggregation and of the maximal level. The aggregation was induced

Table 3. Groups of drugs studied for complement inhibition and platelet aggregation inhibition activity

1. Nonsteroidal antiinflammatory drugs
2. Beta-blocker agents
3. Pyrimido-pyrimidinic compounds
4. Saluretics
5. Drugs active on clotting system

Table 4. In vitro anticomplementary and antiplatelet aggregation activity of several nonsteroidal antiinflammatory agents

C component blocked	Concentration M/H.U.	Drug	Platelet aggregation		
			ADP	Ep.	Coll.
C 1	$2,6 \times 10^{-5}$	Acetylsalicylic acid	Inactive	5×10^{-6} M	5×10^{5} M
C 3	$4,4 \times 10^{-6}$	Flufenamic acid	n.t.	n.t.	n.t.
C 3	$5,5 \times 10^{-6}$	Mefenamic acid	n.t.	5×10^{-5} M	5×10^{5} M
C4 C2 C3	$5,7 \times 10^{-5}$	Indomethacin	10^{-3} M	5×10^{-5} M	5×10^{5} M
C4 C3	$8,4 \times 10^{-5}$	Phenylbutazone	n.t.	5×10^{4} M	5×10^{5} M

by ADP, epinephrine, or collagen according to the technique and the dosage used in previous papers. Every drug was incubated for 30 min at 37°C in the PRP and afterward a second aggregation was induced to test the activity of the substance on this phenomenon. For each drug the minimal concentration needed to actively inhibit the collagen-induced aggregation and the development of the second wave of the ADP and epinephrine-induced aggregation was then evaluated.

The activity on the complement system was studied according to the above-described technique, which is based on the evaluation of the hemolytic activity of a sample of human fresh serum added to an immunohemolytic system, before and after incubation of the serum with the drug. If the result is that the drug actively inhibits activation of the hemolytic reaction, the minimal concentration needed is stated and, successively, by a complex system of recomplementation with a serum sample without a component of the complement, the site of the action of the drug is found. This technique is described elsewhere in detail. The results are reported in Tables 4-8.

We have seen that almost all the drugs which show an antiaggregating effect, are also characterized by an antiinflammatory power, i.e., by an anticomplentary activity. These findings seem to present some interest either from the theoretical or from the practical point of view. Even if the existence of a definite interrelationship between the two processes cannot be accepted,

Table 5. In vitro anticomplementary and antiplatelet aggregation activity of several beta-blocking agents

C component blocked	Concentration M/H.U.	Drug	Platelet aggregation		
			ADP	Ep.	Coll.
C 1	$4,5 \times 10^{-6}$	Propranolol	$0,33 \times 10^{-3}$ M	n.t.	3×10^{-7} M
C 1	$1,5 \times 10^{-5}$	Pindolol	$0,8 \times 10^{-4}$ M	n.t.	15×10^{-7} M
Inactive	-	Practolol	$0,75 \times 10^{-3}$ M	n.t.	$0,8 \times 10^{-4}$ M

Table 6. In vitro anticomplementary and antiplatelet aggregation activity of some pyrimido-pyrimidinic compounds

C component blocked	Concentration M/H.U.	Drug	Platelet aggregation		
			ADP	Ep.	Coll.
C1 C4	1×10^{-5}	Dipyridamole	$0,99 \times 10^{-4}$ M	$0,49 \times 10^{-4}$ M	$2,5 \times 10^{-4}$ M
C1	5×10^{-8}	RA 233	$0,12 \times 10^{-4}$ M	5×10^{-5} M	$2,5 \times 10^{-5}$ M

Table 7. In vitro anticomplementary and antiplatelet aggregation activity of some saluretics

C component blocked	Concentration M/H.U.	Drug	Platelet aggregation		
			ADP	Ep.	Coll.
C1 C4	$8,1 \times 10^{-5}$	Furosemide	$0,9 \times 10^{-6}$ M	$4,5 \times 10^{-7}$ M	$4,5 \times 10^{-7}$ M
C1 C4	$4,7 \times 10^{-5}$	Ethacrinic acid	$3,95 \times 10^{-6}$ M	$1,3 \times 10^{-6}$ M	$1,5 \times 10^{-7}$ M

Table 8. In vitro anticomplementary and antiplatelet aggregation activity of several compounds active on the clotting system

C component blocked	Concentration M/H.U.	Drug	Platelet aggregation		
			ADP	Ep.	Coll.
C 3	$2,2 \times 10^{-5}$	Clofibrinic acid	-	-	-
C 3	$5,5 \times 10^{-4}$	Heparin	$2,9 \times 10^{-2}$ M	n.t.	n.t.
Inactive	-	Streptase	500 U	500 U	500 U
Inactive	-	Nicotinic acid	$5,3 \times 10^{-4}$ M	$2,1 \times 10^{-3}$ M	$5,3 \times 10^{-4}$ M

their contemporaneous participation to the inflammatory reaction appears modified by drugs which lead to decrease the participation of both these factors to the sequential chain.

Now we will emphasize another finding which is summarized in Table 8. Both the streptokinase and the nicotinic acid, which are known as inductors of fibrinolysis, show an evident anti-platelet-aggregating effect, without any inhibitory action on the complement activation. We cannot actually correctly interpret these findings. The problem of the knowledge of the exact relationship between the two chains of activation, the complement and the fibrinolysis, arises which we will try to study in the near future.

To conclude this paper, we summarize our results in the following manner:

1. The platelet aggregation process leads to the consumption of some complement component: in the ADP, epinephrine, and collagen platelet aggregate the presence of some complement component was demonstrated by the istoimmunofluorescent technique. The decrease of the concentration of the third component of the complement system was shown in the supernatant phase of PRP after the platelet aggregation.

2. The dynamic of the platelet aggregation process induced by ADP, epinephrine, or by collagen appears profoundly modified after the activation of the alternate pathway of the complement activation by the inulin or by the bacterial endotoxin.

3. Many drugs known as antiaggregating show also a strong anticomplementary effect.

We think that a hypothesis on the existence of a relationship between the complement activation and the platelet aggregation, at least in some particular situations, can be formulated.

References

Born, G.V.R.: Aggregation of Blood Platelets by ADP and its Reversal. Nature (Lond.) 194, 927 (1962)
Di Perri, T., Auteri, A.: Effect of some pyrimidino-pyrimidinic compounds on the complement; in vitro and in vivo studies. First Meeting of the European Division of Int. Society of Haematology. From Aggregazione Piastrinica, p. 81. Ed.: Boehringer Ingelheim, Milano 1971
Di Perri, T., Auteri, A.: On the anticomplementary activity of some anti-inflammatory drugs. From: Future Trends in Inflammation. Piccin Medical Books. p. 215, 1974
Di Perri, T., Auteri, A.: Attività anticomplementare dell'acido clofibrinoco: ricerche in vivo ed in vitro. Boll. Soc. ital. Biol. sper., in press
Di Perri, T., Auteri, A.: Attività anticomplementare di alcuni farmaci bloccanti beta adrenergici: ricerche in vivo ed in vitro. Boll. Soc. ital. Biol. sper., in press
Di Perri, T., Forconi, S., Auteri, A., Vittoria, A., Laghi Pasine, F., Guercini, F.: Sull'azione antiaggregante piastrinica ed anticomplementare della furosemide e dell'acido etacrinico. Possibilità di un meccanismo d'azione svincolato da quello diuretico nel trattamento dell'insufficienza renale acuta. 14° National Congress of the Italian Society of Nephrology. Bari 1974

Gardner, R., Chater, B.V., Brown, D.L.: The role of Complement in Endotoxin Shock and Disseminated Intravascular Coagulation: Experimental Observations in the Dog. Brit. J. Haemat. 28, 393 (1974)

Goetze, O., Muller-Eberhard, H.J.: The C3-activator system: an alternate pathway of complement activation. J. exp. Med. 134, Supp. 90 (1971)

Henson, P.M., Cochrane, C.G.: Immunological induction of increased vascular permeability: II. Two mechanisms of histamine release from rabbit platelets involving complement. J. exp. Med. 129, 167 (1969)

Jobin, F., Tremblay, F.: Platelet reaction and immune processes. II. The inhibition of platelet aggregation by complement inhibitors. Thromb. Diath. haemorrh. (Stuttg.) 25, 86 (1971)

Jobin, F., Gagnon, F.T.: Platelet reaction and immune processes. IV. The inhibition of complement by pyrazole compounds and other inhibitors of platelet reactions. Canad. J. Microbiol. 16, 63 (1970)

Kane, M.A., May, J.E., Frank, M.M.: Interaction of the classical and alternate complement pathway with endotoxin lipopolysaccharide. J. clin. Invest. 52, 370 (1973)

Mayer, M.M.: Experimental Immunochemistry. Kebat, E.A., Mayer, M.M. (eds.). Springfield: Publ. Chales C. Jhons 1961

Muller-Berghans, G., Lohmann, E.: The role of complement in endotoxin-induced disseminated intravascular coagulation studies in congenitally C6-deficient rabbits. Brit. J. Haematol. 28, 403 (1974)

Mustard, J.F., Packham, M.A.: Factors influencing platelet function: adhesion, release and aggregation. Pharmacol. Rev. 22, 97 (1970)

Nachman, R.L., Weskler, B., Ferris, B.: Increased vascular permeability produced by human platelet granule cationic extract. J. clin. Invest. 49, 274 (1970)

Nachman, R.L., Weskler, B.: The platelet as an inflammatory cell. Ann. N.Y. Acad. Sci. 201, 131 (1972)

Nachman, R.L., Weskler, B., Ferris, B.: Characterization of human platelet vascular permeability enhancing activity. J. clin. Invest. 51, 549 (1972)

Nagayama, M., Zucker, M.B., Beller, F.K.: Effects of variety of endotoxines on human and rabbit platelet function. Thromb. Diath. haemorrh. (Stuttg.) 26, 467 (1972)

O'Brien, J.R.: Effect of antiinflammatory agents on platelets. Lancet, 894 (1968)

Pfueller, S.L., Luscher, E.F.: Studies of the mechanisms of the human platelet release reaction induced by immunologic stimuli. I. Complement-dependent and complement-independent reactions. J. Immunol., Vol. 112, No. 3, March 1974

Pfueller, S.L., Luscher, E.F.: Studies of the mechanisms of the human platelet release reaction induced by immunologic stimuli. II. The effects of Zymosan. J. Immunol., Vol. 112, No. 3, March 1974

Sobel, A., Marcel, G.A., Lagrue, G.: Complement, hémostase et kinines. Interrelations et rôle dans l'inflammation. Sem. Hôp. Paris, No. 8, 549 (1974)

Tomar, R.H., Kolchins, D.: Complement and coagulation, serum beta 1c-beta 1a in disseminated intravascular coagulation. Thromb. Path. Haematol. 27, 389 (1972)

Vittoria, A., Laghi Pasini, F., De Gori, V., Forconi, S., Di Perri, T.: L'azione della furosemide sulla aggregazione piastricina da ADP, adrenalina e collagene. Boll. Soc. ital. Biol. sper., in press

Vittoria, A., Laghi Pasini, F., De Gori, V., Forconi, S., Di Perri, T.: L'azione dell'acido etacrinico sulla aggregazione piastrinica da ADP, adrenalina e collagene. Boll. Soc. ital. Biol. sper., in press

Willoughby, D.A., De Rosa, M.: A unifying concept for inflammation: a new appraisal of some mediators. In: Immunopathology of Inflammation. Fischer, D.K., Houck, J.C. (eds.). Amsterdam: Excerpta Medica 1971

Zimmermann, T.S., Muller-Eberhard, J.J.: Blood coagulation initiation by a complement-mediated pathway. J. exp. Med. $\underline{134}$, 1601 (1971)

Zucker, M.B., Grant, R.A.: Aggregation and release reaction induced in human blood platelets by Zymosan. J. Immunol., Vol. 112, No. 3, March 1974

Discussion and Concluding Remarks

Chairmen's Considerations

BORN: The subject matter of Methodology and Physiology is furthest away from clinical neurology. I think the important thing there is that the basic investigation of platelets is a very interesting biological thing to spend your time on if you are a biologist and, in my case, the fact that platelets really matter or are thought to matter in clinical conditions of the kind in which you are interested, has come as a rather belated bonus.

My own interest and that of many other people in cellular biology is really, you might say, scientific, and although I am a doctor, everything that you find out about the importance of platelets clinically is a rather pleasant surprise. This brings me, however, to believe: the onus, the burden of proof that platelets actually do matter in cerebral ischemias, transient or otherwise, in thromboses, in inflammation, as for instance Dr. Nachmann believes, or in the pathogenesis of atherosclerosis, the burden of proof is always on the people who put these ideas forward. It is very easy to make hypotheses; we can all do that; anybody who has been through medicine can make any number of hypotheses. It is enormously much harder to get evidence which is compatible. You can never say it proves something which is compatible with your hypothesis. And in England the tradition seems to be the other way round, which has its advantages and drawbacks. One tends to do experiments and one is then very cautious with hypotheses. This has also some drawbacks, but it is certainly the tradition in which I have been brought up.

Dr. Nachmann wrote interestingly about the platelets' possible involvement in changes in the vessel wall. This is a nice idea which has been going around and the great thing now again is to get evidence for or against. I am quite certain in my own mind that the evidence for such an effect would be very hard to get, because again, if you think of a blood vessel and the way these little cells occasionally attach themselves to particular places, and if you also think how rapidly the blood flows past, and how rapidly anything that is produced is diluted and removed and antagonized, these kinds of interactions of circulating cells on the wall are quite hard to envisage. This is my own opinion, which is presumably what this is for. But I think it is a very interesting and challenging line of work.

I merely summarized for you the best evidence to my knowledge about the mechanism by which platelets aggregate one to the other and on the basis of this, people subsequently summarize for you various modes of inhibition. The fact that this has some basis in biochemical and pharmacologic logic is gratifying, but of course platelets are at least as complicated as other cells, even though they have no nucleus, they have no DNA, and you know for example the complications with cyclic AMP are quite enormous. We discussed this privately and I am sure most of you are aware of this. And if one says that something acts by rating cyclic AMP in a cell, one really essentially actually says very little indeed.

Professor Abe produced an interesting mathematical model for the way that the optical records show aggregation. In fact he produced two models and his mathematics suggests that one is right and the other is wrong and this is of considerable theoretical interest, but perhaps not of such immediate interest here.

BREDDIN: If we have to conclude on what we can say about our present status in methodology, I would say that we have a number of very good tests to use in the laboratory as far as pharmacologic investigations are concerned and we can be quite happy with this situation. We are not in such a very good situation in the clinical field. If we, for instance, take haemorrhage disorders, we certainly would not use any test which is not specific for the disorder and which is not able to differentiate between normals and patients having this disorder and we are by far not in this situation in the field of thrombosis and thrombotic lesions. Here we can only work with statistic analyses; we can say we have a higher percentage of patients having this whole set of positive tests and this may mean something and we have to prove in some prospective studies if this meaning is correct.

This is the present situation for all our tests. I am rather convinced that this may be changed and that we will have to look into future, that we will have to divide more precise and more suitable tests for this special situation we want to evaluate. We want to know, we want to have a test which more or less precisely has some predictive value and is as precise as possible on the risks of special patients to get thrombosis and of course we know that such a prediction is only possible if there are reactions which are present already beforehand and, for instance, vascular lesions leading to certain changes of platelet behaviour which we can measure beforehand. And at the moment we have to be content with what we have but we should not stop looking for more precise tests and we should always keep in mind that the tests we use at the present for clinical investigations do not fulfill actually the conditions which we demand from such tests.

TAKESHI ABE: First, Professor Zülch examined clinical features of 1000 cases and found that thrombus was the cause four times as much as embolism as a causative factor for the pathogenesis of ischemic cerebrovascular disease. He also noticed that in young patients there was some difference in the instance of embolism, higher in females being influenced by the previous cardiovascular disorders.

Dr. Acheson described mainly the natural history of patients with stroke and TIA and showing that this was not as bad as previously thought. Furthermore she was unable to show any coordination between changes in platelet aggregation and clinical features of cerebrovascular diseases.

Prof. Marshall proposed that amaurosis fugax was mostly due to platelet emboli traveling through branches of retina vessels; and this was confirmed by the fact that fat preparate emboli were also seen in the retinal circulation in patients following heart valve replacement on cardiovascular bypass.

Professor Anthony demonstrated that during migraine attacks platelets lost a considerable amount of their serotonin content but did not lose their ability to take up serotonin when incubated. He also found that the adenosine content was unchanged but migraine plasma seemed to release serotonin from platelets in vitro, suggesting that serotonin loss during migraine was due to the privilege of serotonin releasing factor, which appeared in plasma during the migraine attack.

ACHESON: At the moment, the patients presenting with transient cerebral ischemia is regarded as different in some way or other from the patient presenting with stroke. I would suggest that we should envisage these patients not as two separate populations but as subgroups of the whole. I say this for two reasons. The first is that I have evidence to suggest intracerebral hemorrhage can cause a so-called classical transient cerebral ischemic attack. Secondly, the proportion of patients presenting with stroke and then going on to develop transient cerebral ischemia is too high to be ignored.

My second comment is an appeal for multidisciplinary working. In the 1970s and certainly in the 1980s I feel that team approach is obligatory. By this I mean that the clinician, the neuropathologist, the physicist, the hematologist, the pharmacologist, other experts and other disciplines should work as one team to advance knowledge of cerebrovascular disease.

MEYER: Now, my general impression is that there is now no question of the importance of platelet aggregation in the pathogenesis of cerebrovascular disease and I think this meeting will go down in history, medical history, as establishing a fact.

And the other interesting thing which is of great importance to the clinician and to the patient is that this can be corrected by appropriate therapy, and the influence is that strokes and transient ischemic attacks can be prevented.

And as I look at ulcerated plaques and arteriograms and prescribe antiplatelet therapy and repeat the arteriograms and see the ulcers heal and the emboli disappear, I am beginning to feel no longer that I am a neurologist but a gastroenterologist.

Regarding the papers in this section, I think that Fieschi's paper on the experimental model of cerebral platelet embolism is a considerably important contribution. I think he has established with his team and without any question that the platelet emboli can damage the brain, that it is reversible and this model, obviously, will be used for evaluating various forms of treatment.

I think Isla Williams from Australia also proved that platelet emboli can cause structural damage to the cervical ganglia and neuronal structures.

I think Breda and Swank and Paoletti's papers all showed that the plasma components themselves can influence platelet aggregation, and these in turn are influenced by such things as the diet, hypertension, circulatory shock, cardiopulmonary bypass, a high-fat diet, a low-fat diet, all these things can influence platelet aggregation and embolization.

Coccheri showed that in man, brain trauma with cerebral microthrombosis, so beautifully demonstrated by his electronmicroscopic consultant and associate, established that actual measurements of cerebral arteriovenous differences in man, using a simple platelet count, can show an uptake by the brain of platelets.

Finally, Prencipe and Agnoli showed that in patients with cerebrovascular disease, platelet adhesiveness and platelet consumption are both abnormal and increased and can be modified by platelet inhibitors.

MARSHALL: I do not propose to try and summarize the papers in the section in which Professor Zülch and I were cochairmen. I think to do so would be unnecessary because the work reported in our section was so clearly well designed, carefully executed, and clearly presented that any summary would be superfluous.

What I will do is draw briefly from that section what I personally learned. I think there is a danger, when one starts to work on platelets, that as one takes the blood out of the patient and starts to test the adhesiveness or the aggregation or what have you, one forgets the background from which the platelets have been taken.

The papers by Drs. Lentini, Bouvier, and Ulutin, each in a different way, brought us back to earth in this session stressing that we were considering platelet function in a very complex, pathologic situation which we call cerebrovascular disease.

Dr. Kauchtschischvili carried on the same theme in that man-made disease of smoking, and this seems to me important in keeping our minds into the contexts in which we are supposed to be thinking.

Dr. Cepelak took this theme for me a stage further by setting the story into a more firm biochemical background, but finally Dr. Sterling Meyer and Dr. Bousser did for me, as a neurologist, what was perhaps the most important thing of this section. Somebody said, do we know that platelets have anything whatever to do with cerebrovascular disease?

And certainly, as a neurologist, I would not mind how many platelet aggregates my patients have circulating in the blood providing none of them stuck in the cerebral vessels and caused ischemia. Dr. Bousser showed us very beautifully how the aggregations are a step toward thrombosis embolism, which we do believe is relevant to cerebrovascular disease. And Dr. Stirling Meyer and Dr. Welch in their paper showed us some of the consequences of this phenomenon of embolization: obstruction of vessels on the brain when it occurs.

ZÜLCH: We speak of acute and progressive and stroke in evolution and so on. Is this really, or does this really give entities which are clear and concise and concinct? Well, take my policeman of 20 years: after he has passed his thrombosis he will live perhaps another 50 years. Well, that is different from taking out a bit of cancer where you have a scar. And this scar does not mind everybody when it is under your jacket, you see, and he has got a hemiplegia.

Now, the lady I was referring to, she has a thrombosis of her middle cerebral artery but fortunately all the vessels to the basal ganglia were preserved and she came out with only a neurologic deficit.

Now what is our characteristic for recovery, for a mesuration, or for forming samples? I came into particular great difficulties with regard to this and this is one of the things I particularly learned of this.

BOUVIER: I would like very much to join Dr. Acheson in her comment about the not so great difference between a transient attack and a stroke. As we have known, in heart disease this is about the same problem and most cardiologists do not see that differently: angina pectoris and miocardial infarction. Well, it is finally the same thing but one is permanent.

As for the papers of this section, from Dr. Kadatz we have had a very comprehensive review both of drugs which may influence platelet behaviour in vitro, ex vitro, and in experimental animals, as well the critical review of the techniques and models available. And I am happy to join Dr. Born in agreeing how this kind of work is important.

From the group of Toulouse, Dr. Guiraud and coworkers from their report on the biological and methodologic experience, we have learned as well preliminary information on the results, and I think this is a very important study.

From this paper and the one from Dr. Polli, as from many of the papers in this section, the general rules of preparing a properly randomized trial have been outlined and it is to be hoped that no one will embark into a drug study without remembering those general rules, under penalty of ending the study in blood and tears.

From Dr. Federici's paper we have mostly learned a phantastic degree of logistic complexity which might be necessary to approach ecological problems.

FIELDS: In our particular section there were several remarks which struck me as being particularly important, and one was Professor Kadatz in his very beautiful review of the pharmacologic agents, mentioned, namely, that platelets are presumed to protect vessels from injury. And we do not know how the equilibrium of platelets is maintained in normal flowing blood.

I think one of the very serious problems is to determine whether we are doing something which is beneficial or will adversely affect this equilibrium, and in some way influence the beneficial effect that platelets have in protecting vessels.

Furthermore I think that in many of the papers it was brushcombed very clearly that one should be extremely cautious in extrapolating from in vitro experiments what one would anticipate in the way of effects in vivo in the physiologic properties of platelets.

I also think that we should address ourselves to the question of whether we should take only those persons who have enhanced platelet aggregability to select them for the ones whom we would choose to put into drug trials. This was mentioned several times and I think it is extremely important because one cannot use the laboratory tests as a clue to the beneficial clinical effects. We know that one can alter the platelet function very readily with many agents but yet I would think it would be very risky to interpret these effects as influences of beneficial clinical result.

ULUTIN: The first two papers of the section which I shared the chairmanship with Professor Rascol, were related with the extensive clinical trial of aspirin introduced from the Texas group by Professor Fields. And also sulfinpyrazone from the Canadian group, introduced by Dr. Simard.

They get encouraging results but still too early to come as a final decision in the role of and the treatment of this group of diseases.

The third paper from Professor Di Perri was related with the effects of the antiaggregating drug on the anticomplementary action, and the author showed a close relationship between the activities. This was really an interesting result.

As my conclusion from all these papers, I can summarize as follows: The first: in atherosclerosis the platelet is somewhat different from the normal controls that platelets are more active, they show high glass retention and hyperaggregation. Some contents, such as the platelet antiplasmin is increased and also they are easily released, or we can say it in another way, the percentage of the release by the induced is much higher than compared to the normal.

And also the lifespan, or the half-life of the platelets are decreased. And there are somewhat similarity, at least in the laboratory results, with the chronic low-grade DIC.

When we try to treat it with an antiaggregating drug such as aspirin, dipirodamol, or pretanol carbamid, we see that all these changes or differences from the normal in such cases, turn to the normal, such as adhesion, decrease in glass retention, decrease in sensitivity of ADP, and blockage in release mechanisms, and some of them show a defect, or inhibition, in the carbohydrate metabolism, such as the glucose utilization and lacta formation. Also the lifespan of the platelets is prolonged.

These platelets also show some alteration ultrastructurally. But certainly it is too early to say the real meaning of this. Is this just the effect of this drug in a specific way, related with underlying disease and a real treatment, or is it just a kind of laboratory result?

And I would also like to mention that the laboratory findings do not parallel the clinical findings every time. Sometimes, as I mentioned in my paper, there are very severe clinical pictures with normal laboratory results, or vice versa.

RASCOL: 1. Now it is reasonable to assert that platelet aggregation plays a role in pathogenic mechanism of cerebrovascular disorders.

2. The physicians want to know if an antiaggregating therapy is useful, special in the prevention of stroke.

3. To answer this question we need better information about the natural history of cerebrovascular diseases.

4. When we use an antiaggregating drug we have to know not only its in vitro effectiveness but also its in vivo effects in man after therapeutic dose.

Many drugs in vitro modify the platelet behaviour but at therapeutic doses do not modify the platelet aggregation in man like the Born's method permitted.

So about the problems which concern us, the clinical trials are particularly difficult and the results will be subject to criticism. Nevertheless, we hope with interest for results. Even if they are not positive, they will contribute to a better knowledge of cerebrovascular disease.

Concluding Remarks

C. Fazio

When we decided to organize this symposium, our objective was to try and establish what importance the platelets have in the pathogenesis of cerebrovascular disorders. We have, at least, succeeded partly in our endeavor.

It is, however, opportune to take into consideration two different clinical phases: the preictal one and the acute postictal one.

In a preictal situation the platelet seems to act as a predisposing factor and is, therefore, included among the other "risk factors" already mentioned. In fact, some of the clinical data which has been presented, although criticizable primarily due to a lack of precise methodologic correctness, would appear to demonstrate that in this preictal situation hyperplatelet aggregation exists for threshold stimuli.

It is probably true, however, that for this aggregation tendency to take place the intervention of other factors is required. It is not merely a question of alterations in the vessel walls, which is already well known, but also - as we have heard - of the possibility that an irritated platelet reacts to abnormal plasmatic stimuli.

In this way, pathologic aggregation seems to occur with the formations of thromboemboli which, according to the experimental data presented here, seem in a position to cause cerebral ischemia. The thromboemboli would thus be capable of triggering off acute ischemia.

We, in fact, know that, whatever the pathogenic cause of acute ischemia, there is a circulatory diminution in the focus - there is a stasis and an increase of erythrocyte and platelet aggregation.

In this way the mechanism of platelet thromboemboli seems to be set in motion again. In this case serotonin is released from the platelets and acts on cerebral microcirculation causing venous and arterial spasms, plus an increase in the capillary permeability and consequent vasogenic edema.

All these premises could be of extreme importance in clinical medicine and I believe that we are increasingly authorized to employ and verify clinically the validity of the antiaggregant therapy both in the preictal phase and in the acute postictal phase.

H. Anacker, H. D. Weiß, B. Kramann

Endoscopic Retrograde Pancreaticocholangiography (ERPC)

93 figures. Approx. 140 pages. 1977. ISBN 3-540-08008-2

Contents: Introduction: Preoperative Pancreaticography. – The History of Pancreaticography. – Techniques. – Indications, Contraindications, and Complications. – The Normal Retrograde Pancreaticogram. – The ERPC Roentgenographic Pattern in Changes Due to Age. – The ERPC in Lesions of the Pancreas and the Papilla Duodeni. – ERPC in Diseases of the Biliary Tract. – Correlation of ERPC with Other Roentgenologic and Examinations of the Pancreas. – Conclusion. – **References.** – Subject Index.

This book is the first monograph to deal with the latest method of roentgen pancreatic investigation – endoscopic retrograde pancreaticocholangiography (ERPC). Following an introductory chapter on the historical development of the method, the technical procedure in its various phases is covered in detail. The appearance of the normal pancreatic duct system and its age – related changes serves as a basis for the following chapters on pathology of the pancreatic duct system and extrahepatic biliary tract in various disorders, most importantly chronic pancreatisis and carcinoma of the pancreas. In addition to a description of the purely morphologic changes of the pancreas and biliary tract, special emphasis is placed on the value of ERPC, intrinsically and in comparison with other methods, in large part backed by statistical data. The limitations of ERPC are thus made obvious. Finally, indications are given for each of the many techniques available for application in specific pancreatic disorders.

H. V. Crock, H. Yoshizawa

The Blood Supply of the Vertebral Column and Spinal Cord in Man

120 figures. 44 color plates. XIV, 130 pages. 1977. ISBN 3-211-81402-7

Contents: Origins of Arteries Supply in the Vertebral Column, the Meninges, and Spinal Cord. – The Veins of the Vertebral Column, Spinal Cords, and Vertebral Body. – The Distribution of Arteries within the Vertebral and Spinal Cord.

The book provides the first comprehensive source on the blood supply of the vertebral column and spinal cord. The author's original findings on the venous drainage of the vertebral body (likely to prove to be the anatomic basis of nutrition of the intervertebral disc) as well as some new anatomic findings on the origins of arteries supplying the spinal cord, are presented. The latter findings suggest that the arterial supply of the cord is segmental, though the long established variation in the size of feeding vessels is confirmed. The publication of these findings would seem to have great practical significance for the future management of spinal cord injuries. The authors have attempted to prepare this specialized text and atlas for orthopaedic and neurosurgeons, neurologists, and neurophysiologists and others interested in human microcirculation.

Springer-Verlag Berlin Heidelberg New York

Advances in Cardiopulmonary Resuscitation

The Wolf Creek Conference, 1975

Editor: P. Safar
Associate Editor: J. Elam

92 figures. 256 pages. 1977. ISBN 3-540-90234-1

Contents: The Pre-Arrest Period. – Airway Obstruction and Respiratory Arrest. – Circulatory Arrest. – Drugs in Cardiopulmonary Resuscitation. – Electrocardiography Pacing, Defibrillation. – The Immediate Post-Resuscitative Period. – Special Considerations.

This book is more than the Proceedings of the Wolf-Creek Conference on Advances in Resuscitation. The 46 papers of this volume have been written by 24 authors, of whom were first or second generation pioneers of modern resuscitation, which began in the 1950's. The topics include the pre-arrest period, airway obstruction and respiratory arrest; circulatory arrest; drugs in CPR; electrocardiography-pacing-defibrillation, the immediate post-resuscitative period (with special emphasis on recent advances in brain resuscitation); special considerations (such as massive hemorrhage, near-drowning, educational and legal considerations); and historic vignettes. Some of the papers have primarily teaching value; others bring new, as yet unpublished, data and others are unique because: of unpublished historic facts they convey. The historic vignettes and discussions of other paper include stories and quotes of deceased pioneers in resuscitation, such as Claude Beck and William Kouwenhoven. The 24 clinican-scientists who met for this "think tank" predistributed manuscripts in order to take account of the present developments in CPR, and to have a look into the future. Their edited manuscripts and discussions published in this volume represent a synthesis of recent scientific, clinical and educational advances of respiratory, circulatory and cerebral resuscitation.

Myocardial Failure

Editors: G. Riecker, A. Weber, J. Goodwin
Coeditors: H.-D. Bolte, B. Lüderitz, B. E. Strauer, E. Erdmann

50 tables, approx. 400 pages, 1977. ISBN 3-540-08225-5

International Boehringer Mannheim Symposia

Prompted by the latest scientific findings, this book discusses those aspects of myocardial failure that require further multi- and interdisciplinary investigation in basic and clinical research. Leading scientists and clinicians in biochemistry, physiology, pharmacology, pathology, clinical and experimental cardiology present and correlate data on cardiac function and myocardial failure.

This volume contains the proceedings of the International Symposium on Myocardial Failure, held in Rottach-Egern at Tegernsee (Germany), June 17–19, 1976 under the auspices of the European Society of Cardiology. The papers have been grouped into the following sections: 1) Molecular Basis of Myocardial Function; 2) Sarcoplasmic Reticulum; 3) Membrane-Bound Receptors; 4) New Diagnostic Procedures; 5) Problems of Etiology; 6) Clinical Pharmacology; and finally; 7) Drugs Influencing Myocardial Contractility.

Prices are subject to change without notice

Springer-Verlag Berlin Heidelberg New York

GPSR Compliance
The European Union's (EU) General Product Safety Regulation (GPSR) is a set
of rules that requires consumer products to be safe and our obligations to
ensure this.

If you have any concerns about our products, you can contact us on

ProductSafety@springernature.com

In case Publisher is established outside the EU, the EU authorized
representative is:

Springer Nature Customer Service Center GmbH
Europaplatz 3
69115 Heidelberg, Germany

www.ingramcontent.com/pod-product-compliance
Ingram Content Group UK Ltd.
Pitfield, Milton Keynes, MK11 3LW, UK
UKHW050410240426
12048UKWH00020B/1447